An Introduction to the Event-Related Potential Technique

Second Edition

Steven J. Luck

A Bradford Book
The MIT Press
Cambridge, Massachusetts
London, England

MIT Press books may be purchased at special quantity discounts for business or sales promotional use. For information, please email special_sales@mitpress.mit.edu.

This book was set in Times LT Std by Toppan Best-set Premedia Limited, Hong Kong. Printed and bound in the United States of America.

Library of Congress Cataloging-in-Publication Data
Luck, Steven J. (Steven John), 1963– author.
An introduction to the event-related potential technique / Steven J. Luck—Second edition.
    p. ; cm.
Includes bibliographical references and index.
ISBN 978-0-262-52585-5 (pbk. : alk. paper)
I. Title.
[DNLM: 1. Evoked Potentials. WL 102]
QP376.5
616.8'047547—dc23
2013041747

10 9 8 7 6 5 4

Contents

Online Materials

Online Supplements to Chapters

Supplement to Chapter 3: Linking ERPs with Neural and Cognitive Processes

Online Files

Online Chapters

Preface

I have a love–hate relationship with event-related potentials (ERPs): I love great ERP research, and I hate bad ERP research. My goal in writing this book is to increase the former and decrease the latter. It's as simple as that.

I was fortunate to be trained in Steve Hillyard's lab at the University of California, San Diego, which has a tradition of human electrophysiological research that goes back to some of the first human electroencephalogram (EEG) and ERP recordings in the 1930s. When I wrote the first edition of this book, my goal was to summarize the accumulated body of ERP theory and practice that permeated the Hillyard lab, along with a few ideas of my own, so that this information would be widely accessible to beginning and intermediate ERP researchers.

Just as I was finishing the first edition, I began organizing a summer training workshop called the ERP Boot Camp (see http://erpinfo.org/the-erp-bootcamp). This workshop is now held every summer at the Center for Mind & Brain of the University of California, Davis, and it provides 10 days of intensive training in ERP fundamentals to 35 intermediate-level ERP researchers from around the world. I also started holding Mini ERP Boot Camps at a variety of locations, providing 2 or 3 days of focused training to groups of 10–100 beginning and intermediate ERP researchers. Altogether, I've had an opportunity to provide ERP training to more than 1000 researchers over the past 10 years. The Boot Camps have been a real pleasure, because nothing is more gratifying than teaching something you care about to smart people who are highly motivated to learn it.

The Boot Camps have also taught me a great deal about what new and intermediate ERP researchers want to know, and it gave me the opportunity to develop better ways of explaining ERP methods. My goal for this second edition of the book was to take this experience and use it to update, expand, and clarify the material provided in the first edition. I've organized the material in the way that has proved most effective in the Boot Camps, and I've tried to answer all the questions that I'm typically asked at the various Boot Camps. I hope the end result meets the needs of researchers with varying levels of expertise and a broad range of research interests.

Acknowledgments

Several people have had a direct or indirect impact on the development of the first and second editions of the book, and I would like to thank them for their help.

First, I'd like to thank my early academic mentors. At Reed College, Allen Neuringer single-handedly turned me into a scientist, gave me my first technical skills, demonstrated the importance of creativity in science, and taught me the importance of asking the right questions. At the Oregon Regional Primate Research Center, Martha Neuringer provided my first introduction to the visual system and neuroscience, taught me how to record electrophysiological signals, and showed me how world-class science is done. And back at Reed, Dell Rhodes gave me the opportunity to learn about event-related potentials (ERPs) by putting together an ERP lab from scratch and taught me that learning should extend for a lifetime.

Next, I'd like to thank Steve Hillyard, who is truly a founding father of cognitive neuroscience. Steve taught me innumerable lessons about science, but above all he demonstrated the importance of thinking carefully about every step, from experimental design to writing a manuscript. His fingerprints are on virtually every page of this book. I'd also like to thank everyone who worked with me in the Hillyard lab at the University of California, San Diego, including Jon Hansen, Marty Woldorff, Ron Mangun, Marta Kutas, Cyma Van Petten, Steve Hackley, Hajo Heinze, Vince Clark, Paul Johnston, and Lourdes Anllo-Vento. Working with this group of people was a truly amazing experience. I'd also like to acknowledge the inspiration provided by the late Bob Galambos, who taught me that "you have to get yourself a phenomenon" at the beginning of the research enterprise.

I'd also like to acknowledge the important contributions of my former graduate students at the University of Iowa, who were instrumental in the first edition of this book. Many of the ideas in that edition became crystallized as I taught ERP methods to the "A Team" of Massimo Girelli, Ed Vogel, and Geoff Woodman. I also received many excellent comments and suggestions from Joo-seok Hyun, Weiwei Zhang, Jeff Johnson, Po-Han Lin, and Adam Niese.

A number of graduate students, postdocs, and faculty here at the University of California, Davis, Center for Mind & Brain have been extremely helpful with the second edition, especially the Monday Seminar participants who slogged through drafts of the chapters (including but not limited to Joy Geng, Carly Leonard, Nancy Carlisle, Johanna Kreither, Javier Lopez-Calderon,

Kyle Frankovich, Felix Bacigalupo, Mee-Kyoung Kwon, Ashley Royston, Jesse Bengson, Xiangfei Hong, Beth Stankevich, and Nick DiQuattro). Carly Leonard also played an important role by figuring out how to use independent component analysis (ICA) for artifact correction and providing the example used in chapter 6. I'd also like to thank Javier Lopez-Calderon for his many insights into technical issues and for everything we figured out together while working on ERPLAB Toolbox, our open source package for analyzing ERP data (http://erpinfo.org/erplab). Jaclyn Farrens provided first-rate editorial assistance. Last but not least, I'd like to thank Emily Kappenman, who has been my co-conspirator in countless ERP-related enterprises over the past 7 years, helped me think through many of the issues described in this book, and spent many days reading drafts and providing extremely valuable comments.

I would like to thank David Groppe for his insightful comments about statistical analyses and ICA-based artifact correction.

I would also like to thank my family for providing emotional support and pleasant diversions as I worked on this book. Lisa was, as always, the gravity force at the center of my universe; Alison provided a continuous soundtrack and dance track; and Carter provided peace of mind by making sure that everything was running smoothly in various Minecraft worlds.

Finally, I would like to thank my good friends Paul Myers-Verhage and Paul Hewson for their interesting observations and insightful comments about ERPs, science, and life. I always look forward to talking about Big Ideas with the two of you!

1 A Broad Overview of the Event-Related Potential Technique

Overview, Goals, and Perspective of This Book

The event-related potential (ERP) technique provides a powerful method for exploring the human mind and brain, and the goal of this book is to describe practical methods and underlying concepts that will help you conduct great ERP research.

The first half of this book focuses on essential background information. This first chapter is an overview that is intended for people who are new to ERPs. The second chapter provides a closer look at ERPs, exploring issues that lurk below the surface of almost every ERP study. The third chapter is an overview of the most common and useful ERP components. The fourth chapter describes the design of ERP experiments; its goal is to help you design your own experiments and critically evaluate published research.

The second half of this book provides a detailed explanation of the main steps involved in actually conducting ERP experiments, including recording the electroencephalogram (EEG; chapter 5), rejecting and correcting artifacts (chapter 6), filtering the EEG and ERP waveforms (chapter 7), creating averaged ERPs and conducting time–frequency analyses (chapter 8), measuring amplitudes and latencies (chapter 9), and conducting statistical analyses (chapter 10). The goal is for you to understand how these steps really work so that you can make the best possible choices in how you conduct and analyze your own experiments.

To keep the length and price of this book reasonable, additional chapters and material are provided at a Web site (http://mitpress.mit.edu/luck2e), where they can be accessed by anyone at no cost. This Web site includes supplements to several of the chapters, which provide additional details or advanced materials. It also includes additional chapters on convolution (a simple mathematical procedure that is valuable in understanding ERPs); the relationship between the time and frequency domains; advanced statistical techniques; source localization; how to read, write, and review ERP papers; and how to set up and run an ERP lab.

You should feel free to skip around the book. If you are in the middle of running or analyzing your first ERP experiment, you may want to start with the chapters on recording data and performing basic data analysis steps. But be sure to come back to the first few chapters eventually, because they will help you avoid making common interpretive errors.

I have focused this book on mainstream techniques that are used in my own laboratory and in many other labs around the world. I learned many of these techniques as a graduate student in Steve Hillyard's laboratory at the University of California, San Diego (UCSD), and they reflect a long history of electrophysiological recordings dating back to Hallowell Davis's lab at Harvard in the 1930s. Davis was the mentor of Bob Galambos, who was in turn the mentor of Steve Hillyard. Galambos was actually a subject in the very first ERP experiment in the 1930s, and I got to spend quite a bit of time with him when I was in graduate school. Steve Hillyard inherited the Galambos lab when Bob retired, and he probably did more than any other early ERP researcher to show that ERPs could be used to answer important questions in cognitive neuroscience. Much of this book represents a distillation of 50-plus years' worth of ERP experience that was imparted to me in graduate school.

Although most of the basic ERP recording and analysis procedures described in this book are very conventional, some aspects of my approach to ERP research are different from those of other researchers, reflecting differences in general approaches to science. For example, I believe it is better to record very clean data from a relatively modest number of electrodes rather than to record noisier data from a large number of electrodes. Similarly, I believe that it is better to use a rigorous experimental design and relatively simple analyses rather than to rely on a long series of complex data processing procedures. As you will read in the following chapters, even the simplest processes (e.g., averaging, filtering, artifact rejection) can have unanticipated side effects. Consequently, the more processes that are applied to the data, the further you get from the signal that you actually recorded, and the more likely you are to have artificially induced an effect that does not accurately reflect real brain activity. Of course, some processing is necessary to separate the signal from the noise, but I believe the truth is usually clearest when the data speak for themselves and the experimenter has not tortured the data into confessing whatever he or she wants to hear (see box 1.1).

My views on dense-array recordings and sophisticated data processing techniques are heresy to some ERP researchers, but the vast majority of ERP studies that have had a significant impact on science (i.e., outside the fraternity of ERP researchers) have relied more on clever experimental design than on sophisticated data processing techniques. Even if your plans involve these techniques, this book will give you a very solid background for using them, and you will learn how to apply them wisely.

Chapter Overview

This chapter provides a broad overview of the ERP technique. It is designed to give beginning ERP researchers the big picture of ERP research before we dive into the details. However, even advanced researchers are likely to find some useful information in this chapter.

The remainder of this chapter begins with a brief history of the ERP technique. Two examples of ERP research are then provided to make things more concrete. The next section briefly describes how ERPs are generated in the brain and propagated to the scalp. A more extended

Box 1.1
Treatments and Side Effects

Data processing procedures that attempt to reveal a specific aspect of brain activity by suppressing "noise" in the data are analogous to treatments designed to suppress the symptoms of an underlying medical problem. As any physician can tell you, treatments always have side effects. For example, ibuprofen is a common and useful treatment for headaches and muscle soreness, but it can have negative side effects. According to *Wikipedia*, common adverse effects of ibuprofen include nausea, dyspepsia, gastrointestinal bleeding, raised liver enzymes, diarrhea, epistaxis, headache, dizziness, unexplained rash, salt and fluid retention, and hypertension. These are just the common side effects! Infrequent adverse effects of ibuprofen include esophageal ulceration, hyperkalemia, renal impairment, confusion, bronchospasm, and heart failure. Yes, heart failure!

ERP processing treatments such as filters can also have adverse side effects. According to *Luckipedia*, common adverse effects of filters include distortion of onset times, distortion of offset times, unexplained peaks, and slight dumbness of conclusions. Less frequent adverse effects of filters include artificial oscillations, wildly incorrect conclusions, public humiliation by reviewers, and grant failure.

This does not mean that you should completely avoid filters and other ERP processing procedures. Just as ibuprofen can be used effectively—in small doses—to treat headaches and muscle soreness, mild filtering can help you find real effects without producing major side effects. However, you need to know how to apply data processing techniques in a way that minimizes the side effects, and you need to know how to spot the side effects when they occur so that you do not experience public humiliation by reviewers and grant failure.

example of an ERP experiment is then provided to illustrate the basic steps in conducting and analyzing an ERP experiment. This example is followed by a discussion of two key concepts—oscillations and filtering—and then a more detailed description of the steps involved in collecting and processing ERP data. The final sections describe the advantages and disadvantages of the ERP technique and how it compares with other common techniques.

Note that some basic terminology is defined in the glossary. You should skim through it if you are not sure about the difference between an *evoked potential* and an *event-related potential*, if you don't know how an *SOA* differs from an *ISI* or an *ITI*, or if you are not sure how a *local field potential* differs from a *single-unit recording*.

A Bit of History

I think it's a good idea to know a little bit about the history of a technique, so this section describes the discovery of ERPs in the 1930s and how the use of ERPs has progressed over the ensuing 80-plus years. However, you can skip this section if you don't feel the need for a history lesson.

In 1929, Hans Berger reported a remarkable and controversial set of experiments in which he showed that the electrical activity of the human brain could be measured by placing an electrode on the scalp, amplifying the signal, and plotting the changes in voltage over time (Berger, 1929). This electrical activity is called the electroencephalogram, or EEG. The neurophysiologists of the day were preoccupied with action potentials, and many of them initially believed that the relatively slow and rhythmic brain waves observed by Berger were some sort of artifact. For example, you can get similar-looking waveforms by putting electrodes in a pan of Jello and wiggling it. After a few years, however, human EEG activity was also observed by the respected physiologist Adrian (Adrian & Matthews, 1934), and the details of Berger's observations were confirmed by Jasper and Carmichael (1935) and Gibbs, Davis, and Lennox (1935). These findings led to the acceptance of the EEG as a real phenomenon.

Over the following decades, the EEG proved to be very useful in both scientific and clinical applications. However, the EEG is a very coarse measure of brain activity, and it cannot be used in its raw form to measure most of the highly specific neural processes that are the focus of cognitive neuroscience. This is partly because the EEG represents a mixed-up conglomeration of dozens of different neural sources of activity, making it difficult to isolate individual neuro-cognitive processes. Embedded within the EEG, however, are the neural responses associated with specific sensory, cognitive, and motor events, and it is possible to extract these responses from the overall EEG by means of a simple averaging technique (and more sophisticated techniques, such as time–frequency analyses). These specific responses are called *event-related potentials* to denote the fact that they are electrical *potentials* that are *related* to specific *events*.

As far as I can tell, the first unambiguous sensory ERP recordings from humans were performed in 1935–1936 by Pauline and Hallowell Davis and published a few years later (Davis, 1939; Davis, Davis, Loomis, Harvey, & Hobart, 1939). This was long before computers were available for recording the EEG, but the researchers were able to see clear ERPs on single trials during periods in which the EEG was quiescent (the first computer-averaged ERP waveforms were apparently published by Galambos & Sheatz, 1962). Not much ERP work was done in the 1940s due to World War II, but research picked up again in the 1950s. Most of this research focused on sensory issues, but some of it addressed the effects of top-down factors on sensory responses.

The modern era of ERP research began in 1964, when Grey Walter and his colleagues reported the first cognitive ERP component, which they called the *contingent negative variation*, or CNV (Walter, Cooper, Aldridge, McCallum, & Winter, 1964). On each trial of this study, subjects were presented with a warning signal (e.g., a click) followed 500 or 1000 ms later by a target stimulus (e.g., a series of flashes). In the absence of a task, both the warning signal and the target elicited the sort of sensory ERP response that would be expected for these stimuli. However, if subjects were required to press a button upon detecting the target, a large negative voltage was observed at frontal electrode sites during the period that separated the warning signal and the target. This negative voltage—the CNV—was clearly not just a sensory response. Instead, it appeared to reflect the subject's preparation for the upcoming target. This exciting new finding led many researchers to begin exploring cognitive ERP components (for a review of more recent CNV research, see Brunia, van Boxtel, & Böcker, 2012).

The next major advance was the discovery of the P3 component by Sutton, Braren, Zubin, and John (1965). They created a situation in which the subject could not predict whether the next stimulus would be auditory or visual, and they found that the stimulus elicited a large positive component that peaked around 300 ms poststimulus. They called this the *P300* component (although it is now frequently called *P3*). This component was much smaller when the conditions were changed so that subjects could predict the modality of the stimulus. They described this difference in brain responses to unpredictable versus predictable stimuli in terms of information theory, which was then a very hot topic in cognitive psychology, and this paper generated a huge amount of interest. To get a sense of the impact of this study, I ran a quick Google Scholar search and found more than 27,000 articles that refer to "P3" or "P300" along with "event-related potential." This is an impressive amount of P3-related research. In addition, the Sutton et al. (1965) paper has been cited more than 1150 times. There is no doubt that many millions of dollars have been spent on P3 studies (not to mention the many euros, pounds, yen, yuan, etc.).

During the 15 years after the publication of this paper, a great deal of research was conducted that focused on identifying various cognitive ERP components and developing methods for recording and analyzing ERPs in cognitive experiments. Because people were so excited about being able to record human brain activity related to cognition, ERP papers in this period were regularly published in *Science* and *Nature*. Most of this research was focused on discovering and understanding ERP components rather than using them to address questions of broad scientific interest. I like to call this sort of experimentation "ERPology" because it is simply the study of ERPs.

ERPology experiments do not directly tell us anything important about the mind or brain, but they can be very useful in providing important information that allows us to use ERPs to answer more broadly interesting questions. A great deal of ERPology continues today, resulting in a refinement of our understanding of the components discovered in previous decades and the discovery of new components. Emily Kappenman and I edited a book on ERP components a few years ago that summarizes all of this ERPology (Luck & Kappenman, 2012a).

However, so much of ERP research in the 1970s was focused on ERPology that the ERP technique began to have a bad reputation among many cognitive psychologists and neuroscientists in the late 1970s and early 1980s. As time progressed, however, an increasing proportion of ERP research was focused on answering questions of broad scientific interest, and the reputation of the ERP technique began to improve. ERP research started becoming even more popular in the mid 1980s, due in part to the introduction of inexpensive computers and in part to the general explosion of research in cognitive neuroscience. When positron emission tomography (PET) and then functional magnetic resonance imaging (fMRI) were developed, many ERP researchers thought that ERP research might die away, but exactly the opposite happened; most researchers understand that ERPs provide high-resolution temporal information about the mind and brain that cannot be obtained any other way, and ERP research has flourished rather than withered.

Example 1: The Classic Oddball Paradigm

To introduce the ERP technique, I will begin by describing a simple experiment that was conducted in my laboratory several years ago using a version of the classic *oddball* paradigm (we never published this experiment, but it has been extremely useful over the years as an example). My goal here is to give you a general idea of how a simple ERP experiment works.

As illustrated in figure 1.1, subjects in this experiment viewed sequences consisting of 80% Xs and 20% Os, and they pressed one button for the Xs and another button for the Os. Each letter was presented on a computer monitor for 100 ms, followed by a 1400-ms blank interstimulus interval. While the subject performed this task, we recorded the EEG from several electrodes embedded in an electrode cap. As will be described in detail in chapter 5, EEG recordings typically require one or more *active* sites, along with a *ground* electrode and a *reference* electrode. The EEG from each site was amplified by a factor of 20,000 and then converted into digital form for storage on the *digitization computer*. Whenever a stimulus was presented, *event codes* (also known as *trigger codes*) were sent from the stimulation computer to the EEG digitization computer, where they were stored along with the EEG data (see figure 1.1A).

During each recording session, we viewed the EEG on the digitization computer, but the stimulus-elicited ERP responses were too small to discern within the much larger EEG. Figure 1.1C shows the EEG that was recorded at one electrode site from one of the subjects over a period of 9 s. The EEG waveform shown in the figure was recorded from the Pz electrode site (on the midline over the parietal lobes; see figure 1.1B), where the P3 wave is largest. If you look closely, you can see that there is some consistency in the EEG response to each stimulus, but it is difficult to see exactly what the responses look like. Figure 1.1D is a blowup of one small period of time, showing that the continuous voltage was converted into a discrete set of samples for storage on the computer.

The EEG was recorded concurrently from approximately 20 electrodes in this experiment, and the electrodes were placed according to the International 10/20 System (American Encephalographic Society, 1994a). As shown in figure 1.1B, this system names each electrode site using one or two letters to indicate the general brain region (e.g., Fp for frontal pole, F for frontal, C for central, P for parietal, O for occipital, T for temporal) and a number to indicate the

Figure 1.1
Example ERP experiment using the oddball paradigm. The subject viewed frequent Xs and infrequent Os presented on a computer monitor while the EEG was recorded from several active electrodes in conjunction with ground and reference electrodes (A). The electrodes were placed according to the International 10/20 System (B). Only a midline parietal electrode (Pz) is shown in panel A. The signals from the electrodes were filtered, amplified, and then sent to a digitization computer to be converted from a continuous analog signal into a discrete set of digital samples (D). Event codes were also sent from the stimulus presentation computer to the digitization computer, marking the onset time and identity of each stimulus and response. The raw EEG from the Pz electrode is shown over a period of 9 s (C). Each event code during this period is indicated by an arrow along with an X or an O, indicating the stimulus that was presented. Each rectangle shows a 900-ms epoch of EEG, beginning 100 ms prior to the onset of each stimulus. These epochs were extracted and then lined up with respect to stimulus onset (E), which is treated as 0 ms. Separate averages were then computed for the X and O epochs (F).

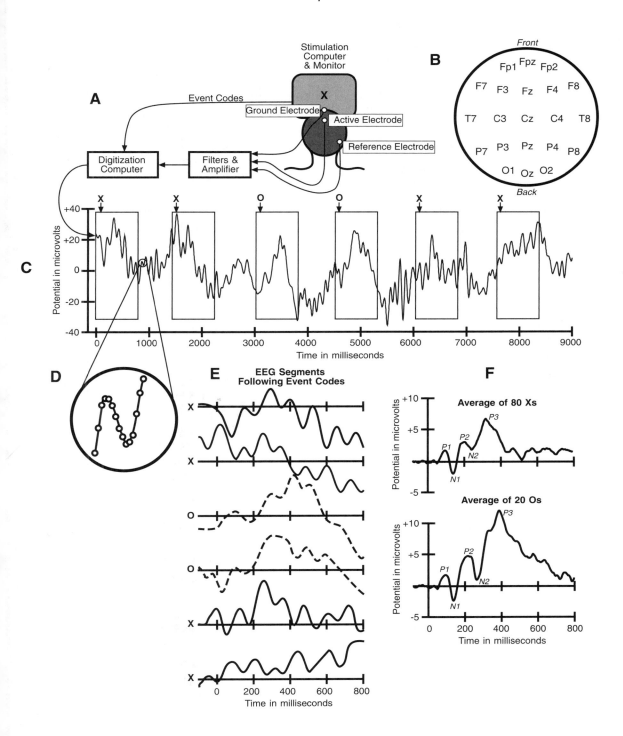

hemisphere (odd for left and even for right) and the distance from the midline (larger numbers mean larger distances). A lowercase z is used to represent the number zero, which indicates that the electrode is on the midline. Thus, F3 lies over frontal cortex to the left of the midline, Fz lies over frontal cortex on the midline, and F4 lies over frontal cortex to the right of the midline (for more details, see figure 5.4 in chapter 5).

At the end of each session, we performed a simple signal-averaging procedure to extract the ERP waveform elicited by the Xs and the ERP waveform elicited by the Os. The basic idea is that the recorded EEG contains the brain's response to the stimulus plus other activity that is unrelated to the stimulus, and we can extract this consistent response by averaging across many trials. To accomplish this, we extracted the segment of EEG surrounding each X and each O (indicated by the rectangles in figure 1.1C) and lined up these EEG segments with respect to the event codes that marked the onset of each stimulus (figure 1.1E). We then simply averaged together the single-trial EEG waveforms, creating one averaged ERP waveform for the Xs and another for the Os at each electrode site (figure 1.1F). For example, the voltage at 24 ms in the averaged X waveform was computed by taking the voltage that was measured 24 ms after each X stimulus and averaging all of these voltages together. Any brain activity that was consistently elicited by the stimulus at that time will remain in the average. However, any voltages that were unrelated to the stimulus will be negative on some trials and positive on other trials and will therefore cancel each other when averaged across many trials.

The resulting averaged ERP waveforms consist of a sequence of positive and negative voltage deflections, which are called *peaks*, *waves*, or *components*. In figure 1.1F, the peaks are labeled *P1*, *N1*, *P2*, *N2*, and *P3*. P and N are traditionally used to indicate positive-going and negative-going peaks, respectively, and the number indicates a peak's position within the waveform (e.g., P2 is the second major positive peak). Alternatively, the number may indicate the latency of the peak in milliseconds (e.g., N170 for a negative peak at 170 ms). If the number is greater than 5, you should assume it is referring to the peak's latency. Components may also be given paradigm- or function-based names, such as the *error-related negativity* (which is observed when the subject makes an error) or the *no-go N2* (which is observed on no-go trials in go/no-go experiments). The labeling conventions for ERP components can be frustrating to new researchers, but they become second nature over time, as discussed in box 1.2. Chapter 3 provides additional details about component naming conventions.

The sequence of ERP peaks reflects the flow of information through the brain, and the voltage at each time point in the ERP waveform reflects brain activity at that precise moment in time. Many of the highest impact ERP studies have made use of this fact to test hypotheses that could not be tested any other way (see chapter 4).

In the early days of ERP research, waveforms were plotted with negative upward and positive downward (largely due to historical accident; see box 1.3). The majority of cognitively oriented ERP researchers now use the traditional mathematical convention of plotting positive upward. However, many excellent researchers still plot negative upward, so it is important to check which

Box 1.2
Component Naming Conventions

ERP component names can be very confusing, but so can words in natural languages (especially languages such as English that draw from many other languages). Just as the English word *head* can refer to a body part, a person who is the director of an organization, or a small room on a boat that has a toilet in it, the ERP term *N1* can refer to at least two different visual components and at least three different auditory components. And just as the English words *finger* and *digit* can refer to the same thing, the ERP terms *ERN* and *Ne* refer to the same ERP component. English words can often be confusing to people who are in the process of learning them, but fluent speakers can usually determine the meaning of a word from its context. Similarly, ERP component names can often be confusing to ERP novices, but expert ERPers can usually determine the meaning from its context.

One source of confusion is that the number following the P or N can either be the ordinal position of the peak in the waveform (e.g., N1 for the first negative peak) or the latency of the peak (e.g., N400 for a peak at 400 ms). I much prefer to use the ordinal position, because a component's latency may vary considerably across experiments, across conditions within an experiment, or even across electrode sites within a condition. This is particularly true of the P3 wave, which almost always peaks well after 300 ms (the P3 wave had a peak latency of around 300 ms in the very first P3 experiment, and the name *P300* has persisted despite the wide range of latencies). Moreover, in language experiments, the P3 wave generally follows the N400 wave, making the term *P300* especially problematic. Consequently, I prefer to use a component's ordinal position in the waveform rather than its latency when naming it. Fortunately, the latency in milliseconds is often approximately 100 times the ordinal position, so that P1 = P100, N2 = N200, and P3 = P300. The one obvious exception to this is the N400 component, which is often the second or third large negative component. For this reason, I can't seem to avoid using the time-based name *N400*.

convention is used in a given ERP waveform plot (and to include this information in your own plots). The waveforms in this book are all plotted with positive upward.

In the experiment shown in figure 1.1, the infrequent O stimuli elicited a much larger P3 wave than the frequent X stimuli. This is exactly what thousands of previous oddball experiments have found (see review by Polich, 2012). If you are just beginning to get involved in ERP research, I would recommend running an oddball experiment like this as your first experiment. It's simple to do, and you can compare your results with a huge number of published experiments.

The averaging process was conducted separately for each electrode site, yielding a separate average ERP waveform for each stimulus type at each electrode site. The P3 wave shown in figure 1.1F was largest at the Pz electrode but could be seen at all 20 electrodes. The P1 wave, in contrast, was largest at lateral occipital electrode sites and was absent at frontal sites. Each ERP component has a distinctive scalp distribution that reflects the location of the patch of cortex in which it was originally generated. As will be discussed in chapter 2 and the online chapter 14, however, it is difficult to determine the exact location of the neural generator source simply by examining the distribution of voltage over the scalp.

Box 1.3
Which Way Is Up?

It is a common convention to plot ERP waveforms with negative voltages upward and positive volt-ages downward. I plotted negative upward in the first part of my career (including the first edition of this book) for the simple reason that this was how things were done when I joined Steve Hillyard's lab at UCSD. I once asked Steve Hillyard's mentor, Bob Galambos, how this convention came about. His answer was simply that this was how things were done when he joined Hal Davis's lab at Harvard in the 1930s (see, e.g., Davis, 1939; Davis et al., 1939). Apparently, this was a common convention among early physiologists. Manny Donchin told me that the early neurophysiologists plotted nega-tive upward, possibly because this allows an action potential to be plotted as an upward-going spike, and this influenced manufacturers of early EEG equipment, such as Grass Instruments. Galambos also mentioned that an attempt was made in the early days of ERP research to get everyone to agree to a uniform positive-up convention, but the whole attempt failed (see Bach, 1998).

I eventually made the switch to positive-up because my primary goal in using ERPs is to make scientific contributions that influence a broad set of researchers (and because Emily Kappenman, then a graduate student, kept reminding me that it was the right thing to do). Almost all scientists outside the ERP world follow the centuries-old convention of the Cartesian coordinate system, in which positive is plotted upward. Plotting with negative upward makes ERP data less approachable for the broader scientific community, and there isn't a good scientific justification for it.

It was quite a bit of work for me to switch from negative-up to positive-up, because I had a huge library of figures and PowerPoint slides that had to be revised. Indeed, I ended up paying a talented undergraduate student, Candace Markley, to go though all my old files and switch the polarity. So I understand why many researchers are reluctant to switch. But it's worth the work in the long run, so I encourage everyone to plot with positive upward, just like the rest of the scientific world.

Example 2: The N170 Component and Face Processing

Now that I've explained a very simple ERP experiment, I'd like to describe a line of research that shows how ERPs can be applied to more interesting questions about the human mind. These experiments focused on the *N170* component, a face-related component that typically peaks around 170 ms after stimulus onset and is largest over ventral areas of visual cortex (figure 1.2). In a typical N170 paradigm, photographs of faces and various types of non-face objects are briefly flashed on a computer monitor, and subjects passively view the stimuli. In the ERP waveforms shown in figure 1.2A, the *X* axis represents time relative to stimulus onset (measured in milliseconds [ms]), and the *Y* axis represents the magnitude of the neural response (in micro-volts [μV]). In the scalp map shown in figure 1.2B, the shading indicates the voltage measured at each electrode site during the time period of the N170 (with interpolated values between the individual electrode sites).

The N170 component is notable because it is larger when the eliciting stimulus is a face compared to when the stimulus is a non-face object, such as an automobile (see review by Rossion & Jacques, 2012). The difference between faces and non-face objects begins approxi-

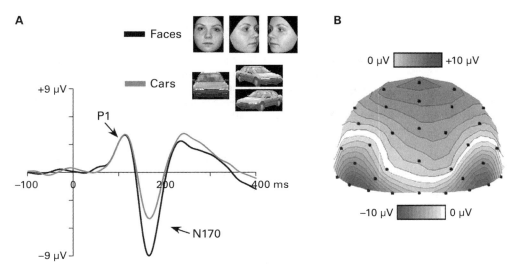

Figure 1.2
Example N170 experiment, including (A) ERP waveforms from an occipito-temporal electrode site (referenced to the average of all electrode sites) and (B) the scalp distribution of the voltage in the N170 latency range. Adapted with permission from Rossion and Jacques (2012). Copyright 2012 Oxford University Press.

mately 150 ms after the onset of the stimulus; this simple fact allows us to conclude that the human brain is able to distinguish between faces and other objects within 150 ms. The scalp distribution helps us to know that this is the same component that is observed in similar studies of the N170, and it suggests that the N170 generator lies in visual cortex (but note that conclusions about generators based on scalp distributions are not usually definitive).

Many researchers have used the N170 to address interesting questions about how faces are processed in the brain. For example, some studies have asked whether face processing is automatic by testing whether the face-elicited N170 is smaller when the faces are ignored. The results of these experiments indicate that face processing is at least partially automatic (Carmel & Bentin, 2002) but can be modulated by attention under some conditions (e.g., when the faces are somewhat difficult to perceive—Sreenivasan, Goldstein, Lustig, Rivas, & Jha, 2009). Other studies have used the N170 to ask whether faces are processed in a specialized face module or whether the same neural process is also used when people process other sorts of complex stimuli for which they have extensive expertise. Consistent with a key role for expertise, these studies have shown that bird experts exhibit an enhanced N170 in response to birds, dog experts exhibit an enhanced N170 in response to dogs, and fingerprint experts exhibit an enhanced N170 in response to fingerprints (Tanaka & Curran, 2001; Busey & Vanderkolk, 2005). Developmental studies have used the N170 to track the development of face processing, showing that face-specific processing is present early in infancy but becomes faster and more sophisticated over the course of development (Coch & Gullick, 2012). Studies of neurodevelopmental disorders

Box 1.4
The Main Virtue of the ERP Technique

Virtually every textbook discussion of cognitive neuroscience techniques notes that the main advantage of the ERP technique is its high temporal resolution and the main disadvantage is its low spatial resolution. Given that this characterization of the ERP technique is so widely accepted, I am constantly amazed at how many studies try to use ERPs to answer questions that require high spatial resolution rather than high temporal resolution. I am also amazed at how many ERP studies use signal processing techniques (e.g., extreme filters) that reduce the temporal precision of the data. It should be obvious that ERPs are most appropriate for answering questions that require high temporal resolution, and I encourage you to think about using ERPs in this way. Some of the studies described in the online supplement to chapter 4 provide excellent examples of how to take advantage of this temporal resolution.

have shown that the N170 is abnormal in children with autism spectrum disorder (Dawson et al., 2002).

The N170 example illustrates the precise temporal resolution of the ERP technique, which is often touted as its main virtue (see box 1.4). ERPs reflect ongoing brain activity with no delay, and an ERP effect observed at 150 ms reflects neural processing that occurred at 150 ms. Consequently, ERPs are especially useful for answering questions about the timing of mental processes. Sometimes this timing information is used explicitly by asking whether two conditions or groups differ in the timing of a given neural response (just as one might ask whether reaction time differs across conditions or groups). In other cases, the timing information is used to determine whether a given experimental manipulation influences sensory activity that occurs shortly after stimulus onset or higher-level cognitive processes that occur hundreds of milliseconds later. For example, ERPs have been used to ask whether attentional manipulations influence early sensory processes or whether they instead influence postperceptual memory and decision processes (see, e.g., Luck & Hillyard, 2000).

Brief Overview of the Neural Origins of ERPs

In almost all cases, ERPs originate as postsynaptic potentials (PSPs), which occur when neurotransmitters bind to receptors, changing the flow of ions across the cell membrane (for more details, see chapter 2 and Buzsáki, Anastassiou, & Koch, 2012). Scalp ERPs are not typically produced by action potentials (except for auditory responses that occur within a few milliseconds of stimulus onset). When PSPs occur at the same time in large numbers of similarly oriented neurons, they summate and are conducted at nearly the speed of light through the brain, meninges, skull, and scalp. Thus, ERPs provide a direct, instantaneous, millisecond-resolution measure of neurotransmission-mediated neural activity. This contrasts with the blood oxygen level–dependent (BOLD) signal in fMRI, which reflects a delayed, secondary consequence of neural activity. Moreover, the close link to neurotransmission makes ERPs potentially valuable as biomarkers in studies of pharmacological treatments.

When a PSP occurs within a single neuron, it creates a tiny electrical dipole (an oriented flow of current). Measurable ERPs can be recorded at the scalp only when the dipoles from many thousands of similarly oriented neurons sum together. If the orientations of the neurons in a given region are not similar to each other, the dipoles will cancel out and will be impossible to detect at a distant electrode. The main neurons that have this property are the pyramidal cells of the cerebral cortex (the primary input–output cells of the cortex). These cells are oriented perpendicular to the cortical surface, and their dipoles therefore add together rather than canceling out. Consequently, scalp-recorded ERPs almost always reflect neurotransmission that occurs in these cortical pyramidal cells. Nonlaminar structures such as the basal ganglia do not typically generate ERPs that can be recorded from the scalp, and interneurons within the cortex are thought to generate little or no scalp ERP activity. Thus, only a fraction of brain activity leads to detectable ERP activity on the scalp.

ERP components can be either positive or negative at a given electrode site. The polarity depends on a combination of several factors, and it is usually impossible to draw strong conclusions from the polarity of an ERP component (see box 2.1 in chapter 2).

When the dipoles from many individual neurons sum together, they can be represented quite accurately with a single *equivalent current dipole* that is the vector sum of the individual dipoles. For the rest of this chapter, the term *dipole* will refer to these aggregates that represent the dipoles from many individual neurons.

The voltage recorded on the surface of the scalp will be positive on one side of the dipole and negative on the other, with a single line of zero voltage separating the positive and negative sides (figure 1.3). The voltage field spreads out through the conductive medium of the brain, and the high resistance of the skull and the low resistance of the overlying scalp lead to further spatial blurring. Thus, the voltage for a single dipole will be fairly broadly distributed over the surface of the scalp, especially for ERPs that are generated in relatively deep cortical structures.

Electrical dipoles are always accompanied by magnetic fields, but the skull is transparent to magnetism, leading to less blurring of the magnetic fields. Consequently, it is sometimes advantageous to record the magnetic signal (the magnetoencephalogram, or MEG) rather than—or in addition to—the electrical signal (the EEG). However, MEG recordings require very expensive equipment and are much less common than EEG recordings.

Example 3: Impaired Cognition in Schizophrenia

This section will provide a more detailed discussion of a specific experiment, in which ERPs were used to study impaired cognition in schizophrenia (Luck et al., 2009). This will serve both to show how ERPs can be used to isolate specific cognitive processes and to provide a concrete example of the steps involved in conducting an ERP experiment.

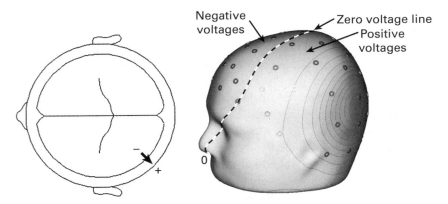

Figure 1.3
Distribution of voltage over the scalp (right) resulting from a single dipole in the brain (left). The dipole is shown in an axial section through a schematic brain, and the positive and negative ends of the dipole are indicated by plus (+) and minus (−) signs, respectively. The scalp distribution shows a strong area of positive voltage right over the positive end of the dipole. This positive voltage gradually declines until it reaches a line of zero voltage, and then weak negative voltages are present on the other side of the head. Images courtesy of J. Bengson.

The goal of this experiment was to ask why behavioral reaction times (RTs) are typically slowed in schizophrenia patients when they perform simple sensorimotor tasks. That is, are RTs slowed in patients because of an impairment in perceptual processes, an impairment in decision processes, or an impairment in response processes? ERPs are ideally suited for answering this question because they provide a direct means of measuring the timing of the processes that occur between a stimulus and a response. On the basis of prior research, we hypothesized that the slowing of RTs in schizophrenia in simple tasks does not result from slowed perception or decision, but instead results from an impairment in the process of determining which response is appropriate once the stimulus has been perceived and categorized (which is called the *response selection* process).

To test this hypothesis, we recorded ERPs from 20 individuals with schizophrenia and 20 healthy control subjects in a modified oddball task. In each 5-min block of trials, we presented a sequence of letters and digits at fixation. Each stimulus was presented for a duration of 200 ms, with a stimulus appearing every 1300–1500 ms (the reason for this particular timing is described near the end of chapter 4). Subjects made a button-press response for each stimulus, pressing with one hand for letters and with the other hand for digits. One of these two categories was rare (20%) and the other was frequent (80%) in any given trial block. Both the category probabilities and the assignment of hands to categories were counterbalanced across trial blocks.

This design allowed us to isolate specific ERP components by means of *difference waves*, in which the ERP waveform elicited by one trial type was subtracted from the ERP waveform elicited by another trial type (much like difference images in fMRI studies). Difference waves are extremely useful in ERP research because they isolate neural processes that are differentially active for two trial types, eliminating the many concurrently active brain processes that do not differ between these trial types. This is important because the different ERP components are ordinarily mixed together, making it difficult to determine exactly which component—and which psychological or neural process—differs across conditions or groups. Difference waves can pull out a subset of the components, making it possible to draw more specific conclusions.

In the current study, rare-minus-frequent difference waves were constructed to isolate the P3 wave, which tells us about the time course of stimulus categorization (i.e., the process of determining whether the stimulus falls into the rare or frequent category). A separate set of difference waves was constructed to isolate the *lateralized readiness potential* (LRP), which reflects the time course of response selection after stimulus categorization (e.g., determining whether the left button or right button is the appropriate response for the current stimulus). The LRP is isolated by subtracting the voltages over the ipsilateral hemisphere (relative to the responding hand) from the voltages over the contralateral hemisphere. We found that RTs were slowed by approximately 60 ms in patients compared to control subjects, and the question was whether this reflected a slowing of perception and categorization (which would produce a delay in the P3 difference wave) or whether it reflected a slowing of postcategorization response selection processes (which would produce a delay in the LRP difference wave). Chapter 3 provides extended discussions of these components and how they can be used to isolate these different processes.

Figure 1.4 shows the ERPs elicited by the rare category, the ERPs elicited by the frequent category, and the rare-minus-frequent difference waves. These are *grand average* waveforms,

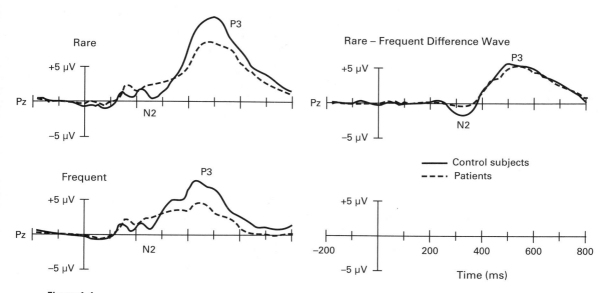

Figure 1.4
Grand average ERP waveforms recorded from schizophrenia patients and healthy control subjects at the Pz electrode site (Luck et al., 2009). ERPs are shown for the rare stimulus category, for the frequent stimulus category, and for the difference between the rare and frequent stimuli.

meaning that average waveforms were first computed across trials for each subject at each electrode site, and then the waveforms at each electrode were averaged across subjects. These grand averages simply make it easier to look at the data (just like graphs of the mean RT across subjects in behavioral experiments).

As in many previous studies, the voltage during the period of the P3 wave (approximately 300–800 ms) was reduced in the schizophrenia group relative to the control group. However, the voltage during this period is the sum of many different components, not just the P3 wave. The rare-minus-frequent difference wave allows us to better isolate the P3 wave and to focus on brain activity that reflects the classification of the stimulus as belonging to the rare or frequent category. Notably, patients exhibited no reduction in the amplitude of the P3 wave in the difference waves (although the preceding N2 was diminished—for similar results, see Potts, O'Donnell, Hirayasu, & McCarley, 2002). The most important finding was that the timing of the P3 was virtually identical in patients and controls, which indicates that patients were able to perceive and categorize these simple stimuli just as fast as controls, even though patient RTs were delayed by 60 ms.

This implies that the slowing of RT reflects an impairment in processes that follow stimulus categorization. Indeed, the LRP—an index of response preparation—was delayed by 75 ms in onset time and diminished by 50% in amplitude for patients compared to controls. Moreover, the degree of amplitude reduction across patients was significantly correlated with the degree of RT slowing. Thus, for a relatively simple perceptual task, the slowed RTs exhibited by the schizophrenia patients appear to result primarily from a slowing of response selection (as evidenced by the later and smaller LRP) rather than a slowing of perception or categorization (as evidenced by no slowing or reduction of the P3).

This example makes two key points. First, it shows how difference waves can be used to isolate specific ERP components that reflect specific processes. Second, it shows how ERPs can be used to precisely assess the timing of specific processes that occur between a stimulus and a response.

Oscillations and Filtering

EEG Oscillations

The brain is constantly active, whether you are awake or asleep and whether or not any distinct stimuli are present. All of this brain activity leads to constant variations in the pattern of PSPs across the billions of neurons in your brain, and this leads to a constantly varying EEG on the scalp. These many different types of brain activity get combined together at the individual scalp electrodes, creating a complicated mixture. One portion of this mixture consists of brief, transient brain responses to internal and external events (i.e., ERPs). Another portion consists of ongoing activity that is not driven by discrete events. Much of this non-event-driven activity is oscillatory in nature, reflecting feedback loops in the brain.

The most prominent such oscillation is the alpha wave, a voltage that goes up and down approximately 10 times per second. This is illustrated in figure 1.5, which shows a 700-ms

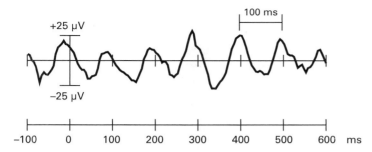

Figure 1.5
Single-trial EEG from an occipital electrode site with large alpha activity. Note that each peak of the alpha oscillation is separated by approximately 100 ms, which tells you that it is occurring at 10 Hz and is therefore an alpha oscillation.

segment of EEG recorded at an occipital scalp site (from 100 ms prior to a stimulus until 600 ms after the stimulus). You can see that the EEG is going up and down repetitively, and you can figure out that the frequency is approximately 10 cycles per second by noting that each cycle lasts approximately 100 ms. Alpha oscillations are usually most prominent over the back of the head and tend to be large when the subject is drowsy or when the subject's eyes are closed. These alpha waves can be either a large signal or a large source of noise, depending on whether you are interested in the processes reflected by the alpha or in some small, transient, stimulus-elicited ERP component that is present at the same scalp sites and is obscured by the alpha waves (see the glossary if you are not sure what I mean by *noise* here).

If you present stimuli at irregular intervals (e.g., every 900–1100 ms), the stimulus will occur at a different point in the alpha cycle (a different *phase*) on each trial, and the alpha oscillations will ordinarily average to nearly zero if you average together a large number of trials (because the voltage at a given poststimulus time point will be positive on some trials and negative on others). However, a stimulus may reset the alpha phase so that the phase after stimulus onset is similar across trials. In this case, considerable alpha may remain in the poststimulus alpha. Some researchers have proposed that ERP components mainly consist of this kind of *phase resetting* of ongoing EEG oscillations (e.g., Makeig et al., 2002). It turns out to be quite difficult to rigorously test this possibility (see review by Bastiaansen, Mazaheri, & Jensen, 2012), but my guess is that only a small proportion of stimulus-locked ERP activity consists of these kinds of oscillations.

A stimulus may also lead to the initiation of a new oscillation, but with a phase that varies from trial to trial. These oscillations will ordinarily cancel out when you create an averaged ERP waveform (because the voltage at a given poststimulus time point will be positive on some trials and negative on others). However, it is possible to perform a *time–frequency* analysis, which extracts the amplitude at a given frequency independent of its phase prior to averaging. This makes it possible to see the time course of stimulus-elicited oscillations (see chapter 8 and the online chapter 12 for details).

EEG oscillations are mainly classified according to frequency bands. In addition to the alpha band (8–13 Hz), there are also delta (<4 Hz), theta (4–8 Hz), beta (13–30 Hz), and gamma (>30 Hz) bands. It is tempting to think that a given frequency band reflects a specific process, but that is not generally true. For example, 8- to 13-Hz oscillations over motor cortex (often called *mu* oscillations) are clearly different from the 8- to 13-Hz alpha oscillations observed over visual cortex.

Fourier Analysis

The EEG typically contains a mixture of multiple simultaneous oscillations at different frequencies. To show you what this mixture looks like in a very simple situation, figure 1.6 shows three individual sine waves and their sum. Although these sine waves are mixed together in the recording, it is possible to determine the amplitudes and frequencies of the individual sine waves. This is achieved by means of *Fourier analysis*, a mathematical process that can compute the amplitudes, frequencies, and phases of the sine waves that sum together to equal the observed waveform (if you need a reminder about what these terms mean, see the *sine wave* entry in the glossary).

The amazing thing about Fourier analysis is that any waveform, no matter how complex, can be reconstructed by summing together a set of sine waves. For example, figure 1.7 shows how

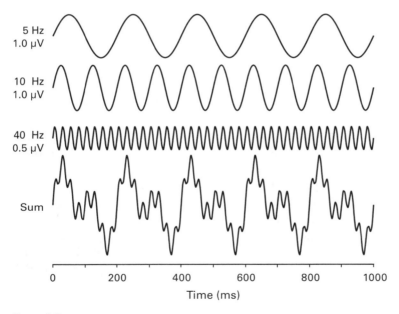

Figure 1.6
Example of the summation of oscillations at different frequencies. Three different sine waves are shown here, along with their sum. The EEG often looks like the sum of several sine waves. The goal of the Fourier transform is to determine the amplitudes, phases, and frequencies of the sine waves that sum together to form a complex waveform.

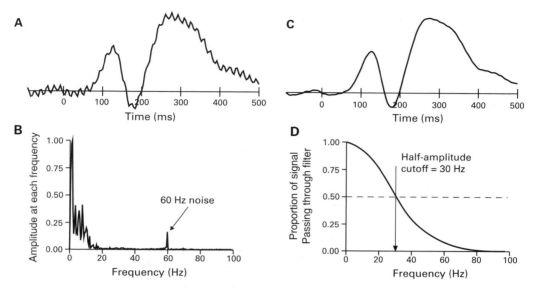

Figure 1.7
(A) ERP waveform containing substantial noise at 60 Hz (which you can determine by counting six peaks in every 100-ms period). (B) Fourier transform of the waveform in panel A, showing the amplitude at each frequency (note that the phase of each frequency is not shown here). (C) Filtered version of the waveform in panel A. (D) Frequency response function of the filter that was used to create the waveform in panel C from the waveform in panel A.

Fourier analysis can be applied to an averaged ERP waveform. Figure 1.7A shows an ERP waveform that contains a lot of "noise" at 60 Hz (artifactual electrical activity picked up from the recording environment that looks like a ripple superimposed on the ERP waveform). Figure 1.7B shows the Fourier transform of this waveform. The X axis of the transformed data is frequency instead of time, and the graph indicates the amplitude at each frequency. Note that most of the amplitude is at frequencies of less than 20 Hz, but there is a fairly substantial amplitude at 60 Hz; this represents the 60-Hz noise oscillation that you can see in the ERP waveform. We could reconstruct the original ERP waveform by taking a sine wave of each frequency and each amplitude shown in figure 1.7B and summing them together (we would also need to know the phase of each frequency, which is not shown here).

Fourier analysis has a fundamental limitation that is not always realized by people who use it: The presence of an amplitude at a given frequency in a Fourier transform does not mean that the original waveform actually *contained* an oscillation at that frequency. It just means that if we wanted to re-create the original waveform by summing together a set of sine waves, we would need to include an oscillating sine wave at that frequency. In some cases, the original waveform really does contain sine waves, such as the 60-Hz noise oscillation shown in the ERP waveform in figure 1.7A. However, the Fourier transform shown in figure 1.7B also shows a lot of activity at 13 Hz, and there is no reason to believe that the brain was actually oscillating at

13 Hz when it generated the ERP waveform shown in figure 1.7A. The activity at 13 Hz in figure 1.7B just means that we would need to use a 13-Hz sine wave of a particular amplitude if we wanted to reconstruct the ERP waveform by adding sine waves together. Chapter 7 will discuss this issue in much greater detail.

Filtering

Filtering is an essential concept in ERP research, and it will arise again and again in the upcoming chapters. Chapter 7 describes filters in detail, but I want to make sure you understand the basics now so that you can understand everything in chapters 2–6. Fortunately, now that you know the basics of the Fourier transform, it's easy for me to explain filtering.

In EEG and ERP research, filters are used to suppress noise that contaminates the data, making it difficult to see the signal of interest. For example, the 60-Hz noise in figure 1.7A would hamper our ability to accurately measure the amplitude and latency of the different components in the waveform. Figure 1.7C shows what the waveform looks like after the high frequencies have been filtered out. The filtered version looks much nicer, doesn't it?

The term *filter* has a general meaning outside of signal processing (e.g., you can have a coffee filter, an air filter, an oil filter, etc.). Filters can be described in terms of what they block and what they do not block. An air filter, for example, may trap particles that are larger than 0.01 mm and allow air and smaller particles to pass through. The filters that are typically applied to EEG and ERP data are usually described in terms of the frequencies that pass through the filter (i.e., the frequencies that are not blocked by the filter and therefore appear in the filter's output). The filter used in figure 1.7 is a *low-pass* filter, which means that it passes low frequencies and attenuates (blocks) high frequencies. It is also possible to use a *high-pass* filter, which attenuates low frequencies and lets higher frequencies pass through. If you apply both a low-pass filter and a high-pass filter at the same time, you will have a *band-pass* filter (i.e., a filter that blocks low and high frequencies, allowing the intermediate frequencies to pass). It's also possible to have a *notch* filter, which filters out one narrow frequency band and passes everything else.

Personally, I find it confusing to describe filters in terms of the frequencies that they pass rather than the frequencies that they block. For example, it's confusing to use the term *low pass* for a filter that blocks high frequencies. However, that's the standard terminology, and we're stuck with it.

For most filters, there is a range of frequencies that "passes through" the filter almost completely (and is therefore present in the filter's output), a range of frequencies that is attenuated almost completely, and a range of frequencies in the middle that is partially attenuated. This is quantified by a filter's *frequency response function*. Figure 1.7D shows the frequency response function for the filter that was used to create the waveform in figure 1.7C. At each frequency, this function tells you the proportion of the signal that will pass through the filter (which is the complement of the proportion that is attenuated by the filter). For example, the frequency response function shown in figure 1.7C has a value of 0.80 at 20 Hz, which means that 80% of the 20-Hz activity will pass through the filter and the remaining 20% will be blocked by the filter. A filter's frequency response function is often summarized by a single number called the *cutoff frequency*. This number typically represents the frequency at which 50% of the signal passes through the filter and 50% is suppressed, and it is therefore called the *half-amplitude cutoff* (see chapter 7 for some important details about cutoff frequencies). The filter shown here has a half-amplitude cutoff at 30 Hz, and it passes less than 10% of the signal at 60 Hz (which is why the 60-Hz noise has been almost completely removed in the filtered waveform). You might find it strange that there is a broad range of frequencies that are partially passed and partially blocked by this filter; the reasons for this will be discussed in chapter 7.

Overview of Basic Steps in an ERP Experiment

This section will provide an overview of the basic steps involved in conducting an ERP experiment, beginning with recording the EEG and finishing with statistical analyses. Each of these topics is covered in more detail in the subsequent chapters, and the goal of this section is to provide a big-picture overview.

Recording the Electroencephalogram (Chapter 5)

Figure 1.1A shows the basic setup of an ERP experiment. The EEG is recorded from electrodes on the scalp, with a conductive gel or liquid between each electrode and the skin to make a stable electrical connection. The electrical potential (voltage) can then be recorded from each electrode, resulting in a separate waveform for each electrode site. This waveform will be a mixture of actual brain activity, biological electrical potentials produced outside of the brain (by the skin, the eyes, the muscles, etc.), and induced electrical activity from external electrical devices that is picked up by the head, the electrodes, or the electrode wires. If precautions are taken to minimize the non-neural potentials, the voltages produced by the brain (the EEG) will be relatively large compared to the non-neural voltages.

The EEG is quite small (usually under 100 microvolts, μV), so the signal from each electrode is usually amplified by a factor of 1,000–100,000. This amplification factor is called the *gain* of the amplifier. A gain of 20,000 was used in the experiment shown in figure 1.1, and a gain of 5000 was used in the experiment shown in figure 1.4. The continuous voltage signal is then turned into a series of discrete digital values for storage on a computer. In most experiments, the voltage is sampled from each channel at a rate of between 200 and 1000 evenly spaced samples per second (i.e., 200–1000 Hz). In the experiment shown in figure 1.1, the EEG was sampled at 250 Hz (one sample every 4 ms).

The EEG is typically recorded from multiple electrodes distributed across the scalp. Different studies use very different numbers of electrodes. For some studies, all of the relevant information can be obtained from five to six electrodes; for others, as many as 256 electrodes are needed. You might think that it's best to record from as many channels as possible, but it becomes more difficult to ensure the quality of the data when you record from a lot of channels (see the online supplement to chapter 5).

Artifact Rejection and Correction (Chapter 6)

There are several common artifacts that are picked up by EEG recordings and require special treatment. The most common of these arise from the eyes. When the eyes blink, a large voltage deflection is observed over much of the head, and this artifact is usually much larger than the ERP signals. Moreover, eyeblinks are sometimes systematically triggered by tasks and may vary across groups or conditions, yielding a systematic distortion of the data. Large potentials are also produced by eye movements, and these potentials can confound experiments that use lateralized stimuli or focus on lateralized ERP responses. Thus, trials containing blinks, eye movements, or other artifacts are typically excluded from the averaged ERP waveforms. In the study shown in figure 1.4, for example, three patients and two controls were excluded from the final analysis because more than 50% of trials were rejected (mainly due to blinks). In the remaining subjects, 23% of trials were rejected on average.

This approach has two shortcomings. First, a fairly large number of trials may need to be rejected, thus reducing the number of trials contributing to the average ERP waveforms. Second, the mental effort involved in suppressing eyeblinks may impair task performance (Ochoa & Polich, 2000). These problems are especially acute in individuals with neurological or psychiatric disorders, who may blink on almost every trial or may perform the task poorly because of the effort devoted to blink suppression. Fortunately, methods have been developed to estimate the artifactual activity and subtract it out, leaving artifact-free EEG data that can be included in the averaged ERP waveforms. Some of these artifact correction techniques are known to make systematic errors in estimating and removing the artifactual activity, but many of these techniques work quite well for blinks and certain other artifacts.

Filtering (Chapter 7)

Filters are usually used to remove very slow voltage changes (<0.01–0.1 Hz) and very fast voltage changes (>15–100 Hz) because scalp-recorded voltages in these frequency ranges are likely to be noise from non-neural sources. Frequencies below 0.1 Hz and above 18.5 Hz were filtered from the waveforms shown in figure 1.4. Filters can dramatically distort the time course of an ERP waveform and can induce artifactual oscillations when the low cutoff is greater than approximately 0.5 Hz or when the high cutoff is less than approximately 10 Hz, so caution is necessary when extreme filters are used. Filters can be applied to the EEG, to the averaged ERPs, or both. The appendix of this book describes the effects of changing the order in which operations such as filtering and averaging are applied to the data.

Computing Average ERP Waveforms (Chapter 8)

ERPs are typically small in comparison with the rest of the EEG activity, and ERPs are usually isolated from the ongoing EEG by a simple averaging procedure. To make this possible, it is necessary to include *event codes* in the EEG recordings that mark the events that happened at specific times, such as the onset of each stimulus (figure 1.1A). These event codes are then used as a time-locking point to extract segments of the EEG surrounding each event (figure 1.1E).

Recall that figure 1.1 shows the EEG recorded over a 9-s period in an oddball task with frequent X stimuli (80%) and infrequent O stimuli (20%). Each rectangle highlights a 900-ms segment of EEG that begins 100 ms before an event code and extends until 800 ms after the event code. The 100-ms period before the event code is used to provide a prestimulus baseline period.

Figure 1.1E shows these same segments of EEG, lined up in time. Stimulus onset is time zero. There is quite a bit of variability in the EEG waveforms from trial to trial, and this variability largely reflects the fact that the EEG is the sum of many different sources of electrical activity in the brain, many of which are not involved in processing the stimulus. To extract the activity that is related to stimulus processing from the unrelated EEG, the EEG segments following each X are averaged together into one waveform, and the EEG segments following each O are averaged together into a different waveform (figure 1.1F). Any brain activity that is not time-locked to the stimulus will be positive at a given latency on some trials and negative at that latency on other trials, and if many trials are averaged together, these voltages will cancel each other out and approach zero. However, any brain activity that is consistently elicited by the stimulus—with approximately the same voltage at a given latency from trial to trial—will remain in the average. Thus, by averaging together many trials of the same type, the brain activity that is consistently time-locked to the stimulus across trials can be extracted from other sources of voltage (including EEG activity that is unrelated to the stimulus and non-neural sources of electrical noise). Other types of events can be used as the time-locking point in the averaging process (e.g., button-press responses, vocalizations, saccadic eye movements, electromyographic activity).

You are probably wondering how many trials must be averaged together for each averaged ERP waveform. This depends on several factors, including the size of the ERP effect being examined, the amplitude of the unrelated EEG activity, and the amplitude of non-neural activity. For large components, such as the P3 wave, very clear results can usually be obtained by averaging together 10–50 trials. For smaller components, such as the P1 wave, it is usually necessary to average together 100–500 trials for each trial type to see reliable differences between groups or conditions. Of course, the number of trials that is required to observe a significant difference will also depend on the number of subjects, the variance across subjects, and the size of the effect. In the experiment shown in figure 1.4, each subject received 256 oddball stimuli and 1024 standard stimuli. This is more trials than would be typical for a P3 study, but it was appropriate given that we were also looking at the much smaller LRP and that we anticipated rejecting a large percentage of trials due to eyeblinks.

Quantification of Amplitudes and Latencies (Chapter 9)

The most common way to quantify the magnitude and timing of a given ERP component is to measure the amplitude and latency of the peak voltage within some time window. For example, to measure the peak of the P3 wave in the data shown in figure 1.4, you might define a measurement window (e.g., 400–700 ms) and find the most positive point in that window. Peak amplitude would be defined as the voltage at this point, and peak latency would be defined as the time of this point. Of course, it is also possible to search for negative peaks, such as the N1 wave.

Finding peaks was the simplest approach to measuring ERPs prior to the advent of inexpensive computers, when a ruler was the only available means of quantifying the waveform (Donchin & Heffley, 1978). This approach is still widely used, but it has several drawbacks, and better methods for quantifying ERP amplitudes and latencies have been developed. For example, the magnitude of a component can be quantified by measuring the mean voltage over a given time window. As discussed in chapter 9, mean amplitude is usually superior to peak amplitude as a measure of a component's magnitude.

A related measure can be used to quantify component latency. Specifically, it is possible to define the midpoint of a component as the point that divides the region under the waveform into two equal-area subregions. This is called the 50% area latency measure, and it was used to quantify the timing of the P3 wave in the data shown in figure 1.4.

Statistical Analysis (Chapter 10)

In most ERP experiments, an averaged ERP waveform is constructed at each electrode site for each subject in each condition. The amplitude or latency of a component of interest is then measured in each one of these waveforms, and these measured values are then entered into a statistical analysis just like any other variable. Thus, the statistical analysis of ERP data is often quite similar to the analysis of traditional behavioral measures.

However, ERP experiments provide extremely rich data sets, usually consisting of several gigabytes of data. This can lead to both the implicit and explicit use of many statistical comparisons in a single study, which can dramatically increase the probability of a Type I error (i.e., concluding that a difference is real when it was actually a result of random variation). The explicit use of multiple comparisons arises when, for example, separate statistical analyses are conducted for each of several different components. The implicit use of multiple comparisons occurs when researchers first look at the waveforms and then decide on the time windows and electrode sites to be used for quantifying component amplitudes and latencies. If a time window is chosen because the difference between conditions is greatest in that time window, then this biases the results in favor of statistical significance, even if the difference was caused by noise. A similar problem arises if the researcher finds the electrode sites with the largest differences between conditions and then uses only those sites for the statistical analyses. With enough electrode sites, it is almost always possible to find a statistically significant difference between two groups or two conditions at a few electrode sites due simply to random noise. When reading papers that describe ERP studies, you should be suspicious if unusual, idiosyncratic, and unjustified electrode sites or measurement windows are selected for the statistical analyses. Fortunately, new statistical methods have been developed that can minimize or eliminate this problem (see online chapter 13).

What Are ERPs Good For?

The ERP technique is the best available technique for answering many important scientific questions, but it is a terrible technique for answering others. To do high-impact ERP research, you need to understand the kinds of questions that ERPs can readily answer. The following paragraphs describe several ways in which ERPs have been successfully used in prior research (for a more extensive discussion, see Kappenman & Luck, 2012). There are certainly other useful ways to apply the ERP technique, but these will provide a good starting point.

Assessing the Time Course of Processing

The most commonly cited virtue of the ERP technique is its temporal resolution (see box 1.4). But this is not merely a matter of being able to reliably measure values of 358 ms versus 359 ms, which can easily be accomplished with reaction time measures, eye tracking measures, cardiac measures, and so forth. The key is that ERPs provide a *continuous* measure of processing, beginning prior to the stimulus and extending past the response. In a behavioral experiment, we get no data during the period between the stimulus and the response, but this is the period when most of the "action" is happening. ERPs give us a measure of the moment-by-moment activity during this period. That is, ERPs show us the "action." ERPs (and other EEG signals) also give us information about the state of the brain prior to the onset of the stimulus, which has an enormous impact on the way that the stimulus is processed (Worden, Foxe, Wang, & Simpson, 2000; Mathewson, Gratton, Fabiani, Beck, & Ro, 2009; Vanrullen, Busch, Drewes, & Dubois, 2011). ERPs also provide information about brain activity that occurs after a response has occurred or after a feedback stimulus has been presented, reflecting executive processes that determine how the brain will operate on subsequent trials (Holroyd & Coles, 2002; Gehring, Liu, Orr, & Carp, 2012).

Determining Which Process Is Influenced by an Experimental Manipulation

What can we do with this wonderful continuous temporal information? A common use is to determine which processes are influenced by a given experimental manipulation. As an example, consider the Stroop paradigm, where subjects must name the color of the ink in which a word is drawn. Subjects are slower when the word is incompatible with the ink color than when the ink color and word are the same (e.g., subjects are slower to say "green" when presented with the word "red" drawn in green ink than when presented with the word "green" drawn in green ink). Do these slowed responses reflect a slowing of perceptual processes or a slowing of response processes? It is difficult to answer this question simply by looking at the behavioral responses, but studies of the P3 wave have been very useful in addressing this issue. Specifically, it has been well documented that the latency of the P3 wave becomes longer when perceptual processes are delayed, but several studies have shown that P3 latency is not delayed on incompatible trials in the Stroop paradigm, indicating that the delays in RT reflect delays in some

Box 1.5
A Neuroimaging Technique?

> Many people include ERPs in the category of neuroimaging techniques, but this doesn't seem right to me. Unambiguous neuroimaging techniques such as fMRI provide an image of the brain, but ERPs do not directly give us an image of the brain. As discussed in online chapter 14, ERPs can be used to create *models* of the distribution of activity over the cortical surface, but the actual image of the brain typically comes from MRI data. It is, of course, possible to plot the distribution of voltage over the scalp, but this makes ERPs a *scalpoimaging* technique rather than a *neuroimaging* technique. ERP waveforms are also images, but they are not neuroimages in any particularly meaningful sense. To avoid overpromising and underdelivering, I prefer to leave the term *neuroimaging* to research that more directly provides an image of the brain.

postperceptual stage (see, e.g., Duncan-Johnson & Kopell, 1981). Thus, ERPs are very useful for determining which stage or stages of processing are influenced (or not influenced) by a given experimental manipulation. Several specific examples are described in the online supplement to chapter 4. I have used ERPs for this purpose in many of my own studies (see especially Vogel, Luck, & Shapiro, 1998).

I would like to stress that the information provided by ERPs is different from, and complementary to, the information provided by neuroimaging techniques (see box 1.5 for a discussion of whether ERPs are a neuroimaging technique). Neuroimaging techniques can isolate different processes to the extent that the different processes are anatomically distinct. It has become very clear, however, that each area of cortex is involved in a great many processes. Thus, finding an effect of an experimental manipulation in primary visual cortex does not guarantee that this effect reflects a modulation of sensory processing; it could instead reflect a working memory representation that was generated 200 ms after stimulus onset and stored in primary visual cortex (Harrison & Tong, 2009; Serences, Ester, Vogel, & Awh, 2009). In this situation, the timing of the effect could tell us whether the effect happened during the initial sensory processing period or at a later point in time.

Identifying Multiple Neurocognitive Processes

In behavioral experiments, it is often parsimonious to invoke a single underlying process to explain changes in behavior that are produced by many different manipulations. However, ERP recordings provide a much richer data set, often making it clear that a given experimental manipulation actually influences several different processes (i.e., several different ERP components) and that a given pattern of behavior might be caused by different mechanisms in different experiments. For example, behavioral studies often treat selective attention as a single mechanism, but different manipulations of attention influence different ERP components (Luck & Hillyard, 2000; Luck & Vecera, 2002). Similarly, different ERP components appear to reflect different mechanisms of memory retrieval (Wilding & Ranganath, 2012).

Covert Measurement of Processing

An important advantage of ERPs over behavioral measures is that ERPs can be used to provide an online measure of processing when a behavioral response is impossible or problematic. This is called the *covert measurement of processing*. In some cases, covert measurement is necessary because the subject is incapable of making a response. For example, ERPs can be recorded from infants who are too young to be instructed to make a response (see review by Coch & Gullick, 2012). ERPs are also used for covert monitoring in people with neurological disorders who are unable to make behavioral responses (Fischer, Luaute, Adeleine, & Morlet, 2004).

Covert monitoring is also useful when normal processing would be distorted by using a task that requires a behavioral response. In attention research, for example, it can be difficult to design a task in which behavioral responses can be obtained for both attended and unattended stimuli——a stimulus isn't really "unattended" if the subject is instructed to respond to it. In contrast, ERPs can easily be used to compare the processing of attended and unattended stimuli without requiring a response to the unattended stimuli. Consequently, ERPs have been used extensively in attention research (see review by Luck & Kappenman, 2012b). In studies of language comprehension, ERPs can be used to assess the processing of a word embedded in the middle of a sentence—at the time the word is presented—rather than relying on a response made at the end of the sentence (see review by Swaab, Ledoux, Camblin, & Boudewyn, 2012).

The ability to measure processing covertly and continuously over the entire course of a task with millisecond-level temporal resolution makes the ERP technique the best available technique for answering many important questions about the human mind. Much of perception, cognition, and emotion unfolds on a timescale of tens or hundreds of milliseconds, and ERPs are particularly valuable for tracking such rapid sequences of mental operations.

A Link to the Brain?

In most cases, ERPs are more valuable for answering questions about the mind than for answering questions about the brain (to the extent that these can really be dissociated). That is, although ERPs are a measure of brain activity, they are usually too coarse to permit specific and definitive conclusions to be drawn about brain circuitry. As an analogy, imagine that you were trying to understand how your computer works by measuring the temperature from sensors placed at a variety of locations on the computer's case. You could learn quite a bit about the general principles by which the computer operated, and you would be able to draw conclusions about some of the major components of the computer's hardware (e.g., the power supply and the hard drive). However, you would never unravel the circuitry of the computer's central processing unit, and you would never decode the computer's program. Similarly, ERPs can occasionally be used to draw strong conclusions about some coarsely defined components of the brain, and they can be used to draw weak conclusions about others. But as discussed in the previous paragraphs, the main advantage of the ERP component is its ability to track the time course of processing, not to measure the operation of specific neural systems.

Box 1.6
ERPs, Desperation, and the Blues

Over the years, I have encountered many cases of people who have tried to use ERPs to answer questions that just can't be answered with this technique. These people desperately want ERPs to be able to answer questions that are better answered with fMRI or single-unit recordings, and this desperation leads them to cast aside their usual critical abilities.

As an analogy, consider this story about the American blues musician Sonny Boy Williamson. In the early 1960s, many young musicians in England were fascinated with American blues music, and they desperately wanted to be able to play it. Sonny Boy Williamson went on a tour of England during this time, and he spent some time jamming with these English musicians, but he was not impressed. According to legend, when he returned to the United States he remarked, "Those English boys want to play the blues so bad—and they DO play it so bad." Whenever I see ERP researchers who try to answer questions about the brain that go beyond the limits of the technique, I always think, "Those ERPers want to study the brain so bad—and they DO study it so bad."

It should be noted that many of these English musicians who played the blues "so bad" became famous rock musicians (e.g., Eric Clapton, Jimmy Page, Jeff Beck). That is, they turned their weakness into a strength by playing a related but distinctly different style of music. By analogy, ERPers should stop trying to be neuroimaging researchers and do their own style of science.

This does not mean that ERPs can never be used to answer questions about the brain. In some cases, the temporal information provided by ERPs can provide at least a coarse answer to such questions. In other cases, we have multiple converging sources of evidence about the neural generator of a given ERP component and can therefore use this component to assess activity in a specific region of cortex (see, e.g., the discussion of the C1 component in chapter 3). Answering questions about the brain with ERPs is therefore possible, but it takes a lot of hard work, cleverness, and careful thought (see box 1.6 for a lighthearted analogy).

People often think it should be possible to combine ERPs with fMRI and thereby obtain both high temporal and high spatial resolution. Although this has sometimes been done, it is much more difficult than most people imagine. The fundamental difficulty is that ERPs and the BOLD signal reflect different aspects of brain activity, and it is quite likely that an experimental manipulation would impact one of these measures without impacting the other. It is even possible to imagine scenarios in which the ERP and fMRI effects would go in opposite directions (Luck, 1999). Consequently, although it may someday be possible, it is not currently possible to directly combine ERP and fMRI data without unjustifiable assumptions.

Biomarkers
ERPs have the potential to be used as biomarkers in medical applications. That is, ERPs can be used to measure aspects of brain function that are impaired in neurological and psychiatric diseases, providing more specific information about an individual patient's brain function than could be obtained from traditional clinical measures (for a detailed discussion, see Luck et al.,

2011). This information could be used to determine whether a new treatment has an impact on the specific brain system that is being targeted. This information could also be used in the clinic to determine which medications are most likely to be effective for a given individual. For example, there is some evidence that the mismatch negativity (MMN) component is a relatively specific measure of PSPs produced by the binding of glutamate to N-methyl-D-aspartate (NMDA) receptors (Javitt, Steinschneider, Schroeder, & Arezzo, 1996; Kreitschmann-Andermahr et al., 2001; Ehrlichman, Maxwell, Majumdar, & Siegel, 2008; Heekeren et al., 2008). The MMN could therefore be used as a biomarker to test whether a new treatment influences NMDA responsiveness or whether a particular patient would benefit from such a treatment.

ERPs have several desirable properties for use as biomarkers: (a) they are directly related to neurotransmission; (b) they are relatively inexpensive and can be recorded relatively easily in clinical settings; (c) they can easily be recorded in animal models (Woodman, 2012); (d) in some cases, they have been shown to be reliable and sensitive measures of individual differences (Mathalon, Ford, & Pfefferbaum, 2000); and (e) they are practical for large ($N > 500$) multisite studies (Hesselbrock, Begleiter, Porjesz, O'Connor, & Bauer, 2001). However, there are also several hurdles that must be overcome for ERPs to be widely used as biomarkers. For example, it is not trivial to develop experimental paradigms that isolate a specific ERP component while also having good measurement reliability. In addition, differences between individuals can reflect "nuisance factors" such as differences in skull thickness and cortical folding patterns, which may make it difficult to use ERPs in clinical settings. Moreover, we do not yet have widely accepted quality assurance metrics that make it possible to demonstrate that valid, low-noise data have been obtained for a given individual. However, these problems are presumably solvable, so ERPs have considerable promise for use as biomarkers in the near future.

What Are ERPs Bad For?

In addition to understanding situations in which ERPs are particularly useful, it is worth considering the shortcomings of the ERP technique and the kinds of questions that cannot be easily answered with ERPs. I have tried to make the limitations as well as the strengths of the ERP technique clear throughout this book, because you need to know the limitations in order to do top-quality research (box 1.7).

The most challenging aspect of ERP research is that the waveforms recorded on the scalp represent the sum of many underlying components, and it is difficult to decompose this mixture into the individual underlying components. This is called the *superposition problem*, because multiple components are superimposed onto the same waveform (see chapter 2 for details). Similarly, it is difficult to determine the neural generator locations of the underlying components. These two problems are the most common impediments to the successful application of the ERP technique. There are many solutions to these two problems, but different solutions are needed in different types of experiments, so it is difficult to provide a simple one-sentence description of when these problems will arise and when they will be solvable. Chapters 2 and 4 and online

Box 1.7
Ugly Little Secrets

I teach two to four ERP Boot Camps each year, including a 10-day boot camp at the University of California, Davis, each summer and a few mini boot camps at universities, industry sites, and conferences. I like to tell boot camp participants that I am going to tell them all of the ugly little secrets involved in ERP research, because they need to know the plain truth if they are going to do ERP research themselves. I say this in a conspiratorial voice, suggesting that they should keep the ugly little secrets to themselves. We don't want fMRI researchers to know our secrets! But we should speak the truth freely among ourselves, so this book presents an unvarnished view of ERPs.

I also tell ERP Boot Camp participants that they will sometimes get depressed when they hear about the limitations of the ERP technique and the problems with some of the analytical approaches that they'd like to use. But there are many strategies that can be used to overcome or sidestep almost every limitation. The key is to fully understand the underlying nature of ERPs and the analytical techniques that are used in ERP research, such as filtering, source localization, and time–frequency analysis. So, if you find yourself getting a little depressed, just keep reading and you will eventually learn how to avoid the limitations of ERPs and run amazing experiments that will bring you fame and fortune.

chapter 14 will describe these problems and the various solutions in more detail. The best solution is often to figure out a clever experimental design in which isolating and localizing a given ERP component is not necessary to distinguish between competing hypotheses (see the discussion of *component-independent experimental designs* in chapter 4).

Another key limitation of the ERP technique is that a given mental or neural process may have no ERP *signature* (i.e., no clear contribution to the scalp-recorded voltage). As will be discussed in chapter 2, scalp ERPs are recordable only when a particular set of biophysical conditions are met, and only a fraction of brain activity meets these conditions. Although there are dozens of distinct ERP components, there are surely hundreds or thousands of distinct brain processes that have no distinct ERP component.

Another limitation arises from the fact that ERPs are small relative to the noise level, and many trials are usually required to accurately measure a given ERP effect. Although some components are large enough to be reliably measured on single trials (mainly the P3 component), it is usually necessary to average between 10 and 500 trials per condition in each subject to achieve sufficient statistical power. This makes it difficult to conduct experiments with very long intervals between stimuli and experiments that require surprising the subjects. In principle, one could increase the number of subjects to make up for a small number of trials per subject, but the time required to prepare the subject usually makes it unrealistic to test more than 50 subjects in a given experiment (and sample sizes of 10–20 are typical). I have frequently started designing an ERP experiment and then given up when I realized that the experiment would require either 10 hours of data collection per subject or 300 subjects.

To use the ERP technique, it is also necessary to have measurable events that can be used as time-locking points. Some imprecision in the timing of the events can be tolerated in many cases (perhaps ±10 ms in a typical cognitive or affective experiment), but ERPs cannot usually be used if the presence or timing of the events is difficult to determine (e.g., when the onset of a stimulus is very gradual).

ERPs are also difficult to use for measuring brain activity that extends beyond a few seconds (e.g., long-term memory consolidation). The main reason for this is that large, slow voltage drifts are present on the scalp due to non-neural factors (e.g., skin potentials), and these drifts add more and more variance to the waveform as time passes after the time-locking point (see figure 8.2D in chapter 8). These slow drifts are ordinarily removed with filters, but this would also remove slow neural effects.

Clean ERPs are difficult to record when subjects make frequent head, mouth, or eye movements. Head movements often cause slight shifts in electrode position, which in turn create large voltage artifacts. Consequently, subjects remain seated in a chair in almost all ERP studies. Mouth movements also create artifacts, especially when the tongue (which contains a powerful dipole) makes contact with the top portion of the mouth. Studies involving speech typically examine the ERPs leading up to the onset of speech, excluding the time period in which the subjects are actually speaking. Like the mouth, the eyes contain a strong dipole, and eye movements lead to large voltage changes on the scalp. Almost all ERP studies therefore require subjects to maintain constant fixation.

The preceding paragraphs describe several of the most common conditions in which ERPs are problematic. This does not mean that ERPs can never be used in these situations; it just means that the challenges will be significant. If you are new to the ERP technique, it is better to avoid these situations. Once you have some experience, you may develop clever ways around these problems, leading to important new discoveries.

Comparison with Other Physiological Measures

Table 1.1 compares the ERP technique with several other physiological recording techniques along four major dimensions: invasiveness, spatial resolution, temporal resolution, and cost. The other classes of techniques that are considered are microelectrode measures (single-unit, multi-unit, and local field potential recordings) and hemodynamic measures (PET and fMRI). ERPs are grouped with event-related magnetic fields (ERMFs), which are the magnetic counterpart of ERPs and are extracted from the MEG (see chapter 2).

Invasiveness

Microelectrode measures (single-unit recordings, multi-unit recordings, and local field potentials) require insertion of an electrode into the brain and are therefore limited to non-human species or human neurosurgery patients. The obvious disadvantage of primate recordings is that

Broad Reviews of the ERP Technique

Coles, M. G. H. (1989). Modern mind-brain reading: Psychophysiology, physiology and cognition. *Psychophysiology, 26,* 251–269.

Coles, M. G. H., Smid, H., Scheffers, M. K., & Otten, L. J. (1995). Mental chronometry and the study of human information processing. In M. D. Rugg & M. G. H. Coles (Eds.), *Electrophysiology of Mind: Event-Related Brain Potentials and Cognition* (pp. 86–131). Oxford: Oxford University Press.

Gaillard, A. W. K. (1988). Problems and paradigms in ERP research. *Biological Psychology, 26,* 91–109.

Hillyard, S. A., & Picton, T. W. (1987). Electrophysiology of cognition. In F. Plum (Ed.), *Handbook of Physiology: Section 1. The Nervous System: Volume 5. Higher Functions of the Brain, Part 2* (pp. 519–584). Bethesda, MD: Waverly Press.

Kappenman, E. S., & Luck, S. J. (2012). ERP components: The ups and downs of brainwave recordings. In S. J. Luck & E. S. Kappenman (Eds.), *The Oxford Handbook of ERP Components* (pp. 3–30). New York: Oxford University Press.

Lindsley, D. B. (1969). Average evoked potentials—achievements, failures and prospects. In E. Donchin & D. B. Lindsley (Eds.), *Average Evoked Potentials: Methods, Results and Evaluations* (pp. 1–43). Washington, DC: U.S. Government Printing Office.

Luck, S. J. (2012). Event-related potentials. In H. Cooper, P. M. Camic, D. L. Long, A. T. Panter, D. Rindskopf, & K. J. Sher (Eds.), *APA Handbook of Research Methods in Psychology: Volume 1, Foundations, Planning, Measures, and Psychometrics* (pp. 523–546). Washington, DC: American Psychological Association.

Picton, T. W., & Stuss, D. T. (1980). The component structure of the human event-related potentials. In H. H. Kornhuber & L. Deecke (Eds.), *Motivation, Motor and Sensory Processes of the Brain, Progress in Brain Research* (pp. 17–49). North-Holland: Elsevier.

Sutton, S. (1969). The specification of psychological variables in average evoked potential experiments. In E. Donchin & D. B. Lindsley (Eds.), *Averaged Evoked Potentials: Methods, Results and Evaluations* (pp. 237–262). Washington, DC: U.S. Government Printing Office.

Vaughan, H. G., Jr. (1969). The relationship of brain activity to scalp recordings of event-related potentials. In E. Donchin & D. B. Lindsley (Eds.), *Average Evoked Potentials: Methods, Results and Evaluations* (pp. 45–75). Washington, DC: U.S. Government Printing Office.

Books on ERPs and Related Topics

Cohen, M. X. (2014). *Analyzing Neural Time Series Data: Theory and Practice.* Cambridge, MA: MIT Press.

Donchin, E., & Lindsley, D. B. (Eds.). (1969). *Average Evoked Potentials, Methods, Results, and Evaluations.* Washington, DC: U.S. Government Printing Office.

Handy, T. C. (Ed.). (2005). *Event-Related Potentials: A Methods Handbook.* Cambridge, MA: MIT Press.

Handy, T. C. (Ed.). (2009). *Brain Signal Analysis: Advances in Neuroelectric and Neuromagnetic Methods.* Cambridge, MA: MIT Press.

Luck, S. J., & Kappenman, E. S. (Eds.). (2012). *The Oxford Handbook of Event-Related Potential Components.* New York: Oxford University Press.

Nunez, P. L., & Srinivasan, R. (2006). *Electric Fields of the Brain, Second Edition.* New York: Oxford University Press.

Picton, T. W. (2011). *Human Auditory Evoked Potentials.* San Diego: Plural Publishing.

Regan, D. (1989). *Human Brain Electrophysiology: Evoked Potentials and Evoked Magnetic Fields in Science and Medicine.* New York: Elsevier.

Rugg, M. D., & Coles, M. G. H. (Eds.). (1995). *Electrophysiology of Mind.* New York: Oxford University Press.

2   A Closer Look at ERPs and ERP Components

Overview

This chapter provides a deeper analysis of the nature of ERPs, with the goal of helping you understand how ERPs are generated in the brain and how the intracranial signals combine together to create the waveforms we record on the surface of the scalp. The peaks that you see in scalp ERP waveforms typically reflect the sum of multiple internal, underlying components, and the conclusions of an experiment often require determining how these underlying components differed across groups or conditions. However, as I mentioned in chapter 1 (and will continue to emphasize throughout this book), it is very challenging to isolate and measure the internal underlying components on the basis of the data that you actually record from the scalp. This is called the *superposition problem*, and this chapter will explain how it arises and why it is so difficult to solve. This can be a little depressing, but don't despair! Chapter 4 describes some strategies you can use to solve this problem (or work around it), along with examples of previous experiments that have successfully used these strategies to provide definitive answers to important scientific questions. Chapter 3 provides a description of the specific ERP components that are most commonly observed in typical experiments.

It is easy to confuse the peaks that we observe in our scalp waveforms with the internal underlying brain components that sum together to create these peaks. In this chapter, I frequently use the phrase *underlying component* to make it clear that I am talking about internal brain activity, not the observed peaks on the scalp.

This chapter will begin with a discussion of basic electrical concepts, followed by a detailed discussion of the neural origins of ERPs and how they mix together in scalp recordings. We will then discuss why it is so difficult to localize the internal generators of scalp ERPs. This will be followed by a description of how the ERPs we record from scalp electrodes can radically misrepresent the underlying components. We will then discuss how difference waves can be used to solve this problem and isolate the underlying components. We will end by considering different ways of defining the term *ERP component*. You might think that it would be more logical to start with the definitions, but they will be much easier to grasp once you've spent some time thinking concretely about the nature of the signals we record from scalp electrodes. The chapter ends with a general discussion of methods that can be used to identify the underlying components that vary across groups or conditions in a given study.

This chapter is designed to provide you with a deep conceptual understanding of ERPs, and it is therefore somewhat abstract. You need this kind of understanding to evaluate previous ERP experiments and to design new ERP experiments. In fact, this chapter includes six specific "rules" for interpreting ERP data. However, if you are in the middle of collecting or analyzing data from an ERP experiment and you need to know the practical details of how to collect clean data and perform appropriate analyses, you might want to skip ahead to chapters 5–10. You can then come back to chapters 2–4 when you are ready to think more deeply about your results and to design new experiments. But don't forget to come back later, because chapter 2 is ultimately the most important chapter in the entire book.

Basic Electrical Concepts

If you are going to measure electrical signals in your research, you need to know a little bit about electricity. In this section, I will provide an overview of the most basic terms and concepts. If you took physics in high school or college, you have already learned these terms and concepts (although many people have told me that electricity was the part of physics they never really understood). If you didn't take physics, then don't worry, because I didn't either. When I was in high school, I took shop classes on electronics rather than physics, because I was more interested in being able to fix my guitar amplifier than I was in learning to predict the velocity of a ball rolling down a hill. Fortunately, the relevant concepts are pretty simple.

Current

Current is the actual flow of electricity (charged particles) through a conductor. It is a measure of the number of charge units (electrons or protons) that flow past a given point in a specific amount of time. Current is measured in amperes, where 1 ampere is equal to 1 coulomb (6.24×10^{18}) of charge units moving past a single point in 1 second.

By convention, physicists and engineers refer to current flowing in the direction of positively charged particles. For example, 1 coulomb of positively charged particles flowing from the left hemisphere to the right hemisphere is viewed as being electrically equivalent to 1 coulomb of negatively charged particles flowing from the right hemisphere to the left hemisphere. Thus, we would talk about the current flowing from the left hemisphere to the right hemisphere in both cases, regardless of whether it is positive charges moving from left to right or negative charges moving from right to left.

When discussing the flow of electricity, it is useful to use the flow of water as an analogy. Electrical current is analogous to the flow of water through a pipe or hose. Just as we can measure the amount of water that passes a given point in a pipe in a given period of time (e.g., 3.6 liters per minute), we can measure the amount of electricity that passes a given point in a conductor in a given period of time (e.g., 3.6 coulombs per second).

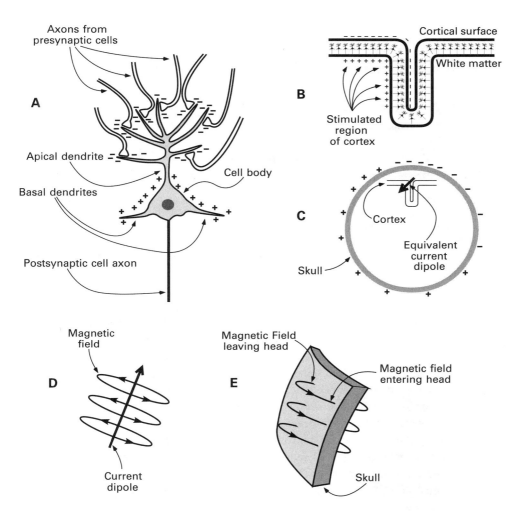

Figure 2.2
Principles of ERP generation. (A) Schematic pyramidal cell during neurotransmission. An excitatory neurotransmitter is released from the presynaptic terminals in the apical dendrite, causing positive ions to flow into this region of the postsynaptic neuron. This creates a net negative extracellular voltage (represented by the "–" symbols) just outside the apical dendrite. To complete the circuit, voltage will flow through the neuron and then exit in the region of the cell body and basal dendrites (represented by the "+" symbols). This flow of current forms a small dipole. The polarity of this dipole would be inverted if an inhibitory neurotransmitter were released rather than an excitatory neurotransmitter. It would also be inverted if the neurotransmission occurred at the cell body or basal dendrites rather than at the apical dendrite. (B) Folded sheet of cortex containing many pyramidal cells. When a region of this sheet is stimulated, the dipoles from the individual neurons summate. (C) The summated dipoles from the individual neurons can be approximated by a single equivalent current dipole, shown here as an arrow. By convention, the arrowhead indicates the positive end of the dipole. The position and orientation of this dipole determine the distribution of positive and negative voltages recorded at the surface of the head. (D) Example of a current dipole with a magnetic field traveling around it. (E) Example of the magnetic field generated by a dipole that lies just inside the surface of the skull. If the dipole is roughly parallel to the surface, the magnetic field can be recorded as it leaves and enters the head; no field can be recorded if the dipole is oriented radially (perpendicular to the surface). Reprinted with permission from Luck and Girelli (1998). Copyright 1998 MIT Press.

Box 2.1
What Does the Polarity of an ERP Component Mean?

I am often asked whether it "means something" if a component is positive or negative. My response is that polarity depends on a combination of four factors:

• Whether the postsynaptic potential is excitatory or inhibitory

• Whether the postsynaptic potential is occurring in the apical dendrite or the basal dendrites and cell body

• The location and orientation of the generator dipole with respect to the active recording electrode

• The location of the reference electrode (which will be discussed in chapter 5)

If you know three of these factors, then the polarity of the ERP can be used to infer the fourth factor. But we don't usually know three of these factors, so the polarity doesn't usually tell us anything. Polarity at a given site is meaningful only insofar as most components will have a constant polarity over a given region of the head, and the polarity of a component can help us determine which component we are seeing. But it cannot ordinarily be used to determine whether the component reflects excitation or inhibition.

the negativity from the next neuron, leading to cancellation. Similarly, if one neuron receives an excitatory neurotransmitter and another receives an inhibitory neurotransmitter, the dipoles of the neurons will be in opposite directions and will cancel. However, if the neurons all have a similar orientation and all receive the same type of input, their dipoles will summate and may be measurable at the scalp. This is much more likely to happen in cortical pyramidal cells than in other cell types or in other brain structures, so ERPs mainly arise from the pyramidal cells.

The summation of the individual dipoles is complicated by the fact that the cortex is not flat, but instead has many folds. Fortunately, however, physicists have demonstrated that the summation of many nearby dipoles is essentially equivalent to a single dipole formed by averaging the orientations of the individual dipoles.[3] This averaged dipole is called an *equivalent current dipole*. It is important to note, however, that whenever the individual dipoles are more than 90° from each other, they will cancel each other to some extent, with complete cancellation at 180°. For example, the orientation of the neurons in the basal ganglia is largely random, making it difficult or impossible to record basal ganglia activity from the scalp.

An important consequence of these facts about ERP generation is that only a fraction of brain processes will produce a scalp ERP "signature." To produce a measurable signal on the scalp, the following conditions must be met:

• Large numbers of neurons must be activated at the same time.

• The individual neurons must have approximately the same orientation.

• The postsynaptic potentials for the majority of the neurons must arise from the same part of the neurons (either the apical dendrite or the cell body and basal dendrites).

• The majority of the neurons must have the same direction of current flow to avoid cancellation.

Volume Conduction

When a dipole is present in a conductive medium such as the brain, current is conducted through that medium until it reaches the surface. This is called *volume conduction* and is illustrated in figure 2.2C. I should note, however, that I don't like the term *volume conduction* very much because it might be taken to imply that we are recording charged particles that pass from the neurons all the way to the scalp electrodes. That isn't the way it works. By analogy, when electricity is generated in a power plant and flows through power lines, the electrons don't go all the way from the power plant to your house. Instead, one electron pushes the next one, which pushes the next one, and so forth. Similarly, when a dipole is active in the brain, you don't have to wait for charged particles to move all the way from the dipole to the surface. Instead, a postsynaptic potential in a set of neurons creates an essentially instantaneous voltage field throughout the entirety of the head, with no meaningful delay. And don't forget that you are measuring voltage, which is the potential for current to flow and not the actual flow of current.

Electricity does not just run directly between the two poles of a dipole in a conductive medium, but instead spreads out across the conductor. Consequently, ERPs do not appear only at an electrode located directly above the dipole but are instead picked up by electrodes located all over the head. The high resistance of the skull causes the voltage to be even more widely distributed. Consequently, the scalp distribution of an ERP component is usually very broad.

Figure 2.2C illustrates a very important fact about ERP scalp distributions. For any dipole location, the voltage will be positive over one portion of the scalp and negative over the remainder of the scalp, with an infinitesimally narrow band of zero that separates the positive and negative portions. In many cases, one of these portions will fall over a part of the head where you don't have any electrodes (e.g., the face or the bottom of the brain), so you might not see both sides of the dipole. In other cases, you will be able to see both the positive and negative sides of the dipoles.[4] The zero band that separates the positive and negative sides of the head will be in a different place for each different dipole, and no single zero band will be present when multiple dipoles are active. Thus, there is no place on the head that consistently has zero voltage (see chapter 5 for additional discussion).

Relationship between Dipoles and Components

An equivalent current dipole represents the summed activity of a large number of nearby neurons. How is this related to the concept of an *ERP component*? We will define the term *ERP component* more carefully later in this chapter, but we can make a simple link to the concept of a dipole at this time. Specifically, when an equivalent current dipole represents the activity of a single functional brain region, this dipole can be considered to be the same thing as an ERP component. The moment-by-moment changes in the magnitude of the dipole (sometimes called the dipole's *source waveform*) constitute the time course of the ERP component. As will be described soon, all of the different dipoles in the brain sum together to give us the complex pattern of positive and negative peaks that we record from our scalp electrodes.

Magnetic Fields

The blurring of voltage caused by the high resistance of the skull can be largely circumvented by recording magnetic fields instead of electrical potentials. As illustrated in figure 2.2D, an electrical dipole is always surrounded by a magnetic field of proportionate strength, and these fields summate in the same manner as voltages. Thus, whenever an ERP is generated, a magnetic field is also generated, running around the ERP dipole. Moreover, the skull is transparent to magnetism, so the magnetic fields are not blurred by the skull,[5] leading to much greater spatial resolution than is possible with electrical potentials. The magnetic equivalent of the EEG is called the *magnetoencephalogram* (MEG), and the magnetic equivalent of an ERP is an *event-related magnetic field* (EMRF).

As illustrated in figure 2.2E, a dipole that is parallel (*tangential*) to the surface of the scalp will be accompanied by a magnetic field that leaves the head on one side of the dipole and enters back again on the other side. If a highly sensitive probe called a SQUID (superconducting quantum interference device) is placed next to the head, it is possible to measure the magnetic field as it leaves and reenters the head. However, if the dipole is perpendicular to the surface of the head (a *radial* dipole), the magnetic field running around the dipole will not leave the head, and it will be "invisible" to the SQUID. For orientations that are between tangential and radial, the strength of the magnetic field that is recorded from outside the head gets weaker and weaker the more radial the dipole is. Similarly, the strength of the extracranial magnetic field becomes very weak for deep dipoles. Thus, MEG recordings are primarily sensitive to superficial, tangential dipoles.

Because magnetic fields are not as widely dispersed as electrical potentials, they can provide more precise localization. However, as will be discussed in online chapter 14, the combination of ERP and ERMF recordings provides even better localization than ERMF recordings alone. Unfortunately, three factors make magnetic recordings very expensive: the SQUID is expensive; the coolant must be continually replenished; and an expensive magnetically shielded recording chamber is necessary to attenuate the earth's magnetic field, which is orders of magnitude larger than the MEG signal. Thus, MEG/ERMF recordings are much less common than EEG/ERP recordings.

The Forward Problem and the Superposition of Components on the Scalp

If I tell you the locations and orientations of a set of dipoles in a brain, along with the shape and conductances of the brain, skull, and scalp, then it would be possible for you to use a relatively simple set of equations to compute the distribution of voltage that would be observed for those dipoles. This is called the *forward problem,* and it is relatively easy to solve. In this section, I will spend some time explaining how the forward problem is solved in a little bit of detail, because it will help you to understand exactly how the ERP components generated in the brain become mixed together in the scalp electrodes. This creates the superposition problem, which I described briefly in chapter 1 and which is often the biggest impediment in ERP studies.

Box 2.2
Four Kinds of Math

ERP waveforms are sequences of numbers, so you can't escape a little bit of math if you want to understand ERPs. If you are a "math person," then you are probably looking forward to seeing some mathematical formalisms as you read the book. But if you are not a math person, the last thing you'll want to see is a lot of equations. There will be some equations in the following chapters, but they will be very simple. And I promise not to write something like "As will be obvious from equation 6.4 ..." because I don't usually find that an equation makes something obvious.

I am not a math person. I managed to get through a year of calculus in my freshman year of college, but that's as far as I got. And I remember almost nothing I learned that year. So this book contains no calculus.

It turns out that you can understand all the key mathematical concepts involved in ERP research without any calculus. In fact, everything I will describe can be boiled down to addition, subtraction, multiplication, and division. If you are not a math person, you can use these four simple mathematical operations to understand things like Fourier analysis, time–frequency analysis, and how filters work. This will require that you spend some time looking at simple equations that combine these four simple operations in interesting ways, but ultimately it's all just arithmetic that anyone can understand.

If you are a math person, I think you will find that boiling things down in this way allows you to find a new appreciation for the logic that underlies the mathematics. And you will probably discover that there are some important things about Fourier analysis, time–frequency analysis, and filtering that you did not fully appreciate before.

Describing the forward problem in detail requires me to introduce a little bit of math, but only a very little bit (see box 2.2).

The forward problem is illustrated in figure 2.3, which shows the ERP waveforms for three hypothetical generator locations. These are called *source waveforms*, and they show the time course of voltage for the components. For any given electrode location on the scalp, a fixed percentage of the source waveform from a given internal generator will propagate to the electrode. That is, X% of the voltage from a given generator will be propagated to a given electrode site, where X has a different value for each combination of internal generator and external scalp electrode. The percentage of the voltage that is propagated will depend on the position and orientation of the generator with respect to the position of the electrode, along with the conductivity of the various tissues within the head. For example, a high percentage of the voltage from a superficial radial dipole will propagate to an electrode on the scalp right over the top of the generator, and a smaller (and inverted) percentage will propagate to an electrode on the opposite side of the head.

It's actually possible to estimate the percentage of propagation from each possible generator location to each electrode site in a given subject by creating a model of the subject's head from a structural MRI scan, combined with normative conductivity values for the different tissues that constitute the head (e.g., brain, skull, scalp). There are software packages that can provide the propagation factors if you provide a structural MRI scan.

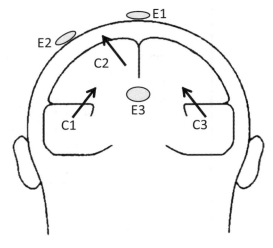

Hypothetical weights from components to electrodes

	Electrode 1	Electrode 2	Electrode 3
Component 1	$w_{1,1} = 1$	$w_{1,2} = 0.25$	$w_{1,3} = 0.5$
Component 2	$w_{2,1} = 0.5$	$w_{2,,2} = 0.5$	$w_{2,3} = -0.5$
Component 3	$w_{3,1} = 1$	$w_{3,2} = 0.75$	$w_{3,3} = 0.5$

Source waveform at the generator location for each component

Observed waveform at each electrode site

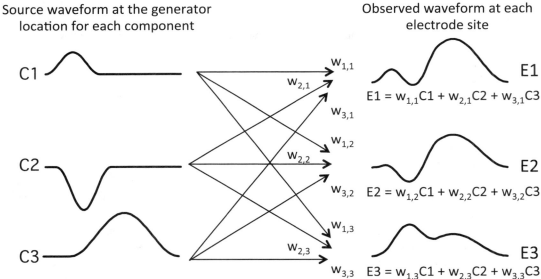

$$E1 = w_{1,1}C1 + w_{2,1}C2 + w_{3,1}C3$$

$$E2 = w_{1,2}C1 + w_{2,2}C2 + w_{3,2}C3$$

$$E3 = w_{1,3}C1 + w_{2,3}C2 + w_{3,3}C3$$

Figure 2.3
Relation between the underlying component waveforms and the observed scalp waveforms. In this example, three components are present (C1, C2, C3), each of which has a source waveform (time course of voltage, shown at the bottom left) and a generator location (represented by the arrows in the head). The contribution of each component waveform to the observed waveform at a given electrode site is determined by a weighting factor that reflects the location and orientation of the generator relative to that electrode, along with the conductivity of the tissues that form the head. The table shows the weighting factors between the three components, and the three electrode sites are given in the table (but note that these are made-up values, not the actual weighting factors from a real head). The observed waveform at a given electrode site (shown at the bottom right) is equal to the sum of each of the component waveforms, multiplied by the weighting factor between each component and that electrode site. The weights are indicated by the w's on the arrows between the component waveforms and the observed waveforms (e.g., $w_{2,3}$ represents the weighting factor between component 2 and electrode 3). Adapted with permission from Kappenman and Luck (2012). Copyright 2012 Oxford University Press.

The proportion of the voltage that is propagated from a given generator location to a given electrode site is called the *weight* between the generator and the electrode. There is a separate weight for each combination of generator location and electrode site. In figure 2.3, the weights are denoted as $w_{x,y}$, where x is the generator location and y is the electrode site. For example, the weight from generator 2 to electrode 1 is denoted $w_{2,1}$. I just made up the weights shown in figure 2.3—they are not the actual weights. For example, real weights would be much less than 1.0, because only a small proportion of the generator voltage makes it all the way out to the scalp electrodes. The real weights would be provided by a software package on the basis of a subject's structural MRI data. Note that the weights are proportions (between −1 and +1), not percentages.

The "fake" weights shown in figure 2.3 make it easy to see how the waveforms at a given electrode site reflect the weighted sum of the internal underlying components. For example, the waveform at electrode E1 is simply the C1 waveform plus 50% of the C2 waveform plus the C3 waveform (because the weights from components C1–C3 to electrode E1 are 1.0, 0.50, and 1.0, respectively). If you think this seems very simple, you're right! The waveform at a given electrode site is always just the weighted sum of all the source waveforms. This simplicity arises from the fact that voltages simply add together in a simple conductor like the human head.

Note that the voltage at a given electrode site is a weighted sum of *all* the underlying components. The weights will be negative on one side of the head and positive on the other, with a narrow band where the weights are zero at the transition between the positive and negative sides of the dipoles. And the weights may be quite small near this band. However, a given electrode site picks up at least some voltage from almost every component in the brain. This means that almost all of the components are mixed together at every electrode site. How many different components are likely to be mixed together in a given experiment? It is difficult to know for sure, but dozens may be present in a typical experiment. For example, evidence for at least 10 different sources was found in the brief period from 50 to 200 ms after the onset of an auditory stimulus in a simple target detection task (Picton et al, 1999), and many more are presumably active when a longer time range is examined and when the subject is engaged in a complex task.

In most cases, we are interested in measuring individual components, not the mixture that we record at a given scalp electrode. Unfortunately, there is no foolproof way to recover the underlying components from the recordings obtained at the scalp. There are many different methods that attempt to recover the underlying components (see box 2.3), but all of them are based on assumptions that are either known to be false or are not known to be true. The mere fact that each of these techniques will arrive at a different solution is an indication that most (or all) of them are incorrect.

As this chapter progresses, we will see why this superposition problem makes life so difficult for ERP researchers. You may find this depressing, but keep in mind that we will eventually discuss a set of strategies for overcoming the superposition problem. And the online supplement to chapter 4 describes some excellent examples of previous studies that have overcome the

Box 2.3
Unmixing the Components

The mixing of underlying components in scalp recordings is such an important problem that many different mathematical procedures have been developed to unmix them. The best known and most widely used procedures are dipole localization methods, principal component analysis, independent component analysis, Fourier analysis, and time-frequency analysis. All of these will be covered in some detail later in the book.

Although these techniques seem very different from each other, they all share a fundamental underlying structure, and it is worth understanding this structure. Specifically, they all assume that the waveforms recorded at the scalp consist of the weighted sum of a set of underlying *basis functions*. However, they make different assumptions about the nature of these basis functions. For example, Fourier analysis assumes that the basis functions are sine waves, but it makes no assumptions about the pattern across the scalp; in contrast, dipole localization methods make no assumptions about the time course of the basis functions, but they assume that the scalp distribution reflects the conductivity of the brain, skull, and scalp. These techniques also differ widely in the mathematical approach that is used to unmix the observed waveforms. However, it is useful to keep in mind that they all assume that the observed waveforms consist of the sum of a relatively small set of basis functions.

superposition problem and made important contributions to our understanding of the mind and brain.

The Challenges of ERP Localization

This section provides a brief discussion of the challenges involved in localizing ERPs. A more complete discussion is provided in online chapter 14.

As noted in the description of the forward problem, it is relatively easy to compute the distribution of voltage on the scalp if you know the locations and orientations of the dipoles (and have a structural MRI scan). However, if I provide you with an observed voltage distribution from the scalp and ask you to tell me the locations and orientations of the dipoles, you will not be able to provide an answer with 100% confidence. This is the *inverse problem*, and it is what mathematicians call an "ill-posed" or "underdetermined" problem. This simply means that there are multiple different sets of dipoles that can perfectly explain a given voltage distribution, and there is no way to tell which is the correct one. In fact, it has been known for more than 150 years that an infinite number of different dipole configurations can produce any given voltage distribution (Helmholtz, 1853; Nunez & Srinivasan, 2006; see also Plonsey, 1963). It is impossible to know with certainty which one of these configurations is the one that is actually responsible for producing the observed voltage distribution (unless you have other sources of information, such as lesion data).

Figure 2.4 illustrates some of the challenges involved in localizing ERPs. When a single dipole is present just under the skull and points directly outward (a superficial, radial dipole), a very

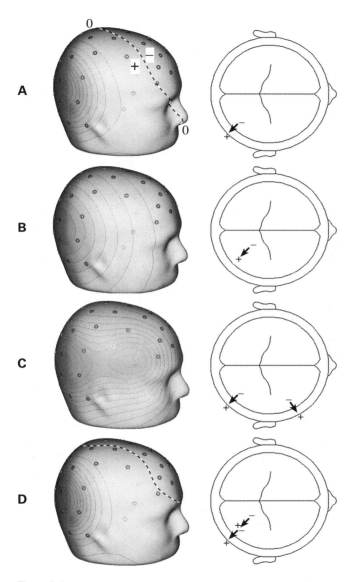

Figure 2.4
Scalp distributions (left) produced by different dipole configurations (right). See the text for descriptions of panels A–D of the figure. Images courtesy of J. Bengson. Reprinted with permission from Luck (2012a). Copyright 2012 American Psychological Association.

focal voltage distribution is observed on the scalp, with the maximum voltage directly over the location of the dipole (figure 2.4A). If this dipole is moved laterally even a small amount (e.g., 1 cm), the voltage distribution will be noticeably different. Consequently, a dipole such as this can be localized quite accurately simply by comparing the observed voltage distribution with the distribution that would be produced by a single hypothetical dipole with various different positions and orientations and selecting the dipole that best fits the observed data. Noise in the data will distort the observed scalp distribution somewhat, but it is still possible to localize a single superficial dipole like the one shown in figure 2.4A with reasonable accuracy as long as the noise is not too large (and assuming you know for certain that it *is* a single superficial dipole). Multiple-dipole configurations also exist that can produce exactly the same distribution of voltage as a single superficial dipole. Thus, we can localize a single superficial dipole with considerable precision (assuming very low levels of noise in the average ERP waveforms), but we cannot know from the scalp distribution that only a single superficial dipole is present.

Now consider the somewhat deeper dipole shown in figure 2.4B. The distribution of voltage over the scalp is broader than that observed with the superficial dipole, and moving the dipole laterally by a given amount won't have as large of an effect on the scalp distribution for the deeper dipole as it would for the superficial dipole. All else being equal, the deeper the dipole, the smaller and more broadly distributed the voltage will be on the scalp, and the less accurately it can be localized. An even more important problem arises in this situation, because it becomes difficult to tell the difference between a deep but focal generator and a superficial generator that extends across a large region of cortex. For example, the error-related negativity has a fairly broad distribution that is consistent with a single deep generator in the anterior cingulate cortex (Dehaene, Posner, & Tucker, 1994), but it is also consistent with the activation of a large patch of superficial cortex.

Figure 2.4C shows a situation in which two dipoles are simultaneously active. Voltages arising from different generators simply sum together, so the scalp distribution of the two simultaneous dipoles is simply equal to the sum of the scalp distributions of the individual dipoles. In this particular example, the two dipoles are both superficial, and they are quite far apart. It would be possible for you to localize the dipoles quite accurately in this situation, assuming that the data were not distorted by much noise and that you already knew that exactly two dipoles were active.

Figure 2.4D shows a different situation in which two dipoles are simultaneously active (the dipoles from figure 2.4A and B). This is a "nightmare scenario" for people who want to localize ERPs, because the scalp distribution of the sum of the two dipoles is nearly indistinguishable from the scalp distribution of the superficial dipole alone. Given even a small amount of noise, it would be impossible to know whether the data were the result of a single superficial dipole or two dipoles that are aligned in this fashion. The problem would still be bad even if the two dipoles were not perfectly aligned. Moreover, imagine an experiment in which these two dipoles were present, and an experimental manipulation led to a change in the amplitude of the deep dipole with no effect in the superficial dipole. It is very likely that an experimental effect of this nature would be attributed to the superficial dipole rather than the deep dipole.

As more and more simultaneous dipoles are added, the likelihood of dipoles aligning in this manner increases, and it becomes more and more difficult to determine how many dipoles are present and to localize them accurately. This problem becomes even worse when the data are noisy. Under these conditions, a set of estimated dipole locations that matches the observed scalp distribution can be quite far from the actual locations.

The examples shown in figure 2.4 are cause for both optimism and pessimism. You should be optimistic because ERPs can be localized accurately if you can be certain that you have a single dipole or that you have two to three dipoles that are not aligned in a way that makes them difficult to localize. But you should also be pessimistic because it will be difficult to localize ERPs unless you *know* that you have a single dipole or that you have two to three dipoles that are not aligned in a way that makes them difficult to localize. Fortunately, there are some things you can do to increase the likelihood that only a single dipole is active (e.g., performing localization on difference waves that isolate a single component).

In many experiments, the number of dipoles could be very large (>10), and localizing ERPs solely on the basis of the observed scalp distribution would be impossible to do with confidence. The only way to localize ERPs in this case is to add external constraints, and this is how existing procedures for localizing ERPs solve the non-uniqueness problem. As will be discussed in detail in online chapter 14, some common procedures allow the user to specify a fixed number of dipoles (Scherg, 1990), whereas other procedures use structural MRI scans and constrain the dipoles to be in the gray matter (Dale & Sereno, 1993; Hämäläinen, Hari, Ilmonieni, Knuutila, & Lounasmaa, 1993) or choose the solution that best minimizes sudden changes from one patch of cortex to the next (Pascual-Marqui, Esslen, Kochi, & Lehmann, 2002). Although each of these methods produces a unique solution, they do not necessarily produce the correct solution. And they may produce quite different solutions, further reducing our confidence that any of them has found the correct solution.

The most significant shortcoming of mathematical procedures for localizing ERPs is that they do not typically provide a well-justified margin of error. That is, they do not indicate the probability that the estimated location falls within some number of millimeters from the actual location. For example, I would like to be able to say that the N2pc component in a particular experiment was generated within 9 mm of the center of the lateral occipital complex and that the probability that this localization is incorrect is less than 0.05. I am aware of no mathematical localization technique that is regularly used to provide this kind of information.[6] Without a margin of error, it is difficult to judge the credibility of a given localization estimate. In most cases, the strongest claim that can be made is that the observed data are consistent with a given generator location.

Although it is usually impossible to definitively localize ERPs solely on the basis of the observed scalp distributions, this does not mean that ERPs can never be localized. Specifically, ERPs can be localized using the general hypothetico-deductive approach that is used throughout science. That is, a hypothesis about the generator location for a given ERP effect leads to a set of predictions, which are then tested by means of experiments. One prediction, of course, is that the observed scalp distribution will be consistent with the hypothesized generator location. However, confirming this prediction is not usually sufficient to have strong confidence that the hypothesis about the generator location is correct. Thus, it is important to test additional predictions. For example, one could test the prediction that damage to the hypothesized generator location eliminates the ERP component. Indeed, researchers initially hypothesized that the P3 component was generated in the hippocampus, and this hypothesis was rejected when experiments demonstrated that the P3 is largely intact in individuals with medial temporal lobe lesions (Polich, 2012). Similarly, one could predict that an fMRI experiment should show activation in the hypothesized generator location under the conditions that produce the ERP component (see, e.g., Hopf et al., 2006). It is also possible to record ERPs from the surface of the cortex in neurosurgery patients, and this can been used to test predictions about ERP generators (see, e.g., Allison, McCarthy, Nobre, Puce, & Belger, 1994). This hypothesis-testing approach has been quite successful in localizing some ERP components.

As was discussed in chapter 1, the main advantages of the ERP technique are its high temporal resolution, its relatively low cost, its noninvasiveness, and its ability to provide a covert and continuous measure of processing. Spatial resolution is simply not one of the strengths of the ERP technique, and it is therefore sensible that this technique should mainly be used in studies designed to take advantage of its strengths and that are not limited by its weaknesses.

Waveform Peaks versus Underlying ERP Components

Now that you understand how the underlying components become mixed together in scalp electrodes—and why it is so difficult to recover the underlying components from the observed scalp waveforms—it is time to consider why this is such a big problem in most ERP studies. In this section, I will show a set of simple artificial waveforms that demonstrate how the superposition problem can easily lead to misinterpretations of ERP data, along with a set of "rules" for avoiding these misinterpretations. In general, the goal of this section is to get you to look beyond the ERP waveforms that you have recorded from the scalp and think about the underlying components that might give rise to these surface waveforms. Even though you cannot usually localize the underlying components, you can learn to make educated guesses about the underlying components from the observed scalp waveforms.

The term *ERP component* will be described more formally later in this chapter. For now, you can just think of a component as electrical activity within a given region of the brain that reflects some neural process occurring in that region, which then propagates to our scalp electrodes.

To understand the underlying "deep structure" of a set of ERP waveforms, the first thing you need to do is understand how the peaks in the observed ERP waveforms can provide a misleading representation of the underlying components. To make this clear, I created some simple simulated components using Excel and then summed them together to create simulated scalp ERP waveforms. The results are shown in figure 2.5. This is a busy and complicated figure, but

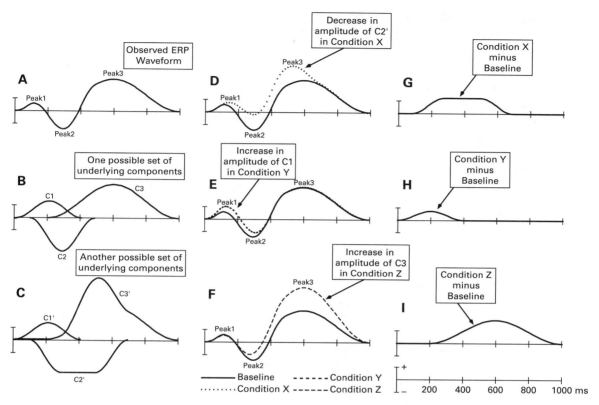

Figure 2.5
Examples of the complications that arise when underlying components sum together to form an observed ERP waveform. Panels B and C show two different sets of underlying components that sum together to produce the observed waveform shown in panel A. Panels D, E, and F show three simulated experiments in which one component differs between a baseline condition and three comparison conditions. In panel D, component C2′ is decreased in condition X compared to the baseline condition. In panel E, component C1 is increased in condition Y compared to the baseline condition, which creates an apparent shift in the latencies of both peak 1 and peak 2. In panel F, component C3 is increased in condition Z relative to the baseline condition, influencing both the amplitude and the latency of peak 2. Panels G, H, and I illustrate difference waves for conditions X, Y, and Z, respectively.

the next several sections will walk you through the different parts. This is probably the most important figure in the whole book, so it's worth taking your time and understanding each panel of the figure.

Voltage Peaks Are Not Special

Our eyes are naturally drawn to the peaks in a waveform, but the peaks often encourage incorrect conclusions about the underlying components. This is illustrated in panels A and B of figure 2.5, which show how three very reasonable components sum together to create the ERP waveform that we observe at the scalp. If you look carefully, you can see how the peaks in the observed waveform do a poor job of representing the underlying components. First, notice that the first positive deflection in the observed waveform (panel A) peaks at 50 ms, whereas the first underlying component (C1 in panel B) peaks at 100 ms. Similarly, the third underlying component (C3 in panel B) begins at 100 ms, but there is no obvious sign of this component in the observed waveform until approximately 200 ms.

It is natural for the human visual system to focus on the peaks in the waveform. However, the points in an ERP waveform where the voltage reaches a maximally positive or maximally negative value do not usually have any particular physiological or psychological meaning, and they are often completely unrelated to the time course of any individual underlying component. The peaks in the observed waveform are a consequence of the mixture of components, and they typically occur at a different time than the peaks of the underlying components. For example, the latency of a given peak in the observed waveform may differ across electrode sites (because of the different weights from the underlying components at the different electrode sites), whereas an underlying component has exactly the same time course at every electrode site (because the voltage propagates instantaneously to all electrodes). Sometimes the peak in the observed waveform occurs at the same time as the peak in an underlying component (e.g., when the underlying component is much larger than the other components). However, even in these cases it is not clear that the peak of activity tells us anything special about the underlying process. That is, theories of cognition or brain processes do not usually say much about when a process peaks. Instead, these theories usually focus on the onset of a process, the duration of the process, or the integrated activity over time (see, e.g., Pashler, 1994; Usher & McClelland, 2001).

This leads to our first and most important "rule" about interpreting ERP data:

Rule 1: Peaks and components are not the same thing. There is nothing special about the point at which the voltage in the observed waveform reaches a local maximum.

In light of this fundamental rule, I am always amazed at how often peak amplitude and peak latency are used to measure the magnitude and timing of ERP components. These measures often provide a highly distorted view of the amplitude and timing of the underlying components. Chapter 9 will describe better techniques for quantifying ERP data that do not rely on peaks.

Peak Shapes Are Not the Same as Component Shapes

Panel C of figure 2.5 shows another set of underlying components that also sum together to equal the ERP waveform shown in panel A. These three components are labeled C1′, C2′, and C3′. From the single observed waveform in panel A, there is no way to tell whether it was actually created from components C1, C2, and C3 in panel B or from C1′, C2′, and C3′ (or from some other set of underlying components). There are infinitely many sets of underlying components that could sum together to produce the ERP waveform in panel A (or any other ERP waveform).

If the observed waveform in panel A actually consisted of the sum of C1′, C2′, and C3′, the peaks in the observed panel would be a very poor indicator of the shapes of the underlying components. For example, the relatively short duration of peak 2 in panel A bears little resemblance to the long duration of component C2′ in panel C. If you saw a waveform like panel A in an actual experiment, you probably wouldn't guess that a very long-duration negative component like C2′ was a major contributor to the waveform.

This leads to our second rule:

Rule 2: It is impossible to estimate the time course or peak latency of an underlying ERP component by looking at a single ERP waveform—there may be no obvious relationship between the shape of a local part of the waveform and the underlying components.

In a little bit, we will talk about how difference waves can sometimes be used to determine the shape of an underlying ERP component. You can also gain a lot of information by looking at the waveforms from multiple scalp sites because the time course of a given component's contribution to one electrode site will be the same as the time course of its contribution to the other electrode sites (because ERPs propagate to all electrodes instantaneously).

The Time Course and Scalp Distribution of an Effect

Panels D–F of figure 2.5 show three simulated experiments in which a "baseline" condition is compared with three other conditions, named X, Y, and Z. The waveform for the baseline condition is the same as that shown in panel A, which could be the sum of either components C1, C2, and C3 from panel B or components C1′, C2′, and C3′ from panel C. I am calling this a "baseline" condition only because I am using it for comparison with conditions X, Y, and Z.

In the simulated experiment shown in panel D, we are assuming that the observed waveform for the baseline condition is the sum of components C1′, C2′, and C3′ and that component C2′ is decreased by 50% in condition X relative to the baseline condition. However, if you just saw the waveforms in this panel, and you didn't know the shapes of the underlying components, you might be tempted to conclude that at least two different components differ between the baseline condition and condition X. That is, it looks like condition X has a decrease in the amplitude of a negative component peaking at 300 ms and an increase in the amplitude of a positive component peaking at 500 ms. You might even conclude that condition X had a larger positive component peaking at 100 ms. This would be a very misleading set of conclusions about the underlying components because the effect actually consists entirely of a decrease in the amplitude of a single, broad, negative component (C2′). These incorrect conclusions arise from our natural inclination to assume that the peaks in the observed waveform have a simple and direct relationship to the unknown underlying components. This leads us to conclude that the increased amplitude in peak 3 in panel D reflects an increase in the amplitude of a long-latency positive component, when in fact it is a result in of a decrease in the amplitude of an intermediate-latency negative component.

Panel E shows a simulated experiment comparing the baseline condition with condition Y. This time, we're assuming that the underlying components are C1, C2, and C3 (rather than C1′, C2′, and C3′). Condition Y is the same as the baseline condition except for an increase in the amplitude of component C1. However, if you saw these observed waveforms, you might be tempted to conclude that condition Y has an increase in the amplitude of an early positive component peaking at 100 ms followed by a decrease in a negative component that peaks at 300 ms.

Similarly, panel F shows a simulation in which the amplitude of component C3 is increased in condition Z compared to the baseline condition. Although only this late positive component differs across conditions, the observed waveform for condition Z appears to exhibit a reduction of a negative component at 300 ms as well. However, this is just a consequence of the fact that component C3 overlaps in time with component C2.

In panels D–F, a change in the amplitude of a single component caused a change in the amplitude of multiple peaks, making it likely that the researcher will falsely conclude that multiple underlying components differ across conditions. I have seen this general pattern of waveforms countless times, and people often misinterpret the effects in exactly this way. This leads to our third rule:

Rule 3: An effect during the time period of a particular peak may not reflect a modulation of the underlying component that is usually associated with that peak.

Difference Waves as a Solution

Difference waves can often be used to provide a better estimate of the time course and scalp distribution of the underlying components. Difference waves are a very simple concept: You simply take the voltage at each time point in one ERP waveform and subtract the voltage from a different ERP waveform at the corresponding time points. The result gives you the difference in voltage between the two waveforms at each time point. In an oddball experiment, for example, you can subtract the frequent-trial waveform from the rare-trial waveform (rare-minus-frequent), and the resulting difference wave will show the difference in voltage between these trial types at each point in time (see, e.g., figure 1.4 in chapter 1). The difference wave from a given electrode site is analogous to a difference image in an fMRI experiment, except that differences are computed at each time point in an ERP difference wave and at each voxel in an fMRI difference image.

Panels G–I of figure 2.5 show difference waves from the three simulated experiments shown in panels D–F. The baseline waveform has been subtracted from the waveforms for conditions

X, Y, and Z. These difference waves are very helpful in showing us the nature of the experimental effects. They make the time course of the effects very clear, showing that they are different from the time course of the peaks in the observed waveforms. Indeed, under these conditions—in which the experimental effects consist entirely of a change in the amplitude of a single underlying component—the difference waves reveal the time course and scalp distribution of the underlying component that differs across conditions. That is, if you compare the difference wave in panel G to the underlying components in panel C, you will see that the difference wave has exactly the same time course as component C2′. And because the difference waves in these examples subtract away everything except the one component that differs across conditions, the difference waves will have the same topography as the underlying component.

You cannot be guaranteed that a difference wave will contain only a single component. For example, the rare-minus-frequent difference waves from the oddball paradigm that were discussed in chapter 1 (see figure 1.4) contained both an N2 wave and a P3 wave. However, there is a good chance that you will have a single component in your difference waves if you compare conditions that are very subtly different.

How can you know if a broad effect in a difference wave consists of a change in a single underlying component or a sequence of two or more components? First, you can look at the scalp distribution of the difference wave, collapsed across the entire duration of the effect, to see if the scalp distribution is too complex to be explained by a single dipole. Second, you can compare the scalp distributions of the early and later parts of the difference wave. If they are the same, then you probably have a single component; if they are different, then you have at least two components. Difference waves are discussed in more detail later in this chapter.

Interactions between Amplitude and Latency

Although the amplitude and latency of an underlying component are conceptually independent, they can become confounded when the underlying components become mixed together at the scalp. For example, panel F of figure 2.5 illustrates the effect of increasing the amplitude of component C3. Because component C3 overlaps with peak 2, this change in the amplitude of component C3 causes a shift of approximately 20 ms in the peak latency of peak 2 in condition Z relative to the baseline condition. This leads to our next rule:

Rule 4: Differences in peak amplitude do not necessarily correspond with differences in component size, and differences in peak latency do not necessarily correspond with changes in component timing.

This is a somewhat depressing rule, because the most obvious virtue of the ERP technique is its temporal resolution, and it is unpleasant to realize that estimates of component timing can be distorted by changes in the amplitude of temporally overlapping components. Keep in mind, however, that this sort of temporal distortion mainly happens when a large change in a large component overlaps with a smaller component, leading to a small latency change in the smaller component. If you think carefully about a given effect and spend some time looking at the difference waves, you should be able to determine whether the latency differences could plausibly be a result of amplitude changes (or vice versa). In addition, chapters 3 and 4 will discuss how this problem can be completely avoided by looking at the onset time of a difference wave.

Distortions Caused by Averaging

In the vast majority of ERP experiments, the ERP waveforms are isolated from the background EEG by means of signal averaging. It's tempting to think of averaging as a process that simply attenuates the nonspecific EEG, allowing us to see what the single-trial ERP waveforms look like. However, to the extent that the single-trial waveform varies from trial to trial, the averaged ERP may provide a distorted view of the single-trial waveforms, particularly when component latencies vary from trial to trial. This is illustrated in figure 2.6. Panel A illustrates three single-trial ERP waveforms (without any EEG noise), with significant latency differences across trials, and panel B shows the average of those three single-trial waveforms. The averaged waveform differs from the single-trial waveforms in two significant ways. First, it is smaller in peak amplitude. Second, it is more spread out in time. As a result, even though the waveform in panel B is the average of the waveforms in panel A, the onset time of the averaged waveform in panel B reflects the onset time of the earliest single-trial waveform and not the average onset time. In other words, the onset time in the average waveform is not equal to the average of the single-trial onset times (see chapter 8 and online chapter 11 for additional discussion). This leads to our next rule:

Rule 5: Never assume that an averaged ERP waveform directly represents the individual waveforms that were averaged together. In particular, the onset and offset times in the averaged waveform represent the earliest onsets and latest offsets from the individual trials or individual subjects that contribute to the average.

Fortunately, it is often possible to measure ERPs in a way that avoids the distortions created by the averaging process. For example, the area under the curve in the averaged waveform shown in figure 2.6A is equal to the average of the area under the single-trial curves in figure 2.6B. Similarly, it is possible to find the time point that divides the area into two equal halves, and this can be a better measurement of latency than peak measures. These methods are described in detail in chapter 9.

Comparisons across Experiments

It can be very difficult to determine whether an ERP effect in one experiment reflects the same underlying component as an ERP effect in another experiment. For example, imagine that you've conducted an experiment in which you recorded ERPs from bilingual speakers of two languages, Xtrinqua and Kirbish, focusing on the N400 component elicited by words in these two languages. You found that the N400 was larger (more negative) for proper nouns in Xtrinqua than for proper nouns in Kirbish. You therefore conclude that processing proper nouns requires more work in Xtrinqua than in Kirbish. However, this assumes that the effect you have observed actually reflects a change in the same N400 component that was observed in prior studies of semantic

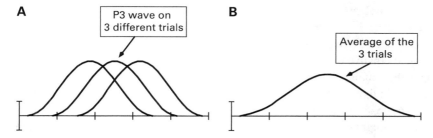

Figure 2.6
Illustration of the distortions in an averaged ERP waveform that arise from trial-to-trial latency variation. (A) A single-trial component at three different latencies, representing trial-to-trial variations in timing. (B) The average of these three single-trial waveforms. The average waveform is broader than the single-trial waveforms and has a smaller peak amplitude. In addition, the onset time and offset time of the average waveform reflect the earliest onset and latest offset times, not the average onset and offset times.

integration. Perhaps you are instead seeing a smaller P3 rather than a larger N400. Because the P3 is related to the amount of cognitive resources devoted to processing a stimulus, this would suggest that it is actually easier to process Xtrinqua proper nouns than Kirbish proper nouns. This is nearly the opposite of the conclusion that you would draw if the effect was an increase in N400 amplitude rather than a decrease in P3 amplitude. This type of problem has come up many times in the language ERP literature (see review by Swaab, Ledoux, Camblin, & Boudewyn, 2012), and analogous problems arise in other ERP domains. Scalp distribution can sometimes be used to rule out a given component, but this only works when the scalp distributions of two components are easily distinguished (which is not the case for P3 and N400).

This leads to our final rule of ERP interpretation:

Rule 6: An ERP effect observed in one experiment may not reflect the same underlying brain activity as an effect of the same polarity and timing in previous experiments.

Solutions to this problem will be described in the final section of this chapter.

Some General Comments about the Component Structure of ERP Waveforms

The six rules of ERP interpretation presented in this chapter have been violated in a very large number of published ERP experiments (including some of my own papers!). Violations of these rules significantly undermine the strength of the conclusions that can be drawn from these experiments, so you should look for violations when you read ERP papers. If you are relatively new to the ERP technique, it would be worth looking back through old ERP papers that you've previously read and identifying any violations of these rules. This will be good practice, and it may cause you to reevaluate your conclusions about some prior findings.

This section of the book can be depressing, because it focuses on the most difficult challenges that face ERP researchers. But don't despair! Chapter 4 provides some strategies for overcoming these challenges.

Also, many of these problems are not nearly as bad when you consider the waveforms recorded at multiple scalp sites. Because the ERP waveforms at different scalp sites reflect exactly the same components, but scaled by different weighting factors (as shown in figure 2.3), you can rule out many alternative explanations by looking at the data from multiple electrodes. Consider, for example, the simulated experiment shown in figure 2.7 (which is based on the C1′, C2′, and C3′ components in figure 2.5). In this imaginary experiment, ERPs were recorded from the Fz, Cz, and Pz electrode sites in condition A and condition B. If you just saw the data from the Cz electrode site, you might conclude that the P3 wave was larger and earlier for condition B than for condition A. However, you can rule out this conclusion by looking at the scalp distribution of the waveforms (both the parent waveforms and the difference waves). From the whole set of data, you can see that the P3 wave is largest at Pz, a little smaller at Cz, and quite a bit smaller at Fz (which is the usual pattern). The experimental effect, however, is largest at Fz, smaller at Cz, and near zero at Pz, and it extends from approximately 150 to 450 ms. Both the timing and the scalp distribution of the experimental effect are inconsistent with a modulation of the P3 wave and are more consistent with a frontal N2 effect. You should also note that the scalp distribution of the effect is the same throughout the time course of the effect, suggesting that it really does consist of a single component that extends from approximately 150 to 450 ms, rather than consisting of separate early and late effects. Thus, a careful examination of the data from multiple electrode sites, including the difference waves, will allow you to avoid making naïve mistakes in interpreting your data. Moreover, as you gain more experience with ERPs, you will be able to do a better job of figuring out the underlying components (see box 2.4).

Although I am not a big fan of mathematical techniques for determining the underlying components in ERP experiments, they can certainly be useful for generating plausible hypotheses and for showing that some hypotheses are inconsistent with the data. For example, if you were to perform source localization on the difference waves shown in figure 2.7 (with many more electrode sites, of course), this would make it clear that the effect is not consistent with the likely generators of the P3 wave. Source localization would also make it clear that a single generator source could account for both the early and late portions of the effect. Thus, even though you should be skeptical about the three-dimensional coordinates of the effect that would be provided by source localization techniques, these techniques are still valuable for testing more general hypotheses about the component structure of your data. This is especially true if you combine source localization with the use of difference waves (see box 2.5 for an example of my own rather lame attempt at using math to understand the underlying component structure of an experiment).

Difference Waves as a Tool for Isolating Components

In this section, I want to spend some additional time discussing the value of difference waves in addressing the problems caused by the mixing of components in our observed ERP waveforms. As shown in figures 2.5 and 2.7, difference waves can sometimes reveal the time course

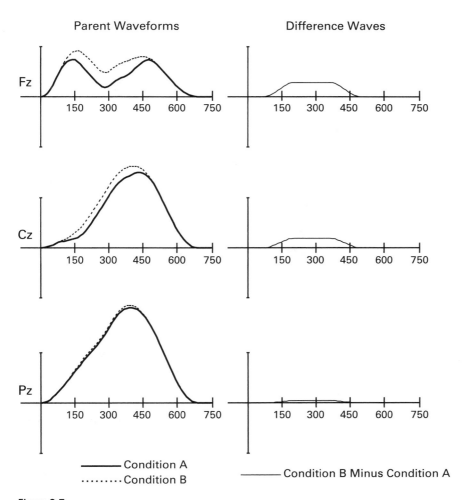

Figure 2.7
Simulated experiment in which three underlying components (similar to components C1′, C2′, and C3′ in figure 2.5) propagate to electrode sites Fz, Cz, and Pz with different weighting factors. The amplitude of a negative-going component (C2′) is reduced in condition B compared to condition A. Because the weighting from this component is greatest at Fz, intermediate at Cz, and smallest at Pz, the difference between conditions A and B is greatest at Fz, intermediate at Cz, and smallest at Pz. This can be seen in the parent waveforms from the individual conditions (left column) and in the difference waves (right column). Note that because the difference waves were formed by subtracting condition A from condition B rather than vice versa, the difference waves show a positive deflection rather than a negative deflection.

Box 2.4
Experience Matters

It appears that people can learn to make good inferences about the underlying components when examining ERP data. Many years ago, Art Kramer conducted a study in which he created artificial ERPs from a set of realistic underlying components (Kramer, 1985). He then gave these waveforms (from multiple conditions at multiple electrode sites) to eight graduate students and two research associates who had between 1 and 10 years of experience with ERPs. There was a strong effect of experience: the experienced ERP researchers were able to figure out the nature of the underlying components, whereas the people who were new to ERPs could not reliably determine the underlying component structure of the data. Of course, there are limits to this conclusion. It was based on a small sample of researchers from one particular lab who were shown one type of data, and there is no guarantee that a given experienced researcher will always (or even usually) reach a correct interpretation. When writing a journal article, you couldn't cite your years of experience with ERPs as evidence that your interpretation of the underlying component structure is correct. However, once you gain enough experience with ERPs (and learn to apply everything you have read in this chapter), you should be able to figure out the likely pattern of underlying ERP components from a given experiment with some confidence.

and scalp distribution of an underlying ERP component. A well-constructed difference wave (i.e., one based on a well-controlled and relatively subtle experimental manipulation) will always contain fewer components than the *parent* waveforms from which the difference wave was constructed. And fewer components means less opportunity for confusion due to the mixing of components.

Given that I'm a big fan of difference waves, it's a bit ironic that the first published use of difference waves to assess the component structure of an experiment were in a paper by Risto Näätänen and his colleagues (Näätänen, Gaillard, & Mantysalo, 1978) that was designed to question the interpretation of a previous study by my graduate mentor, Steve Hillyard (Hillyard, Hink, Schwent, & Picton, 1973). However, Jon Hansen and Steve Hillyard responded with a paper in which they embraced the difference wave approach and fought back with their own difference waves (Hansen & Hillyard, 1980).

The logic behind difference waves is very simple. As shown in figure 2.3, the ERP waveform at a given electrode site is the weighted sum of a set of underlying source waveforms (recall that *source waveform* is just another term for the time course of an ERP component at the generator site). If two conditions differ in some of the source waveforms and not in others, then the difference wave created by subtracting one condition from the other will eliminate the source waveforms that are identical in the two conditions, making it possible to isolate the components that differ. This is completely reasonable because ERPs from different sources simply sum together.

As described earlier in this chapter, figures 2.5 and 2.7 show how difference waves can help reveal the time course of an effect. To see how difference waves can help reveal the scalp

Box 2.5
My Own Experience with Mathematical Approaches

When I was making the simulated waveforms shown in figure 2.7, I realized that these waveforms looked a lot like the data from one of the first experiments I conducted in graduate school (Luck, Heinze, Mangun, & Hillyard, 1990). In this experiment, subjects attended either to the left or the right side of a bilateral visual display, and we wanted to see if the P1 and N1 components would be larger over the hemisphere contralateral to the attended side (relative to the ipsilateral hemisphere). The waveforms are shown in the illustration that follows.

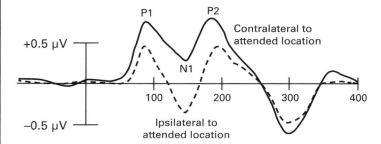

As you can see, we found that the waveform was more positive over the contralateral hemisphere from approximately 80 to 200 ms, including the time ranges of the P1, N1, and P2 components. From visual inspection of the data, it was not clear whether this broad attention effect was a single component or a modulation of separate P1 and P2 components. Because I thought of myself as a statistics hotshot at that time, I decided that I would use principal component analysis (PCA) to answer this question. PCA produced nice results indicating that the attention effect consisted of modulations of separate P1 and P2 components. However, the more I thought about the results, the more I realized that they could be an artifact of PCA assumptions. Also, Steve Hillyard was not as impressed by fancy mathematical techniques as I was (for reasons that I now appreciate). Our compromise was to put the PCA results into an appendix in the published paper. Fortunately, the conclusions of the paper did not hinge on the PCA results.

distribution of an effect, consider the N170 component shown in figure 2.8. As discussed in chapter 1, the N170 is larger in response to pictures of faces than pictures of non-face objects such as cars (for a review, see Rossion & Jacques, 2012). If you simply measure the scalp distribution at 170 ms for trials on which faces were presented, the distribution will reflect face-specific processing plus all of the nonspecific activity that is present at 170 ms. However, if you first construct face-minus-nonface difference waves and then measure the scalp distribution, you will see the topography of face-specific brain activity, uncontaminated by the nonspecific activity.

Bruno Rossion's lab conducted a study in which they examined the development of the N170 (Kuefner, de Heering, Jacques, Palmero-Soler, & Rossion, 2010), and this study showed the importance of measuring scalp distributions from difference waves. In this study, ERPs were recorded from children between 4 and 17 years of age while they viewed images of faces,

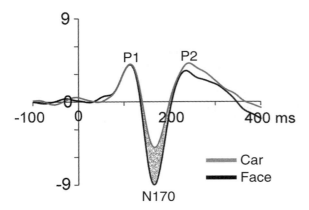

Figure 2.8
ERPs elicited over visual cortex by car and face stimuli, with the N170 effect highlighted by the shaded region. Adapted with permission from Rossion and Jacques (2012). Copyright 2012 Oxford University Press.

scrambled faces, cars, and scrambled cars. When the scalp distribution was measured at the time of the N170 component from the ERPs elicited by faces, the scalp distribution changed markedly over the course of development. Previous studies had found similar effects and concluded that the neuroanatomy of face processing changed during this period of development. However, these changes in scalp distribution could instead be a result of distortion from other ERP components that are also present at 170 ms and that change in relative amplitude over the course of development. To test this, Kuefner et al. constructed difference waves in which they subtracted the ERPs elicited by scrambled faces from the ERPs elicited by intact faces, removing any activity that was not face specific. The scalp distribution of the resulting difference wave was virtually identical from 4 to 17 years, indicating that the face-specific processing had the same neuroanatomical locus across this developmental period.

Difference waves can also help you avoid a visual illusion that sometimes causes people to misperceive ERP effects. Figure 2.9 shows a simulated experiment in which this illusion arises. In this experiment, the difference between condition A and condition B is exactly the same at the Fz and Cz electrode sites. However, because this effect is superimposed on the steeply rising edge of the P3 wave at the Cz site, the effect looks smaller at Cz than at Fz. The difference wave shows that the effect is exactly the same at these sites. The illusion arises because our visual system tends to focus on the distance between two curves along an axis that is approximately perpendicular to the curves (e.g., the horizontal difference between the condition A and condition B waveforms at the Cz site). However, the difference in amplitude at a given time point is the vertical distance between the two waveforms at that time point. The steeper the slope of two overlaid waveforms, the more our visual system focuses on the horizontal difference between the waveforms rather than the vertical distance, and this leads us to underestimate the true

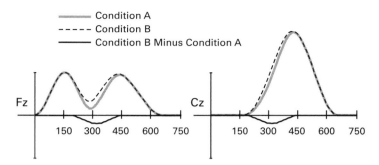

Figure 2.9
Visual illusion in which the effect of an experimental manipulation appears to be smaller when it is superimposed on a steeply sloped ERP wave. In this example, the difference between conditions A and B is exactly the same size at the Fz and Cz scalp sites (as shown by the difference waves). However, because this difference is superimposed on a steeply sloped P3 wave at the Cz site, but not at the Fz site, the effect looks smaller at Cz than at Fz when judged by looking at the parent waveforms rather than by looking at the difference waves.

difference in amplitude between the two conditions. Once you look at the difference waves in figure 2.9, you can see that the difference between conditions A and B is exactly the same at Fz and Cz in this experiment. This is yet one more advantage of difference waves.

Although difference waves are an incredibly useful way to isolate the underlying components from the mixture of components in the parent waveforms, they do have some limitations. First, you are making implicit assumptions about the nature of the difference between conditions A and B when you choose whether to compute A minus B or to compute B minus A. Consider, for example, the simulated experiment shown in figure 2.7. The difference waves in this experiment make it look as if a frontal positive component is larger in condition B than in condition A, but the effect was actually created (in my simulations) by decreasing the amplitude of a frontal negative component. Thus, you need to be careful about the implicit assumptions you are making about the direction of the effect when you subtract one waveform from another.

Another important issue is that the difference wave will always be noisier than the parent waveforms. To understand why this is true, consider what happens when you add two waveforms rather than subtract two waveforms. When you add two waveforms, A and B, some of the noise in A cancels out noise in B, but mostly the noise increases rather than canceling, and the overall noise in the summed waveform increases (although it doesn't double). It turns out that exactly the same thing happens when you subtract one waveform from the other instead of adding the two waveforms. The reason for this is that the noise at a given point in the waveform can be either positive-going or negative-going, so it doesn't matter whether you add or subtract pure noise (on average). The end result will be a noisier waveform. However, this is not usually a problem in practice. If you do a statistical analysis on the difference waves, the p value may be exactly the same as if you did the analysis on the parent waveforms (this is true for linear measures, such as mean amplitude, as will be discussed in chapter 9). For example, if you have waveforms for conditions A and B in two groups of subjects, you could do a two-way analysis of variance (ANOVA) on the mean amplitude from 400 to 600 ms with factors of condition and group. Alternatively, you could measure the mean amplitude from 400 to 600 ms in difference waves formed by subtracting the condition B waveforms from the condition A waveforms and then do a t test comparing these values for the two groups. The p value from this t test will be identical to the p value from the group × condition interaction in the ANOVA.

What Is an ERP Component?

A Conceptual Definition

We are now ready to provide a formal definition of *ERP component*. In the early days of ERP research, a component was defined primarily on the basis of its polarity, latency, and general scalp distribution. For example, the P3a and P3b components were differentiated on the basis of the earlier peak latency and more frontal distribution of the P3a component relative to the P3b component (Squires, Squires, & Hillyard, 1975). However, polarity, latency, and scalp distribution are superficial features that don't really capture the essence of a component. For example, the peak latency of the P3b component may vary by hundreds of milliseconds depending on the difficulty of the target–nontarget discrimination (Johnson, 1986), and the scalp distribution of the auditory N1 wave depends on the pitch of the eliciting stimulus in a manner that corresponds with the tonotopic map of auditory cortex (Bertrand, Perrin, & Pernier, 1991). Even polarity may vary: the C1 wave, which is generated in primary visual cortex, is negative for upper-field stimuli and positive for lower-field stimuli due to the folding pattern of this cortical area (Clark, Fan, & Hillyard, 1995). The difficulty of using latency, scalp distribution, and polarity to identify components becomes even greater when multiple components are mixed together at a given electrode site. For example, the N2 elicited by an oddball is superimposed on top of P2 and P3 waves, so the overall voltage during the N2 time range is often positive.

I do not mean to imply that you should not use latency, scalp distribution, and polarity to help identify ERP components. In fact, these factors are extremely useful in determining whether a given experimental effect reflects, for example, a P3a or P3b component. However, these factors do not *define* an ERP component; they are merely superficial consequences of the nature of the component. Instead, a component should be defined by the factors that are intrinsic to the component. For example, an influential discussion of ERP components by Manny Donchin, Walter Ritter, and Cheyne McCallum noted that, "an ERP component is a subsegment of the ERP whose activity represents a functionally distinct neuronal aggregate" (Donchin, Ritter, & McCallum, 1978, p. 353). My own, very similar definition is this:

Conceptually, an ERP component is a scalp-recorded neural signal that is generated in a specific neuroanatomical module when a specific computational operation is performed.

By this definition, a component may occur at different times under different conditions, as long as it arises from the same module and represents the same computational operation (e.g., the encoding of an item into working memory in a given brain area may occur at different delays after the onset of a stimulus because of differences in the amount of time required to identify the stimulus and decide that it is worth storing in working memory). The scalp distribution and polarity of a component may also vary according to this definition, because the same cognitive function may occur in different parts of a cortical module under different conditions (e.g., when a visual stimulus occurs at different locations and therefore stimulates different portions of a topographically mapped area of visual cortex). It is logically possible for two different cortical areas to accomplish exactly the same cognitive process, but this probably occurs only rarely and would lead to a very different pattern of voltages, and so this would not usually be considered a single ERP component (except in the case of mirror-image regions in the left and right hemispheres).

Although this definition captures the essence of an ERP component, it is completely useless as an operational definition. That is, because we are recording the mixed signals from multiple generators at every electrode site, we have no way of directly determining the contribution of each neuroanatomical module to the observed waveform. And we don't ordinarily know the computational operation reflected by a scalp voltage. So we cannot use this definition to determine which component is being influenced by a given experimental manipulation or to determine whether the same component is active in multiple experiments. However, this definition is useful insofar as it helps to formalize what we believe is going on under the skull.

You should also keep in mind that many well-known ERP components (e.g., N400, mismatch negativity, P3b) are probably not individual components according to this definition. That is, multiple brain areas that are carrying out similar but somewhat different computations likely contribute to these scalp-recorded voltage deflections (see, e.g., the chapters on these components in Luck & Kappenman, 2012b). As research progresses, we gain the ability to subdivide these multicomponent conglomerations into their parts. In some cases, this leads to new names for the individual components. For example, the "N2 component" that was identified in early ERP studies turned out to consist of a set of many different individual components that can be separated by means of various experimental manipulations, and these individual components (sometimes called *subcomponents*) were given new names (e.g., *anterior N2*, *N2c*, *N2pc*). In some cases, these subcomponents have been divided even further (see, e.g., Luck, 2012b). In other cases, we still use a single name to refer to something that likely consists of multiple similar components, especially when we don't yet have good methods for isolating the individual subcomponents (e.g., *N400* is still used even though there are probably at least two neurally distinct subcomponents that contribute to it).

An Operational Definition

Manny Donchin provided an alternative definition that is more like an operational definition: "A component is a set of potential changes that can be shown to be functionally related to an experimental variable or to a combination of experimental variables" (Donchin et al., 1978, p. 353). They summarized this idea more compactly by saying that, "an ERP component is a source of *controlled*, observable variability" (Donchin et al., 1978, p. 354). This is very similar to the idea, illustrated in figures 2.5 and 2.7, that difference waves can be used to isolate ERP components. Donchin et al. noted that electrode site is one key source of variability, which corresponds to the idea that changes in the scalp distribution of a difference wave between the early and late parts of the effect mean that different components are active in the early and late parts of the effect. Thus, by this definition, any consistent difference between the waveforms in different experimental conditions is an ERP component. This is a clear and easily applied operational definition of *ERP component*.

However, there are two shortcomings of this operational definition. First, it ignores the key idea that an ERP component is generated by a specific neuroanatomical module (or "neural aggregate" to use the terminology of Donchin et al., 1978). That is, even if the scalp distribution of an experimental effect did not vary over time or over the conditions of a particular experiment, the scalp distribution might clearly indicate that more than one dipole was active (e.g., if the scalp distribution did not follow a simple dipolar pattern). In such a case, we would want to conclude that multiple components were active, but tightly interlinked. And we could reasonably assume that these components could be shown to be functionally distinct in future experiments.

Second, Donchin's operational definition focuses exclusively on *experimental* (controlled) manipulations, completely ignoring variations that are spontaneous or correlational in nature. To take an obvious example of why this is an unnecessary limitation, consider a case in which a patient group and a control group differ in terms of the voltage between 150 and 250 ms over lateral frontal electrode sites. In this case, there must be at least one ERP component (according to my conceptual definition) that varies between these groups. This is just as convincing as an experimentally controlled within-subject difference in the voltage between 150 and 250 ms over lateral frontal electrode sites. Subtler examples are also important to consider. For example, if you look at the distribution of voltage over the scalp from moment to moment as the EEG spontaneously varies, it is possible to used advanced statistical techniques (e.g., principal component analysis, independent component analysis) to isolate systematic patterns of covariation across electrode sites that reflect spontaneous fluctuations in the amplitude of underlying ERP components.

When taking these factors into consideration, we can update Donchin's operational definition as follows:

An ERP component can be operationally defined as a set of voltage changes that are consistent with a single neural generator site and that systematically vary in amplitude across conditions, time, individuals, and so forth. That is, an ERP component is a source of systematic and reliable variability in an ERP data set.

This should not be taken to imply that we cannot isolate a component until we have determined its generator. If the scalp distribution of a putative component has a complex structure that is inconsistent with a single dipole, then this should not be considered a single component. But if it has a simple structure that is consistent with a single dipole (or with a pair of mirror-symmetric dipoles in the left and right hemispheres), then we can provisionally accept it as a single component.

How Can We Identify Specific ERP Components?

Now that we have both a conceptual definition and an operational definition for ERP components, we can consider in more detail a question that has come up many times in this chapter: when we see differences in the ERPs between different conditions or different groups in a given study, how can we determine which underlying components are responsible for these differences?

This is still not an easy question to answer, because our operational definition applies to a single data set and cannot be used to determine if the components (sources of consistent variability) that we see in one experiment are the same as the components that were observed in another experiment. Scalp distributions can be helpful, especially when derived from difference waves to eliminate the contribution of other overlapping components. If the scalp distribution of the effect in the new experiment is the same as the scalp distribution observed in previous experiments, then this is one piece of evidence in favor of it being the same component; if the scalp distribution is different, then this is evidence against the hypothesis. In practice, this can provide strong evidence *against* the hypothesis that two effects reflect the same component, but it cannot provide strong evidence *in favor* of the hypothesis. First, it is often impossible to provide direct quantitative comparisons of scalp distributions across experiments (e.g., when you are comparing your data with published data from another lab), and this will make it difficult to rule out the possibility that the scalp distribution in the new experiment is subtly but significantly different from the scalp distribution in a previous experiment. Second, concluding that two scalp distributions are the same requires accepting the null hypothesis, and conclusions of this sort are almost always weak. Source localization techniques can also provide evidence about whether generator sources are the same for ERP effects observed in different experiments, but this is essentially equivalent to asking whether the scalp distributions are the same and therefore has the same limitations.

We can also ask whether the component that we isolate in one experiment shows the same pattern of interaction with other factors as the component that was isolated in prior experiments. For example, if we believe we are seeing a P3b effect, we can ask whether it interacts with the probability of the stimulus category (see Luck et al., 2009; and see Experiment 4 in Vogel, Luck, & Shapiro, 1998). This can be a useful adjunct to comparisons of scalp distribution, but it requires a very solid theory of the interactions that should occur. Some of the clearest cases involve combinations of experimental factors and scalp distribution; as will be described in chapter 3, the N2pc, contralateral delay activity (CDA), and LRP components are isolated by computing a difference wave in which the waveforms recorded over the ipsilateral hemisphere (relative to an attended location or a response hand) are subtracted from the waveforms recorded over the contralateral hemisphere. These components can be very easily isolated from other ERP components, and they have been particularly useful for testing precise hypotheses about attention, working memory, and response selection.

In general, the difficult problem of identifying ERP components is solved in the same way that every other difficult problem in cognitive neuroscience is solved: with converging evidence. If you make a slight change in an experimental paradigm and you find the same effects as before (in terms of latency and scalp distribution), it is very likely that you have isolated the same underlying ERP components. If you then make a very large change in the paradigm, and now you find a very different latency or a somewhat different scalp distribution, you have to do some further research to determine whether you are still looking at the same component. If the latency is quite different, you can look for other evidence that the timing of the relevant process has changed. If the scalp distribution is a little different, you can look at whether the scalp distribution changes over the time course of the effect, suggesting that you now have two components contributing to the effect. If the scalp distribution is approximately the same as in your prior research, you can run additional experiments to see if the effect interacts with other manipulations in a way that is predicted by your hypothesis about the nature of the underlying component. You could also ask whether subject-to-subject differences in the scalp distribution of your new experimental effect are correlated with subject-to-subject differences in your previous effect.

The bottom line is that it is more difficult to isolate and identify specific ERP components than people typically realize, but this challenge can be overcome if you think carefully about the problem and come up with clever ways to test your hypotheses. Chapter 4 will describe a variety of strategies for overcoming this challenge, including circumventing the problem entirely by designing experiments that do not depend on identifying a specific ERP component.

Suggestions for Further Reading

Donchin, E. (1981). Surprise! … Surprise? *Psychophysiology*, *18*, 493–513.

Donchin, E., Ritter, W., & McCallum, W. C. (1978). Cognitive psychophysiology: The endogenous components of the ERP. In E. Callaway, P. Tueting, & S. H. Koslow (Eds.), *Event-Related Brain Potentials in Man* (pp. 349–441). New York: Academic Press.

Kappenman, E. S., & Luck, S. J. (2012). ERP components: The ups and downs of brainwave recordings. In S. J. Luck & E. S. Kappenman (Eds.), *The Oxford Handbook of ERP Components* (pp. 3–30). New York: Oxford University Press.

Makeig, S., & Onton, J. (2012). ERP features and EEG dynamics: An ICA perspective. In S. J. Luck & E. S. Kappenman (Eds.), *The Oxford Handbook of ERP Components* (pp. 51–86). New York: Oxford University Press.

Näätänen, R., & Picton, T. (1987). The N1 wave of the human electric and magnetic response to sound: A review and an analysis of the component structure. *Psychophysiology*, *24*, 375–425.

Picton, T. W., & Stuss, D. T. (1980). The component structure of the human event-related potentials. In H. H. Kornhuber & L. Deecke (Eds.), *Motivation, Motor and Sensory Processes of the Brain, Progress in Brain Research* (pp. 17–49). North-Holland: Elsevier.

3 Overview of Common ERP Components

Overview

This chapter provides an overview of the ERP components that are most commonly encountered in cognitive, affective, and clinical neuroscience research. Some of the individual components could justify an entire chapter, but I will cover just the basics. For more extensive descriptions, see the edited volume on ERP components that Emily Kappenman and I organized (Luck & Kappenman, 2012b).

ERP components are classically divided into three main categories: (1) *exogenous* sensory components that are obligatorily triggered by the presence of a stimulus (but may be modulated to some degree by top-down processes); (2) *endogenous* components that reflect neural processes that are entirely task-dependent; and (3) *motor* components that accompany the preparation and execution of a given motor response. Although these terms provide a useful first approximation for dividing up the various ERP components, the category boundaries are not always clear. For example, is the lateralized readiness potential—which reflects response preparation whether or not a response is actually emitted—an endogenous component or a motor component? Consequently, you should not worry too much about the boundaries between these terms.

It is important for you to learn about all the major ERP components, even the ones that don't seem relevant to your area of research. As an analogy, cardiologists need to know the anatomy and physiology of all the major systems of the human body because the heart influences the whole body and the whole body influences the heart. Similarly, even if your main interest is language, you need to know about ERP components related to all the major systems of the brain. This is important because components from other domains will influence language-elicited ERP waveforms, and also because the components from these other domains may end up being directly useful to you for testing hypotheses about language. As chapter 4 describes in more detail, some of the highest-impact ERP studies have "hijacked" a component from one domain and used it to study a different domain.

One of the central issues that must be addressed for a given component is the nature of the psychological or neural processes that it reflects. Indeed, the conceptual definition of the term *ERP component* that I presented in chapter 2 requires that a component be related to a particular computation. It turns out that this is a difficult issue to address, and an online supplement to this chapter will consider the general question of whether and how it is possible to assess the mental or neural function represented by a given component.

This chapter will describe the various ERP components in an order that roughly corresponds with their time of occurrence. We will begin with the contingent negative variation (CNV), which precedes a target stimulus, and then move on to the major visual and auditory sensory ERP components, with a brief discussion of sensory components from other modalities. We will then turn to the various components that contribute to the overall N2 wave, including the auditory mismatch negativity and the visual N2pc component (along with the distractor positivity [P_D] and contralateral delay activity [CDA] components, which are closely related to N2pc). This will be followed by a discussion of the P3 family of components and then a discussion of components related to language, long-term memory, emotion, error processing, and motor responses. We will then discuss steady-state ERPs, which are oscillatory neural responses that are evoked by rapidly oscillating stimuli. The final section will describe how four major ERP components were originally discovered. Before discussing individual components, however, it is necessary to say a few words about the naming conventions for ERP components (as briefly mentioned in chapter 1).

Naming Conventions

Unfortunately, the naming of ERP components is often inconsistent and sometimes ill-conceived. The most common convention is to begin with a P or N to indicate that the component is either positive-going or negative-going, respectively. This is followed by a number indicating the peak latency of the waveform (e.g., N400 for a negative component peaking at 400 ms) or the ordinal position of the peak within the waveform (e.g., P2 for the second major positive peak). This seems like a purely descriptive, theory-free approach, but it is not usually used this way. For example, the term *P300* was coined because it was positive and peaked at 300 ms when it was first discovered (Sutton, Braren, Zubin, & John, 1965). In most studies, however, the same functional brain activity typically peaks between 350 and 600 ms, but this component is still often labeled P300. Many investigators therefore prefer to use a number that represents the ordinal position of the component in the waveform (e.g., P3 instead of P300). This can still be confusing. For example, the first major peak for a visual stimulus is the P1 wave, which is observed over posterior electrode sites with a peak latency of approximately 100 ms. This component is not typically visible at anterior scalp sites, where the first major positive peak occurs at approximately 200 ms. This anterior positive peak at 200 ms is typically labeled *P2*, because it is the second major positive peak overall, even though it is the first positive peak in the waveform recorded at the anterior electrode sites.

Another problem is that a given label may refer to a completely different component when different sensory modalities are considered. For example, the auditory P1 wave bears no special relationship to the visual P1 wave. However, later components are largely modality-independent, and the labels for these components typically refer to the same brain activity whether the stimuli are auditory or visual. For example, N400 refers to the same brain activity whether the eliciting stimulus is auditory or visual.

Although the conventions for naming ERP components can be very confusing to novices, experts usually have no trouble understanding exactly what is meant by these names. This is just like the problem of learning words in natural languages: two words that mean different things may sound exactly the same (homophones); two different words may have the same meaning (synonyms); and a given word may be used either literally or metaphorically. This is certainly an impediment to learning both natural languages and ERP terminology, but it is not an insurmountable problem, and in both cases some work is needed to master the vocabulary.

ERP components are sometimes given more functional names, such as the *syntactic positive shift* (which is observed when the subject detects a syntactic error in a sentence) or the *error-related negativity* (which is observed when the subject makes an obviously incorrect behavioral response). These names are often easier to remember, but they can become problematic when subsequent research shows that the same component can be observed under other conditions. For example, some investigators have argued that the error-related negativity is not directly related to the commission of an error and is present (although smaller) even when the correct response is made (Yeung, Cohen, & Botvinick, 2004).

The Contingent Negative Variation and Stimulus-Preceding Negativity

As you may recall from the brief description in chapter 1, the contingent negative variation (CNV) is a broad negative deflection between a warning stimulus and a target stimulus (Walter, Cooper, Aldridge, McCallum, & Winter, 1964). When the period between the warning and target stimuli is lengthened to several seconds, it is possible to see that the CNV actually consists of a negativity after the warning stimulus, a return to baseline, and then a negativity preceding the target stimulus (Loveless & Sanford, 1975; Rohrbaugh, Syndulko, & Lindsley, 1976). The first negative phase may reflect processing of the warning stimulus, and the second negative phase may reflect the readiness potential that occurs as the subject prepares to respond to the target (for a more nuanced account, see the comprehensive review by Brunia, van Boxtel, & Böcker, 2012).

A related ERP component is the stimulus-preceding negativity (SPN; see review by van Boxtel & Böcker, 2004). The SPN is a negativity that grows as the subject anticipates the occurrence of an information-bearing stimulus, such as a feedback tone, irrespective of whether an overt response is required for this stimulus. The SPN can be one of the subcomponents of the CNV.

It is important for you to know about these components because they can create confounds in your experiments. For example, if the interval between stimuli is relatively long in an experiment, an SPN will occur prior to each stimulus, and this can confound the results if the degree of anticipation differs across conditions.

Visual Sensory Responses

C1

Although the first major visual ERP component is typically the P1 wave, the P1 is sometimes preceded by a component called *C1*, which is largest at posterior midline electrode sites. Unlike most other components, it is not labeled with a P or an N because its polarity can vary. The C1 wave appears to be generated in area V1 (primary visual cortex), which in humans is folded into the calcarine fissure (figure 3.1). The part of area V1 that codes the lower visual field is on the upper bank of the fissure, and the part that codes the upper visual field is on the lower bank. In both cases, the positive side of the dipole points to the cortical surface, but the cortical surface

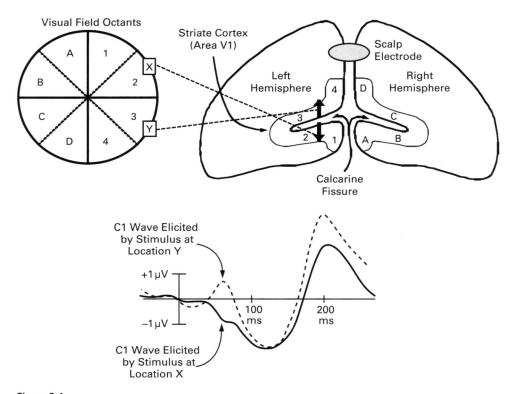

Figure 3.1
Relationship between stimulus position and C1 polarity. If a subject is staring at the center of the circle shown on the left side of the figure, the visual field can be divided into eight octants. Each octant projects to a different region of striate cortex, which wraps around the calcarine fissure as shown on the right side of the figure. If a stimulus is presented at location X in octant 2, just above the horizontal meridian, this will generate a dipole with a positive end that points downward and a negative end that points upward. This will lead to a negative potential recorded at a scalp electrode located on the midline (e.g., at the Oz or Pz electrode locations). If a stimulus is instead presented at location Y in octant 3, just below the horizontal meridian, the dipole will be located on the other side of the calcarine fissure, and the positive end will therefore point upward. This will lead to a positive potential at the scalp electrode.

is flipped for the upper bank of the fissure relative to the lower bank. As a result, the voltage recorded from an electrode above the calcarine fissure is positive for stimuli in the lower visual field and negative for stimuli in the upper visual field (Jeffreys & Axford, 1972; Clark, Fan, & Hillyard, 1995). When the C1 wave is positive, it sums together with the P1 component, creating a single positive-going wave. Consequently, a distinct C1 wave is not usually observed unless upper-field stimuli are used to generate a negative C1 wave (which can easily be distinguished from the positive P1 wave). The C1 wave typically onsets 40–60 ms poststimulus and peaks 80–100 ms poststimulus, and it is highly sensitive to basic visual stimulus parameters, such as contrast and spatial frequency.

The C1 wave provides my favorite example of the localization of an ERP component because the evidence does not just come from mathematical source localization techniques. Instead, the evidence comes from the confirmation of several predictions that derive from the hypothesis that C1 is generated in area V1. The most important of these predictions is that the C1 wave should invert in polarity for upper versus lower field stimuli, which is based on the known anatomy and physiology of visual cortex. There is no other area of visual cortex that is folded in a manner that would lead to this inversion. Moreover, the polarity actually flips slightly below the horizontal midline, which fits perfectly with the details of the mapping of the visual field onto area V1 (as discussed by Clark et al., 1995). The hypothesis that C1 is generated in area V1 also predicts that it should onset before any other visual component, and this hypothesis is also confirmed by the C1 onset latency. This hypothesis also predicts that the scalp distribution of the C1 wave should be consistent with a generator near the location of the calcarine fissure, which has also been verified (Clark et al., 1995; Di Russo, Martinez, Sereno, Pitzalis, & Hillyard, 2002). Thus, strong converging evidence supports the hypothesis that C1 is generated in area V1.

P1 and N1

The C1 wave is followed by the P1 wave, which is largest at lateral occipital electrode sites and typically onsets 60–90 ms poststimulus with a peak between 100 and 130 ms. Note, however, that P1 onset time can be difficult to assess because of overlap with the C1 wave. In addition, P1 latency will vary substantially depending on stimulus contrast. A few studies have attempted to localize the P1 wave by means of mathematical modeling procedures, sometimes combined with co-localization with fMRI effects, and these studies suggest that the early portion of the P1 wave arises from dorsal extrastriate cortex (in the middle occipital gyrus), whereas a later portion arises more ventrally from the fusiform gyrus (see Di Russo et al., 2002). Note, however, that many areas across the cortex are activated within the first 100 ms after the onset of a visual stimulus, and many of these areas presumably contribute to the voltages recorded in the C1 and P1 latency range (Foxe & Simpson, 2002). Like the C1 wave, the P1 wave is sensitive to variations in stimulus parameters, as would be expected given its likely origins in extrastriate visual cortex. The P1 wave is also modulated by selective attention (see review by Hillyard, Vogel, & Luck, 1998; Luck, Woodman, & Vogel, 2000) and by the subject's state of arousal (Vogel &

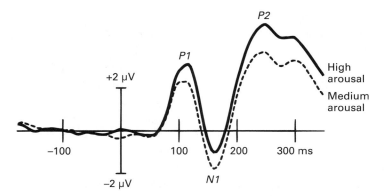

Figure 3.2
ERPs elicited by foveal visual stimuli under conditions of medium versus high arousal (from the study of Vogel & Luck, 2000). Note that arousal influences the entire waveform, beginning with the P1 wave.

Luck, 2000). Other top-down variables do not appear to reliably influence the P1 wave. In particular, P1 amplitude is not sensitive to whether a stimulus matches a task-defined target category (Hillyard & Münte, 1984).

The effect of arousal is illustrated in figure 3.2. In this experiment (Vogel & Luck, 2000), subjects pressed a button whenever they detected a stimulus (a foveal array of five letters). In the *medium arousal* condition, subjects were simply given our usual instructions for RT tasks; namely, "to respond as quickly as possible." In the *high arousal* condition, subjects were encouraged to respond even faster. They were given feedback at the end of each block of trials that consisted of their mean RT for that block and a message stating that they must respond even faster in the next block. This manipulation was effective: mean RT was 292 ms in the medium arousal condition and 223 ms in the high arousal condition. As shown in figure 3.2, the amplitude of the P1 wave was increased in the high arousal condition relative to the low arousal condition. However, the difference between conditions persisted well after 300 ms, influencing the amplitude over the entire waveform. It is not clear which underlying components were affected later in the waveform, but it is clear that arousal can influence the measured value at virtually any point in the ERP waveform. It is therefore important to control for arousal in ERP experiments. For example, if two conditions are tested in separate trial blocks, and one condition is more difficult than the other, this may lead to arousal differences that confound the ERP measurements (see chapter 4 for additional discussion of this issue).

The P1 wave is followed by the N1 wave. There are several visual N1 *subcomponents*. That is, there are several distinct components that sum together to form the N1 peak. These components are not necessarily functionally related, and they are called *subcomponents* only because they contribute to the same visually salient deflection in the waveform. The earliest N1 subcomponent peaks 100–150 ms poststimulus at anterior electrode sites, and there appear to be at least two posterior N1 components that peak 150–200 ms poststimulus, one arising from parietal

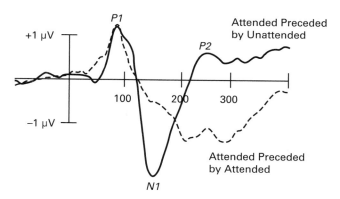

Figure 3.3
ERPs elicited by lateralized visual stimuli at an attended location that were preceded either by a stimulus at the same attended location or at an unattended location in the opposite visual field (from the study of Luck et al., 1990). Note that amplitude of the N1 wave is greatly reduced when the previous stimulus was at the same location.

cortex and another arising from lateral occipital cortex. Many studies have shown that all three N1 subcomponents are influenced by spatial attention (see reviews by Mangun, 1995; Hillyard et al., 1998). In addition, the lateral occipital N1 subcomponent appears to be larger when subjects are performing discrimination tasks than when they are performing detection tasks, which has led to the proposal that this subcomponent reflects some sort of discriminative processing (Ritter, Simson, Vaughan, & Friedman, 1979; Vogel & Luck, 2000; Hopf, Vogel, Woodman, Heinze, & Luck, 2002).

The visual N1 component is highly *refractory*. That is, if a stimulus at a particular location is preceded by another stimulus at the same location at a short delay, the response to the second stimulus is greatly reduced. This is illustrated in figure 3.3 (from the study of Luck, Heinze, Mangun, & Hillyard, 1990). The figure shows the ERP recorded at temporal-occipital electrodes in response to an attended stimulus that was preceded by either an attended stimulus at the same location or an unattended stimulus at a different location. Although the P1 wave was approximately equivalent for these two trial types, the N1 wave was dramatically reduced when the attended stimulus was preceded by an attended stimulus in the same location. Many ERP components show effects such as this under some conditions, and this creates confounds in many experiments (to read about how this confounded some of my own experiments, see box 4.5 in chapter 4).

N170 and the Vertex Positive Potential

Jeffreys (1989) compared the responses to faces and non-face stimuli, and he found a difference from 150 to 200 ms at central midline sites that he named the *vertex positive potential* (VPP; note that the Cz electrode site at the very top of the head is sometimes called the *vertex* site). Jeffreys noted that this effect inverted in polarity at more lateral sites, but he did not have any recordings from electrode sites over inferotemporal cortex. More recent studies using a broader range of electrode sites have found that faces elicit a more negative potential than non-face stimuli at lateral occipital electrode sites, especially over the right hemisphere, with a peak at approximately 170 ms (Bentin, Allison, Puce, Perez, & McCarthy, 1996; Rossion et al., 1999). This effect is typically called the *N170* wave (see figure 1.2 in chapter 1). It is likely that the N170 and the VPP are just the opposite sides of the same dipole (see review by Rossion & Jacques, 2012); the relative sizes of the N170 and VPP depend on what reference location is used (see chapter 5 for an in-depth discussion of reference electrodes). Note that the N170 is one of the subcomponents of the N1 wave. As was discussed in chapter 2, the voltage measured at 170 ms reflects the sum of the N170 component and other face-insensitive components, and you should be careful not to assume that the voltage elicited by a face in this time range solely reflects the face-sensitive N170 component.

A key question is whether the N170 effect (the difference between face and non-face stimuli) truly reflects face-specific processing or whether it reflects low-level properties that just happen to be more prominent in faces. Several lines of evidence now provide strong evidence that this effect is truly related to face perception (reviewed by Rossion & Jacques, 2012). For example, research by the late, great Shlomo Bentin showed that simple stimuli, which are not ordinarily perceived as faces, will elicit a larger N170 when subjects are primed to perceive these stimuli as faces (Bentin & Golland, 2002; Bentin, Sagiv, Mecklinger, Friederici, & von Cramon, 2002).

Much fMRI research has asked whether the same neural mechanisms that are ordinarily used to process faces are also used to process other complex stimuli for which the perceiver is an expert (e.g., cars for people who are car experts). The N170 has been used to address the same issue. Although ERPs lack the anatomical specificity of fMRI, their greater temporal resolution makes it possible to ask whether the timing of the processing is the same for faces and non-face objects. This is important because fMRI effects that are the same for faces and non-face objects might reflect processes that occur long after perception is complete. By showing that both faces and non-face objects elicit the same negative-going component at 170 ms, we can have more confidence that these effects reflect perception rather than some later postperceptual process. Supporting this possibility, initial studies demonstrated that bird and dog experts exhibit enhanced N170 responses for birds and dogs, respectively (Tanaka & Curran, 2001). But how can we know that this effect reflects the same neural circuits that produce the N170 for faces? A very clever set of experiments by Bruno Rossion and his colleagues answered this question by showing that the simultaneous presentation of the object of expertise and a face will reduce the N170 elicited by the face, which is typically taken as a sign of competition for the same neural circuits (Rossion, Kung, & Tarr, 2004; Rossion, Collins, Goffaux, & Curran, 2007).

Several studies have attempted to localize the N170 and its magnetic counterpart, the M170 (reviewed by Rossion & Jacques, 2012). Different studies have come up with substantially different estimates of the generator location, leading Rossion and Jacques (2012) to conjecture that several different anatomical areas (e.g., the fusiform face area and the occipital face area) contribute to the scalp N170/M170.

P2

A distinct P2 wave follows the N1 wave at anterior and central scalp sites. This component is larger for stimuli containing target features, and this effect is enhanced when the targets are relatively infrequent (see Luck & Hillyard, 1994a). For example, figure 3.4B shows data recorded from the Cz electrode site in a visual oddball paradigm, in which the oddballs elicited a larger P2, N2, and P3 relative to the standards. Although both P2 and P3 are larger for oddball stimuli in some paradigms, the P2 effect occurs only when the target is defined by fairly simple stimulus features, whereas the P3 effect can occur for arbitrarily complex target categories. At posterior

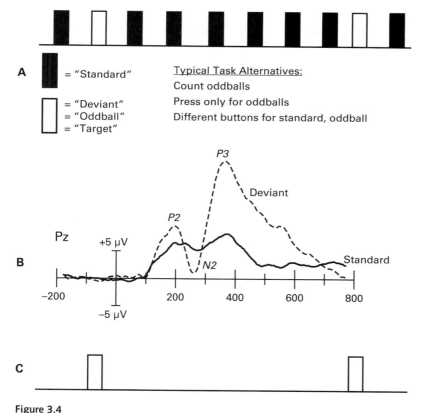

Figure 3.4
(A) Stimulus sequence in a typical oddball experiment, in which 80% of the stimuli are standards and 20% are targets (also known as oddballs or deviants). (B) Typical ERPs elicited at the Pz electrode site by standards and deviants in a visual oddball paradigm. (C) Target-only paradigm, in which the targets are presented at the same time as in the oddball paradigm, but without the standards. The P3 in this paradigm is nearly identical to the P3 in the traditional paradigm.

sites, the P2 wave is often difficult to distinguish from the overlapping N1, N2, and P3 waves. Consequently, not as much is known about the posterior P2 wave.

Auditory Sensory Responses

Figure 3.5 shows the typical ERP components evoked by the presentation of an auditory stimulus (see reviews by Picton, 2011; Pratt, 2012). If the stimulus has a sudden onset (such as a click), a distinctive set of peaks can be seen over the first 10 ms that reflect the flow of information from the cochlea through the brainstem and into the thalamus. These *auditory brainstem responses* (ABRs) are typically labeled with Roman numerals (waves I–VI). They are highly automatic and can be used to assess the integrity of the auditory pathways, especially in infants.

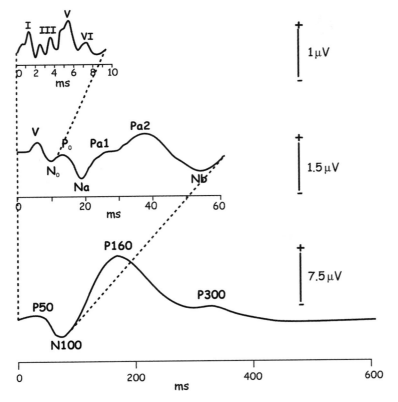

Figure 3.5
Typical sequence of auditory sensory components. The waveform elicited by a click stimulus is shown over different time ranges with different filter settings to highlight the auditory brainstem responses (top), the midlatency responses (middle), and the long-latency responses (bottom). Adapted with permission from Pratt (2012). Copyright 2012 Oxford University Press.

When my children were born, they were both given ABR screening tests, and it was gratifying to see that a variant of my main research technique is commonly used in a clinical application.

The ABRs are followed by the *midlatency responses* (MLRs; defined as responses between 10 and 50 ms), which probably arise at least in part from the medial geniculate nucleus and the primary auditory cortex. Attention has its first reliable effects in the midlatency range, but I don't know of any other cognitive variables that influence auditory activity in this time range.

The MLRs are followed by the *long-latency responses*, which typically begin with the P50 (aka P1), N100 (aka N1), and P160 (aka P2). The phrase *long-latency response* is a bit confusing, because these are relatively short latencies compared to high-level cognitive components, such as P3 and N400. However, the transmission of information along the auditory pathway is very fast, and 100 ms is relatively late from the perspective of auditory sensory processing. The long-latency auditory responses can be strongly influenced by high-level factors, such as attention and arousal.

Like the visual N1 wave, the midlatency and long-latency auditory responses become much smaller when the interval between successive stimuli decreases, with refractory periods that may exceed 1000 ms (this is true for sensory components in other modalities as well). Thus, when evaluating an ERP study, it is important to assess whether a difference between groups or conditions might be confounded by differences in the interstimulus interval.

Like the visual N1 wave, the auditory N1 wave has several distinct subcomponents (see review by Näätänen & Picton, 1987). These include (1) a frontocentral component that peaks around 75 ms and appears to be generated in the auditory cortex on the dorsal surface of the temporal lobes, (2) a vertex-maximum potential of unknown origin that peaks around 100 ms, and (3) a more laterally distributed component that peaks around 150 ms and appears to be generated in the superior temporal gyrus. Further fractionation of the auditory N1 wave is possible (see, e.g., Alcaini, Giard, Thevenet, & Pernier, 1994). The N1 wave is sensitive to attention: Although some attention effects in the N1 latency range reflect the addition of an endogenous component, the N1 wave itself (or at least some subcomponents) can be influenced by attention (Woldorff et al., 1993).

Figure 3.5 illustrates a common feature of ERP waveforms, namely, that the peaks tend to be narrow early in the waveform and become progressively broader later in the waveform. The ABR peaks, for example, last only 1–2 ms each, whereas the MLR peaks last 10–20 ms each, and some of the "cognitive" peaks such as P3 may last for several hundred milliseconds. This is not a coincidence. In almost any physical system, timing precision relative to a starting point becomes less and less precise as time progresses. This simply reflects the accumulation of timing errors. For example, if there is a 1-ms variance in timing a 10-ms interval, and we want to time a 70-ms interval, then the 1-ms variance associated with each of the seven successive 10-ms intervals will add together, and there will be a 7-ms variance in the timing of the 70-ms interval. Consequently, the variance across trials in the starting time of a given ERP component will tend to be larger for later components. This trial-to-trial variance will cause the component to become temporally "smeared" when the individual trials are averaged together (chapter 8 and online chapter 11 will provide a more precise way of describing and understanding the exact nature of this smearing, using the mathematical concept of *convolution*).

Somatosensory, Olfactory, and Gustatory Responses

The vast majority of cognitive ERP experiments use auditory or visual stimuli, so I will provide only a brief mention of components from other modalities. The response to a somatosensory stimulus (see review by Pratt, 2012) begins with one of the rare ERP components that reflects action potentials rather than postsynaptic potentials, arising from the peripheral nerves. This *N10* response is followed by a set of subcortical components (ca. 10–20 ms) and a set of short- and medium-latency cortical components (ca. 20–100 ms). An N1 wave is then observed at approximately 150 ms, followed by a P2 wave at approximately 200 ms.

It is difficult to record olfactory and gustatory ERP responses, largely because it is difficult to deliver precisely timed, sudden-onset stimuli in these modalities (which is necessary when averaged ERP waveforms are computed). However, these potentials can be recorded when appropriate stimulation devices are used (see, e.g., Wada, 1999; Ikui, 2002; Morgan & Murphy, 2010; Singh, Iannilli, & Hummel, 2011).

The N2 Family

Many clearly different components have been identified in the time range of the second major negative peak (for reviews, see Näätänen & Picton, 1986; Folstein & Van Petten, 2008; Luck, 2012b). Early reports of N2 components typically came from oddball experiments, in which infrequently occurring targets (aka oddballs or deviants) are interspersed among frequently occurring standards (see figure 3.4A; see also the discussion of the oddball paradigm near the beginning of chapter 1).

Subdivision into N2a, N2b, and N2c Subcomponents
As described by Näätänen and Picton (1986), a repetitive, nontarget stimulus will elicit an N2 deflection that can be thought of as the *basic N2* (although it doubtless contains several subcomponents). If deviant stimuli are occasionally presented within the repetitive train (e.g., in an oddball paradigm), a larger amplitude is observed in the N2 latency range for the deviants. Early research (summarized by Pritchard, Shappell, & Brandt, 1991) suggested that this effect could be divided into three subcomponents—termed *N2a*, *N2b*, and *N2c*—on the basis of manipulations of attention and stimulus modality. The N2a is an automatic effect that occurs for auditory mismatches, even if they are task-irrelevant. This effect is more commonly known as the *mismatch negativity* (MMN). Task-irrelevant visual oddballs can also elicit a negative-going wave from approximately 100 to 200 ms (Czigler, Balazs, & Winkler, 2002), but the early portion of this effect appears to reflect temporal rareness rather than mismatch detection per se (Kenemans,

Jong, & Verbaten, 2003). Thus, the MMN appears to be a mainly auditory phenomenon. The MMN will be discussed in more detail in a later section.

If the deviants are task-relevant, then a somewhat later N2 effect is observed, largest over central sites for auditory stimuli and over posterior sites for visual stimuli (Simson, Vaughan, & Ritter, 1977). These anterior and posterior effects were labeled *N2b* and *N2c*, respectively. More recent research has shown that an anterior N2 effect can also be observed for visual stimuli under some conditions (as described in the next section). The terms *N2b* and *N2c* have therefore been largely replaced by the terms *anterior N2* and *posterior N2*.

Anterior N2

There are several anterior N2 (N2b) effects that have been repeatedly reported in the literature (see the excellent review by Folstein & Van Petten, 2008). One common effect is related to response inhibition. Figure 3.6, for example, shows an example from the go/no-go paradigm (Bruin & Wijers, 2002). In this paradigm, subjects make a manual response for one stimulus and withhold this response for the other stimulus. When the no-go stimulus is presented, this produces conflict between the go response and the no-go response, especially when the go stimulus is more common than the no-go stimulus. Accordingly, the no-go N2 is largest when the go stimulus is more common than the no-go stimulus.

The anterior N2 is also sensitive to mismatch, but only when the stimuli are attended. That is, subjects must be performing some kind of task with the stimuli, even if it does not explicitly involve discriminating mismatches. For example, Suwazono, Machado, and Knight (2000) presented subjects with 70% standard stimuli (simple shapes), 20% targets (triangles), and 10% novel nontarget stimuli (color photographs). Even though the task did not require subjects to discriminate between the simple standards and the novel stimuli, the novel stimuli elicited an enhanced anterior N2. Similarly, Luck and Hillyard (1994a) presented subjects with homogeneous arrays

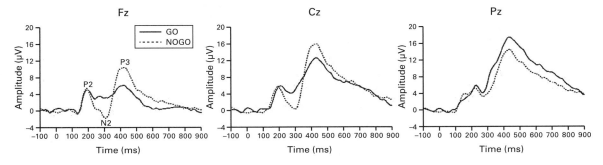

Figure 3.6
Grand average ERPs from the go/no-go study of Bruin and Wijers (2002). In this study, the letters M and W were presented at fixation, and the subject was instructed to make a finger-lift response for one stimulus but not the other (counterbalanced across trial blocks). The probabilities of the go and no-go stimulus were .25/.75, .50/.50, or .75/.25. The data shown here overlay the waveforms for go stimuli when they had a probability of .25 and no-go stimuli when they had a probability of .25. Adapted with permission from Bruin and Wijers (2002).

of colored rectangles or arrays that contained a "pop-out" item, which could be deviant in color, shape, or size. One of the three deviant types was the target, but both the target and nontarget deviants elicited a larger anterior N2 than the homogeneous arrays. Moreover, this effect was relatively independent of the probability of the deviants. However, it disappeared when the subjects were not attending to any of the three deviants. Thus, subjects must be attending to the stimulus arrays for this effect to be observed, but they do not need to be attending to the specific mismatch dimension.

Several anterior N2 effects are related to conflict between competing response alternatives. The no-go N2, for example, may reflect competition between the go response and the no-go "response." An increased anterior N2 is also seen in the Eriksen flankers task. In this task, subjects make one of two responses to a central stimulus, and simultaneous task-irrelevant flanking stimuli may be assigned to the same response as the target (*compatible* trials) or to the opposite response (*incompatible* trials). The anterior N2 is typically larger for incompatible trials than for compatible trials (Folstein & Van Petten, 2008), presumably because the flanking stimuli activate the incorrect response to some extent, which then conflicts with the activation of the target response (e.g., Gratton, Coles, Sirevaag, Eriksen, & Donchin, 1988; Gehring, Gratton, Coles, & Donchin, 1992; Yeung et al., 2004). In fact, Yeung et al. (2004) proposed that the anterior N2 is actually the same as the error-related negativity (described later); as the activation of the incorrect response increases, the likelihood that the incorrect response is actually emitted also increases, so this brain activity will be larger when subjects actually make an incorrect response. A larger anterior negativity is also seen on incompatible trials relative to compatible trials in the Stroop paradigm (West & Alain, 1999, 2000; Liotti, Woldorff, Perez, & Mayberg, 2000). This effect is considerably later than the anterior N2 effect in the flankers paradigm, but it is possible that it nonetheless reflects the same process.

The anterior N2 component is also observed in the stop-signal paradigm. In this paradigm, the subject tries to make a speeded response to a visual target, but on some trials a tone occurs a few hundred milliseconds after the target, indicating that the response should be withheld. The anterior N2 is larger when subjects are unsuccessful at withholding the response than when they are successful (Pliszka, Liotti, & Woldorff, 2000; van Boxtel, van der Molen, Jennings, & Brunia, 2001), consistent with the idea that conflict should be greatest when go representation is so strong that it cannot be overcome by the stop representation.

An anterior N2 effect is also observed for feedback indicating whether a response was correct or incorrect. This is called the *feedback-related negativity* (FRN), and it is typically larger for negative feedback than for positive feedback (Holroyd & Coles, 2002; Nieuwenhuis, Yeung, Holroyd, Schurger, & Cohen, 2004; Hajcak, Holroyd, Moser, & Simons, 2005). Again, this can be considered a case of conflict between the response that was made by the subject and the correct response that was indicated by the feedback.

It is tempting to assume that all of these anterior N2 effects reflect the same underlying component, generated in medial frontal cortex (possibly the anterior cingulate cortex). However, fMRI data indicate that several distinct frontal brain regions exhibit activity related to the variables that influence the anterior N2 (see, e.g., Ridderinkhof, Ullsperger, Crone, & Nieuwenhuis, 2004). It is therefore quite plausible that multiple ERP components contribute to the anterior N2, just as multiple ERP components may contribute to the N170.

Posterior N2

The typical posterior N2 (N2c) is very much like the P3 wave, insofar as it is seen for task-relevant targets and is larger for rare targets than for frequent targets. Renault, Ragot, Lesevre, and Redmond (1982) proposed that this component reflects the process of categorizing a stimulus, because the duration of the component (measured from the target-minus-standard difference wave) depends on the difficulty of the categorization. However, increasing the difficulty of categorization also increases the onset latency of the P3 wave, and this may artificially change the apparent duration of the N2c component. Thus, the functional significance of the posterior N2 component is not clear.

The N2pc (N2-posterior-contralateral) component occurs during approximately the same time interval as the N2c component. The N2pc component is observed at posterior scalp sites contralateral to an attended object and reflects some aspect of the focusing of attention. The N2c and N2pc components can be distinguished from each other because the N2c is highly sensitive to the probability of the target whereas the N2pc is not (Luck & Hillyard, 1994a). The N2pc component will be discussed in greater detail in a later section of this chapter.

The Mismatch Negativity

The mismatch negativity (MMN) is a relatively automatic response to an auditory stimulus that differs from the preceding stimuli. It typically peaks between 160 and 220 ms, with a fronto-central midline scalp maximum.

Figure 3.7 shows the results of a typical MMN experiment (Näätänen & Kreegipuu, 2012). Tones were presented at a rate of approximately 1 per second while the subjects read a book and ignored the tones. Most (80%) of the tones were 1000-Hz standards, and the rest (20%) were slightly higher-pitched tones (1004, 1008, 1016, or 1032 Hz). The brain cannot easily discriminate between 1000, 1004, and 1008 Hz, so there was little or no difference in the ERP waveforms elicited by these three tones. However, the 1016- and 1032-Hz deviants elicited a greater negativity than the 1000-Hz standards from approximately 100 to 220 ms, and this greater negativity is the MMN. This MMN is often isolated from the rest of the ERP waveform with a deviant-minus-standard difference wave (shown in the right column of figure 3.7).

The most widely accepted theory of the MMN proposes that it reflects the comparison of a short-lived memory trace of the standards with the current stimulus (see review by Näätänen & Kreegipuu, 2012). However, the MMN effect is often contaminated by other brain activity. As was described earlier in this chapter, the auditory N1 wave is highly refractory. When the same pitch is presented many times, the neurons responding to this pitch will become less responsive, leading to a smaller N1 wave for this pitch. When a different pitch is presented, a somewhat

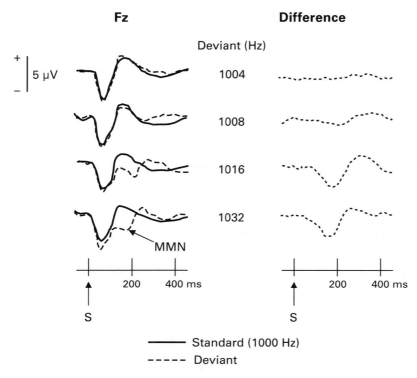

Figure 3.7
Example mismatch negativity results. While the subjects read a book, brief tones were presented at a rate of approximately 1 tone per second, with 80% standard tones (1000 Hz) and 20% deviant tones (1004, 1008, 1016, or 1032 Hz). The mismatch negativity is isolated by difference waves (right column) in which the ERP elicited by the standard stimulus is subtracted from the ERP elicited by the deviant stimulus. Adapted with permission from Näätänen and Kreegipuu (2012). Copyright 2012 Oxford University Press.

different population of neurons will be stimulated, and these neurons will produce a larger response than the neurons that code the standard pitch. This N1 difference between standards and deviants may therefore contribute to the MMN effect.

Three pieces of evidence indicate that, although this type of refractory confound can contribute to the MMN, a portion of the MMN cannot be explained in this manner. Specifically, an MMN can be elicited by tones that deviate from the standard by having a lower intensity (e.g., Woldorff, Hackley, & Hillyard, 1991; Woldorff, Hillyard, Gallen, Hampson, & Bloom, 1998), by tones occurring at an unexpectedly early time (Ford & Hillyard, 1981), and by the omission of a stimulus (Rüsseler, Altenmuller, Nager, Kohlmetz, & Munte, 2001). None of these results could be explained by a release from refractoriness when the deviant is presented. Studies that wish to isolate the "true" MMN from N1 refractory effects should use one of these procedures.

The MMN is often called *preattentive* and *automatic* because it is observed even if subjects are not using the stimulus stream for a task (e.g., if they are reading a book while the stimuli are being presented). However, the MMN can be eliminated for stimuli presented in one ear if the subjects focus attention very strongly on competing stimuli in the other ear (Woldorff et al., 1991). The problem with the terms *preattentive* and *automatic* is that attention is a complex set of cognitive processes that operate at different stages of processing under different conditions (see Luck & Vecera, 2002; Luck & Kappenman, 2012a), making it difficult to draw a clear line between *preattentive* and *attentive* activity or between *automatic* and *controlled* processing. It is best to steer clear of these terms when describing the MMN and stick to the facts. That is, you might describe the MMN as, "a response that is sufficiently automatic that it can be observed when the subject is reading a book."

Because of its relatively high degree of automaticity, the MMN has been very useful as a means of assessing processing in individuals who cannot easily make behavioral responses, such as preverbal infants (Csepe, 1995; Trainor et al., 2003) and people who are comatose (Fischer, Luaute, Adeleine, & Morlet, 2004). For example, the MMN can be used to assess infants' sensitivity to various linguistic contrasts (Dehaene-Lambertz & Baillet, 1998; Cheour, Leppanen, & Kraus, 2000).

As reviewed by Näätänen and Kreegipuu (2012), at least two cortical generator sources contribute to the MMN. One is located in the auditory cortex of the supratemporal plane, just anterior to the N1 generator site. This generator leads to a negativity over the front of the head and a positivity at inferior posterior electrodes (when a nose reference site is used), and it is thought to reflect the mismatch detection process itself (see chapter 5 for more on the effects of reference electrode placement). The second is thought to arise from the prefrontal cortex and to reflect processes involved in shifting attention to the deviant sound. There is some intriguing evidence that the MMN specifically reflects the flow of current through NMDA receptor-mediated ion channels (Javitt, Steinschneider, Schroeder, & Arezzo, 1996; Umbricht et al., 2000; Kreitschmann-Andermahr et al., 2001; Ehrlichman, Maxwell, Majumdar, & Siegel, 2008; Heekeren et al., 2008; Tikhonravov et al., 2008), but some contradictory evidence has also been reported (Oranje et al., 2000).

N2pc, Distractor Positivity, and Contralateral Delay Activity

In this section, we will consider three high-level visual components that are isolated from the rest of the ERP waveform by measuring the difference in voltage between electrodes contralateral and ipsilateral to an object that is being attended (N2pc), an object that is being suppressed (distractor positivity; P_D), or an object that is being maintained in working memory (contralateral delay activity; CDA). The ability to effectively isolate these components from other components has made them extremely valuable for answering specific questions about high-level visual processes.

N2pc

As described earlier in this chapter, the posterior N2 contains a subcomponent that is called *N2-posterior-contralateral* (N2pc) because it is larger at contralateral sites than at ipsilateral sites relative to the location of an attended visual object (see review by Luck, 2012b).

Consider, for example, the experiment shown in figure 3.8 (from the study of Luck et al., 2006). Each stimulus array contained one red square, one green square, and a large number of black distractor squares. The locations of the individual squares varied at random from trial to trial, with the constraint that the two colored items were always on opposite sides of the display. The subject was instructed to attend either to red or to green at the beginning of each trial block

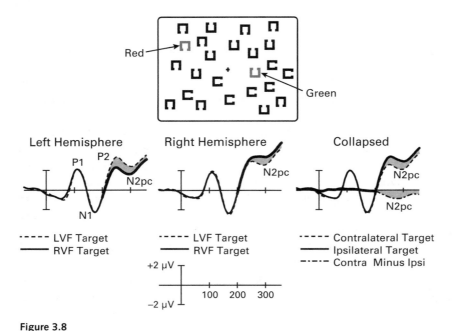

Figure 3.8
Typical N2pc paradigm and grand average ERP waveforms from posterior occipito-temporal electrode sites (from the study of Luck et al., 2006). To avoid any possibility of physical stimulus confounds, each stimulus array contained a distinctly colored item on each side, and one of these two colors was designated as the target color in the instruction screen for each trial block. Thus, the same stimulus array could be used to induce subjects to focus on either the left or right visual field, depending on which color was defined as the target. The locations of the items varied at random from trial to trial, except that the two pop-out colors were always on opposite sides. The subject was required to press one of two buttons to indicate whether the gap in the target item was at the top or the bottom of the square. The voltage in the N2 latency range over the left hemisphere was more negative when the target was in the right visual field than when it was in the left visual field, and the voltage over the right hemisphere was more negative when the target was in the left visual field than when it was in the right visual field. The data were collapsed into an ipsilateral waveform (left hemisphere/left target averaged with right hemisphere/right target) and a contralateral waveform (left hemisphere/right target averaged with right hemisphere/left target). The N2pc is defined as the difference between these contralateral and ipsilateral waveforms (shown as the shaded region), which was made explicit by constructing a contralateral-minus-ipsilateral difference wave. Adapted with permission from Luck (2012b). Copyright 2012 Oxford University Press.

and to press one of two buttons on each trial to indicate whether the attended-color object contained a gap on its top or a gap on its bottom. Because either red or green could be the target color (depending on the instructions given to the subject at the beginning of the trial), the same physical stimulus arrays could be used for trials that required focusing attention on the left (e.g., attend-red for the array shown in figure 3.8) or on the right (e.g., attend-green for the array shown in figure 3.8). This was important for avoiding a variety of stimulus confounds.

The N2pc component consists of a greater negativity when the attended item is contralateral to the recording electrode than when the attended item is ipsilateral (see the end of the chapter for the story of how N2pc was discovered). It typically occurs during the time range of the N2 wave (200–300 ms) and is observed at posterior scalp sites over visual cortex, with a maximum voltage near the PO7 and PO8 electrodes. In figure 3.8, the N2pc can be seen over the left hemisphere as a more negative voltage for targets in the right visual field (RVF) than for targets in the left visual field (LVF), and it can also be seen over the right hemisphere as a more negative voltage for LVF targets than for RVF targets. To avoid overall differences between LVF and RVF targets and overall differences between the left and right hemispheres, it is useful to create a collapsed contralateral waveform (the average of RVF for the left hemisphere and LVF for the right hemisphere) and a collapsed ipsilateral waveform (the average of LVF for the left hemisphere and RVF for the right hemisphere). The difference between these contralateral and ipsilateral waveforms is used to isolate the N2pc component from other overlapping ERP components. The rightmost portion of figure 3.8 shows the contralateral-minus-ipsilateral difference wave, illustrating how this difference wave eliminates the other components (e.g., P1 and N1). If you study the auditory modality, you might be interested to know that Marissa Gamble and I discovered an auditory analog of the N2pc component, which we called the *N2ac* component because it was observed over anterior contralateral electrode sites (Gamble & Luck, 2011). This is shown in figure 10.6 in chapter 10.

The N2pc component is useful for determining whether attention has been covertly directed to a given object and for assessing the time course of attentional orienting. For example, although an N2pc is typically observed for both simple feature targets and more complex conjunction targets, it is larger for conjunction targets than for feature targets (Luck, Girelli, McDermott, & Ford, 1997), and it can be eliminated for feature targets but not for conjunction targets when subjects are performing an attention-demanding secondary task (Luck & Ford, 1998). Similarly, N2pc can be used to determine whether attention is automatically captured by salient but irrelevant objects (Eimer & Kiss, 2008; Lien, Ruthruff, Goodin, & Remington, 2008; Sawaki & Luck, 2010). N2pc has also been used to show that masked, subliminal objects can nonetheless attract attention (Woodman & Luck, 2003a). The timing of the N2pc component has been used to show that objects associated with reward trigger faster shifts of attention (Kiss, Driver, & Eimer, 2009), that attention shifts serially from object to object in some visual search tasks (Woodman & Luck, 1999, 2003b), and that people with schizophrenia can shift attention just as quickly as healthy control subjects under some conditions (Luck et al., 2006).

Combined MEG/ERP/fMRI studies have shown that the topography of the N2pc component is consistent with a generator source in area V4 and the lateral occipital complex (LOC) (Hopf et al., 2000, 2006). Notably, the V4 source may be present only when the scale of the competition between the target and the distractors is fine enough to occur within the receptive fields of individual V4 neurons; when the scale of distraction is coarse, the N2pc may arise solely from the LOC (Hopf et al., 2006). This stretches the conceptual definition of the term *ERP component* provided in chapter 2. That is, the V4 and LOC sources may reflect the same computational operation (suppression of a competing distractor), but applied to different scales of information in distinct but nearby regions of cortex. Do these sources therefore reflect different components or should they be treated as the same component? If it turns out that we can reliably distinguish between these different sources and be certain that they reflect the same computational operation, I would be so happy that I wouldn't care about picky little definitional issues!

I originally hypothesized that the N2pc reflects the suppression of the distractors surrounding the attended object (Luck & Hillyard, 1994b), and subsequent research has confirmed that it is sensitive to the proximity of the distractors (Luck et al., 1997; Hopf et al., 2006). However, a clever study by Clayton Hickey, Vince Di Lollo, and John McDonald provided strong evidence that N2pc does not directly reflect distractor suppression (Hickey, Di Lollo, & McDonald, 2009; see box 3.1 for a story about this study). They presented arrays containing only one target and one distractor, and they borrowed a trick that Geoff Woodman and I previously developed for

Box 3.1
Honor

You might notice that the experimental design shown in figure 3.9 contains a low-level sensory imbalance, with a lateralized stimulus on one side of the screen but not the other, which could have contributed to the contralateral-minus-ipsilateral differences found by Hickey et al. (2009). To minimize this problem, Hickey et al. used colored stimuli that were the same luminance as the background, which reduces the sensory ERP amplitudes. I was one of the reviewers when this paper was submitted for publication to the *Journal of Cognitive Neuroscience*, and in my review I noted that it was still possible—although unlikely—that lateralized sensory responses were contaminating the N2pc and P_D effects. The study was done in the lab of John McDonald, who had trained as a postdoc in Steve Hillyard's lab several years after I left the lab, and I knew him fairly well. As a result, I had a little bit of fun in my review (which I signed), noting that, "I view this as a matter of honor more than a matter of publishability." The authors rose to the challenge and ran a control experiment ruling out contributions from sensory lateralizations, and this control experiment ended up providing additional evidence about the link between the P_D component and distractor suppression. Honor is important!

Several years later, Clayton Hickey was asked to review a paper that Risa Sawaki and I had submitted to *Visual Cognition*. He pointed out a potential alternative explanation of our results, and he brought up the issue of "honor" that I had raised in my review of his paper. He even quoted from my review of his paper. Risa and I were able to argue convincingly that this alternative explanation was unlikely, but it was quite amusing to see my own review used against me.

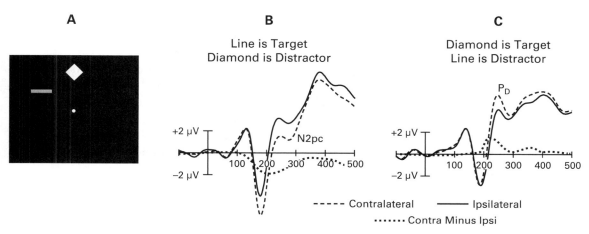

Figure 3.9
Example stimuli and grand average ERP waveforms from the study of Hickey et al. (2009). (A) Each array contained a bright green square or diamond along with a short or long red line that was isoluminant with the background. The red stimulus elicited almost no lateralized sensory activity, making it possible to attribute any later lateralized activity to top-down processing. (B) When the red line was the target, a contralateral negativity (N2pc) was elicited by the red item. (C) When the green diamond was the target, the ERP was more positive contralateral to the red distractor item than ipsilateral to this item. This contralateral positivity is called the distractor positivity (P_D). Waveforms courtesy of John McDonald.

isolating the N2pc to one of two items (Woodman & Luck, 2003b). Specifically, one of the two items was presented on the vertical midline, and the other was on the left or right side (figure 3.9A). There is no contralateral or ipsilateral side for the item on the vertical midline, so any brain activity elicited by this item should cancel out in a difference wave constructed by subtracting the waveform ipsilateral to the other item from the waveform contralateral to the other item.

When the target was lateralized and the distractor was on the midline, Hickey et al. (2009) found that the voltage was more negative over the hemisphere contralateral to the target than over the ipsilateral hemisphere (figure 3.9B). That is, they found the typical N2pc pattern, even though there were no lateralized distractors. This is pretty strong evidence that the N2pc is not directly related to the processing of the distractor but instead reflects processing of the target. Moreover, when the distractor was lateralized and the target was on the midline, the voltage was actually more positive and not more negative over the contralateral hemisphere relative to the ipsilateral hemisphere (figure 3.9C). Hickey et al. termed this positivity the *distractor positivity* (P_D). If the N2pc component directly reflected suppression of the distractor, then a negativity rather than a positivity should have been observed contralateral to the distractor.

What, then, does the N2pc component reflect? It is clearly related to the focusing of attention onto an object (for a discussion of the evidence for this claim, see Luck, 2012b), and its amplitude is influenced by the presence of nearby distractors, although it does not reflect processing of these distractors. The best current explanation is that it reflects the allocation of some kind of limited-capacity process to one or more relevant objects (Ester, Drew, Klee, Vogel, & Awh, 2012), but in a way that is influenced by the presence of nearby distractors. Note that it is entirely possible that the N2pc does not reflect a single specific process, but instead reflects multiple processes that are applied to an attended stimulus and that are stronger in the contralateral hemisphere than in the ipsilateral hemisphere. That is, the N2pc may reflect multiple *consequences* of focusing attention onto a lateralized object.

By now, you should be noticing a theme: A given ERP effect may reflect multiple related processes (multiple face-related processes for the N170 component, multiple control-related processes for the anterior N2 component, and multiple attention-related perceptual processes for the N2pc component). Given the enormous complexity of the human brain, a difference wave based on a fairly subtle experimental manipulation (e.g., ipsilateral versus contralateral target location) may include activity from several processes that are implemented in nearby regions of cortex during the same general time period.

Distractor Positivity

As shown in figure 3.9C, the distractor positivity (P_D) component consists of a more positive voltage over the hemisphere contralateral to a distractor than over the ipsilateral hemisphere. Three findings from that study suggest that it reflects some kind of suppressive or inhibitory process that is applied to the distractor. First, it is lateralized with respect to the distractor, not with respect to the target, which indicates that the P_D reflects a process that operates on the distractor. It seems very likely that a process that operates on a distractor will be inhibitory in nature. Second, the polarity of the P_D component is opposite to that of the N2pc component, but the scalp distributions are quite similar. If we assume that the N2pc represents some kind of excitation (because it is applied to the target), then a reversal from an excitatory postsynaptic potential to an inhibitory postsynaptic potential in the same population of neurons would give us the opposite polarity. This is only weak evidence that the P_D represents an inhibitory process because it rests on difficult-to-test assumptions, but it is at least consistent with an inhibitory process. Third, Hickey et al. showed that the P_D was eliminated if subjects were required only to detect the presence of the target item rather than to discriminate its identity, which presumably reduced the need to actively suppress the distractor.

Additional evidence that the P_D reflects a suppressive process comes from subsequent studies conducted by Risa Sawaki in my lab. These studies showed that distractor items presented within a bilateral array will elicit a P_D rather than an N2pc if the distractor is highly salient (e.g., a color pop-out) or partially matches a working memory representation or target template (Sawaki & Luck, 2010, 2011; Sawaki, Geng, & Luck, 2012). A similar effect was reported by Eimer and Kiss (2008), but this was before Hickey et al. (2009) isolated the P_D component and proposed that it reflects distractor suppression, so Eimer and Kiss didn't realize they were seeing an inhibition-related component. Risa has also shown that P_D amplitude may be correlated with behavioral measures of attentional capture, with a larger P_D associated with less capture (Sawaki et al., 2012). Thus, there is growing evidence that the P_D reflects a process that is involved in

Box 3.2
Ignorance Is Not Always Bliss

Ever since the first N2pc experiments that I conducted in graduate school, I noticed that the N2pc was often followed by a contralateral positivity. For example, you can see this positivity following the N2pc component in figure 2 of Luck and Hillyard (1994b). This positivity was present in some experiments more than in others, but I could not decipher any consistency in the pattern of results. I decided that it was going to be nearly impossible to understand this positivity because it was partially canceled by the N2pc, making it difficult to measure separately from the N2pc. Because I didn't want to deal with this positivity, I started publishing N2pc studies in which the figures ended at approximately 300 ms so that people couldn't see the positivity that followed N2pc (see, e.g., figure 3.8 in this chapter and figure 5 in Luck et al., 1997).

When Risa Sawaki started getting interested in the P_D, she proposed that this positivity that followed the N2pc was the same as the P_D component; that is, the N2pc reflects the focusing of attention onto an object, and the P_D reflects the cancellation of attention after perception of the object was complete. I explained to Risa my skepticism about being able to figure out this subsequent positivity, but she ignored my advice. As I mentioned in box 3.1, it is sometimes good to ignore your mentor when he or she tells you that something can't be done, and Risa demonstrated this principle by running several experiments that explored the P_D that follows the N2pc component, leading to a very nice paper in the *Journal of Neuroscience* (Sawaki et al., 2012). This required developing a new way of measuring and analyzing the N2pc and P_D components to minimize the problem of cancellation (which you can read about in the paper).

the suppression of potentially distracting visual objects. Notably, an N2pc component is often followed by a P_D component, which appears to reflect a cancellation or resetting of attention after perception is complete (Sawaki et al., 2012) (box 3.2).

Contralateral Delay Activity and Working Memory

Ed Vogel and Maro Machizawa conducted a beautiful set of experiments extending the basic N2pc approach into the domain of working memory (Vogel & Machizawa, 2004). Previous studies by Dan Ruchkin and others had described a sustained negative voltage during the maintenance period of working memory tasks, which they called the negative slow wave (NSW; see review by Perez & Vogel, 2012). The NSW increases in amplitude as the memory load increases, and it has a frontal distribution for verbal memory tasks and a temporo-parietal distribution for visual memory tasks. Although the NSW has been a useful component, it faces two problems as a specific measure of working memory. First, to demonstrate that a neural measure actually reflects stored working memory representations per se, it is necessary to show that the measure is closely related to the amount of information that is stored in working memory. A given measure may increase as the load increases simply because the task is becoming more difficult, and not because of changes in the amount of information being stored in memory. Second, it is important to be able to isolate a memory-related component from other components that might be present during the delay interval, such as anticipation-related components that may be present as subjects wait for the cue to report the stimuli at the end of the trial (see the section on the contingent negative variation and stimulus-preceding negativity later in this chapter). Vogel and Machizawa solved both of these problems by means of a contralateral-minus-ipsi-lateral difference wave (as described near the end of the chapter in the section on the discovery of the CDA).

The basic paradigm of Vogel and Machizawa (2004) is shown in figure 3.10A. Each trial began with a *cue arrow*, which told the subjects which side should be remembered on that trial. The cue was followed by a *sample array*, which contained one to four colored squares on each side of the fixation point. Subjects were supposed to store the items on the cued side of the array in memory, completely ignoring the items on the uncued side. At the end of the trial, a *test array* was presented, and subjects reported whether one of the items on the cued side had changed

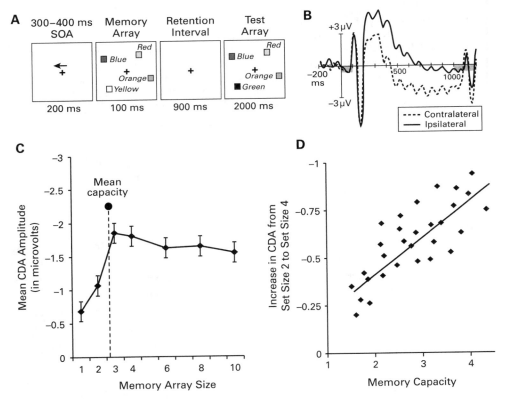

Figure 3.10
ERP results from Vogel and Machizawa (2004) showing the contralateral delay activity (CDA). (A) Experimental design. (B) Grand average ERP waveforms, time-locked to the sample stimulus and extending through the delay interval. (C) Mean CDA amplitude as a function of memory set size. (D) Correlation between subject's visual working memory capacity, measured behaviorally, and the magnitude of the increase in CDA between set sizes 2 and 4. Adapted with permission from Perez and Vogel (2012). Copyright 2012 Oxford University Press.

color between the sample and test arrays. The colors on the uncued side always remained the same between the sample and test arrays, giving subjects a strong incentive to remember the items on the cued side but not the items on the uncued side.

Figure 3.10B shows the ERP waveforms recorded over posterior scalp sites contralateral and ipsilateral to the cued side, time locked to the sample array and extending through the delay interval until the time of the test array. The voltage was more negative over the contralateral hemisphere than over the ipsilateral hemisphere from approximately 200 ms after the onset of the sample array until the onset of the test array. The early portion of this contralateral negativity presumably consists of an N2pc component, but the N2pc is not ordinarily sustained for more than a few hundred milliseconds. Thus, the sustained activity during the later portion of the delay interval represents an additional ERP component, which Vogel and Machizawa (2004) called *contralateral delay activity* (CDA). Although both the N2pc and CDA are contralateral negativities over posterior scalp sites, the CDA has a more parietal distribution than the N2pc, and MEG studies suggest that the CDA is generated in posterior parietal cortex (Robitaille, Grimault, & Jolicoeur, 2009), whereas the N2pc appears to be generated in ventral occipito-temporal cortex (Hopf et al., 2000, 2006).

Figure 3.10C shows that CDA amplitude is closely tied to working memory capacity. Average storage capacity in this experiment, measured behaviorally, was slightly under three items, and the CDA reached an asymptote at a set size of three items. This demonstrates that the CDA is related to memory capacity per se and does not just increase as the task becomes more difficult. Moreover, individual differences in working memory capacity, as measured behaviorally, were tightly correlated with the set size at which CDA amplitude reached asymptote. Specifically, figure 3.10D shows that a given subject's memory capacity was correlated with the change in CDA amplitude between set size 2 and set size 4. This tight correlation between the CDA and behaviorally measured capacity provides excellent evidence that the CDA is closely related to the actual maintenance of information in working memory. This also shows how ERPs can be extremely useful in assessing individual differences (see also Vogel, McCollough, & Machizawa, 2005; Leonard et al., 2012).

The P3 Family

Varieties of P3 Components

There are several distinguishable ERP components in the time range of the P3 wave (for a review of the P3 wave, see Polich, 2012). The first major distinction was made by Squires, Squires, and Hillyard (1975), who distinguished between a frontally maximal P3a component and a parietally maximal P3b component. Both were elicited by unpredictable, infrequent changes in the stimuli, but the P3b component was present only when these changes were task-relevant. When ERP researchers (including myself) refer to *the P3 component* or *the P300 component*, they almost always mean *the P3b component* (in fact, I have already used the term *P3* to refer to the P3b component several times in this book).

Other studies have shown that an unexpected, unusual, or surprising task-irrelevant stimulus within an attended stimulus train will elicit a frontal P3-like response (e.g., Courchesne, Hillyard, & Galambos, 1975; Soltani & Knight, 2000; Polich & Comerchero, 2003), but it is not clear whether this response is related to the P3a component as originally described by Squires et al. (1975). For example, Verleger, Jaskowski, and Waushckuhn (1994) provided evidence that the P3b component is observed for targets that are infrequent but are in some sense expected or awaited, whereas the frontal P3 wave is elicited by stimuli that are truly unexpected or surprising. However, it is not clear that this frontal P3 is as automatic as the P3a observed by Squires et al. (1975).

Theories of Functional Significance

Given the thousands of published P3 experiments, you might think that we would have a very thorough understanding of the P3 wave. But you'd be wrong! We know a great deal about the effects of various manipulations on P3 amplitude and latency, but there is no clear consensus about what neural or cognitive processes are reflected by the P3 wave. The most widely cited theory was developed by Donchin (1981), who proposed that the P3 wave is related to a process he called "context updating." This is often interpreted to mean that the P3 wave reflects the updating of working memory, but this is not what Donchin meant. He likes to point out that he never used the phrase *working memory*, and if you read his paper (which I strongly encourage), he uses the word *context* to mean something very different from working memory (and very different from the way that *context* is used by people like Jonathan Cohen). For Donchin, *context* representations are broad representations of the overall state of the environment, not specific representations of individual items or tasks.

The evidence supporting the context updating theory of the P3 wave is pretty sparse, and very few direct tests have been performed in the time since Donchin's original description of the theory. If you are interested in the P3 wave, you should probably read Donchin's original paper (Donchin, 1981), Verleger's extensive critique of this theory (Verleger, 1988), and the response of Donchin and Coles to this critique (Donchin & Coles, 1988). In my own laboratory's research on attention, we have frequently assumed that the P3 wave reflects working memory updating, and this has led to a variety of very sensible results (e.g., Luck, 1998a; Vogel, Luck, & Shapiro, 1998; Vogel & Luck, 2002). But this assumption certainly carries some risk, so you should be careful in making assumptions about the meaning of the P3 wave.

Another key concept, again raised by Donchin (1981), is that the process reflected by the P3 wave is *strategic* rather than *tactical*. A tactical response is something that is done to deal with the current situation (e.g., when a pilot suddenly banks to the left to avoid a flock of birds). A strategic response is something that is done to prepare for the future (e.g., when a pilot chooses a different route to avoid turbulence that is predicted to occur in an hour). Donchin proposed that the P3 wave reflects a strategic process rather than a tactical process because it frequently occurs too late to have an impact on the behavioral response. Moreover, Donchin argued that the amplitude of the P3 elicited by a stimulus is predictive of later memory for that stimulus (for a review of memory-related ERP effects, see Wilding & Ranganath, 2012).

Effects of Probability

Although we do not know exactly what the P3 wave means, we do know what factors influence its amplitude and latency (for extensive reviews of the early P3 literature, see Pritchard, 1981; Johnson, 1986; for more recent reviews, see Picton, 1992; Polich & Kok, 1995; Polich, 2004, 2012). The hallmark of the P3 wave is its sensitivity to target probability. As shown in great detail by Duncan-Johnson and Donchin (1977), P3 amplitude gets larger as target probability gets smaller. However, it is not just the overall probability that matters; local probability also matters, because the P3 wave elicited by a target becomes larger when it has been preceded by more and more nontargets.

A crucial detail is that the probability of a given physical stimulus is not the relevant factor. Instead, it is the *probability of the task-defined stimulus category* that matters. For example, a classic experiment by Marta Kutas showed that if subjects are asked to press a button when detecting male names embedded in a sequence containing male and female names, the amplitude of the P3 wave will depend on the relative proportions of male and female names in the sequence even though each individual name appears only once (see Kutas, McCarthy, & Donchin, 1977). Similarly, if the target is the letter E, occurring on 10% of trials, and the nontargets are selected at random from the other letters of the alphabet, the target will elicit a very large P3 wave even though the target letter is approximately four times more probable than any individual nontarget letter (see Vogel et al., 1998).

For decades, people assumed that the amplitude of the P3 depends on the *sequential probability* of the stimulus category (i.e., where the probability is the number of stimuli in a particular category divided by the total number of stimuli). However, more recent studies indicate that P3 amplitude is largely dependent on the *temporal probability* of the stimulus category (i.e., where probability is the number of stimuli in a particular category divided by the time period over which the stimuli are presented) (Polich, 2012). These two factors are typically confounded. For example, figure 3.4A shows a series of stimuli in which the target stimulus is both sequentially rare (because only two of the 10 stimuli are targets) and temporally rare (because only two target stimuli are presented over a fairly long period of time). If you simply leave out the standards, but maintain the same timing of the targets (as in figure 3.4C), the P3 elicited by the targets is about the same as it would be with the standards included, even though the sequential probability of the targets is now 100% (Polich, Eischen, & Collins, 1994; Katayama & Polich, 1996).

P3, Resource Allocation, and Task Difficulty

P3 amplitude is larger when subjects devote more effort to a task, leading to the proposal that P3 amplitude can be used as a measure of resource allocation (see, e.g., Isreal, Chesney, Wickens, & Donchin, 1980). However, P3 amplitude is smaller when the subject is uncertain of whether a given stimulus was a target or nontarget. Thus, if a task is made more difficult, this might increase P3 amplitude by encouraging subjects to devote more effort to the task, but it might decrease P3 amplitude by making subjects less certain of the category of a given stimulus. Consequently, there is no simple rule for determining whether the P3 will get larger or smaller for more difficult tasks. Johnson (1984, 1986) proposed that the variables of probability (P), uncertainty (U), and resource allocation (R) combine to influence P3 amplitude in the following manner: P3 amplitude $= U \times (P + R)$.

P3 Latency and Stimulus Categorization

The fact that P3 amplitude depends on the probability of the task-defined category of a stimulus has an important but often-overlooked consequence. Specifically, it is logically necessary that a difference in P3 amplitude between the rare and frequent trials means that the subject has begun to categorize the stimulus according to the rules of the task by the time the difference is present. As an example, consider figure 3.11, which provides a cartoon version of the Kutas et al. (1977) experiment in which subjects were instructed to categorize names as male or female. The category of female names was rare and the category of male names was frequent, but each individual name occurred only once. In this cartoon, the difference between the two categories deviated from 0 μV at approximately 300 ms. This provides essentially bullet-proof evidence that the brain had begun to determine the category of the names by 300 ms (with the caveat, described in chapter 2, that the onset of a difference will reflect the trials with earliest onset).

Because a difference in P3 amplitude between rare and frequent stimuli cannot occur until the brain has categorized the stimulus as rare or frequent, any manipulation that postpones stimulus categorization (including increasing the time required for low-level sensory processing or higher-level categorization) must necessarily increase the onset time of P3 probability effect. That is, anything that increases the amount of time required for the brain to determine whether a given stimulus falls into the rare category or the frequent category will necessarily increase

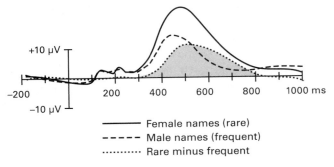

Figure 3.11
Cartoon version of the ERP waveforms from the study of Kutas et al. (1977). These waveforms are not intended to represent the actual waveforms reported by Kutas et al., but instead show the general principle involved in using a rare-minus-frequent difference wave to assess the timing of stimulus categorization.

the onset time of the difference in brain activity between rare and frequent trials. Not only is this logical, it has been confirmed in countless studies. Kutas et al. (1977) framed this general concept in terms of the idea—which was common then but is no longer widely accepted—that the P3 on rare trials reflects surprise, and they noted that, "before a stimulus can surprise it must be identified." They therefore proposed that P3 latency for rare stimuli reflects *stimulus evaluation time*. The Donchin lab mostly focused on the latency of the P3 peak rather than P3 onset time, and they focused on the ERPs elicited by the rare stimuli rather than focusing on the rare-minus-frequent difference wave. Consequently, their notion that P3 peak latency on rare trials reflects stimulus evaluation time is not quite as precise as the statement that the onset of the difference between rare and frequent trials cannot logically occur until after the brain has begun to determine whether the stimulus falls into the rare category or the frequent category. But these two ways of thinking about P3 latency are very closely related.

When rare-minus-frequent difference waves are used, very strong conclusions can be drawn about whether a given experimental manipulation influenced the amount of time required to categorize a stimulus. An example of this is provided in figure 3.12, which shows the task and results from a study using the psychological refractory period paradigm (Luck, 1998b). In this task, subjects saw two targets on each trial, separated by a variable SOA, and they made speeded responses to both targets. At short SOAs, the brain will still be busy processing the first target when the second target appears, and this delays the response to the second target. Thus, we found that the response to the second target was delayed by hundreds of milliseconds at short SOAs compared to long SOAs. Hal Pashler has theorized that this delay for the second target reflects a postponement of *response selection*, the process of determining which response should be made once the stimulus has been identified (see review by Pashler, 1994). Earlier processes, such as the perception and categorization of the stimulus, should not be delayed at short SOAs. If this theory is correct, then the amount of time required to perceive and categorize the second target should be relatively unaffected by differences in the SOA. The experiment shown in figure 3.12 tested this hypothesis by having one version of the second target be rare and the other be frequent, making it possible to construct a rare-minus-frequent difference wave for the second target. The P3 latency in this difference wave was slowed by only 51 ms at the shortest SOA compared to the longest SOA, which was less than 25% of the total RT slowing seen behaviorally. This result indicated that the RT slowing was mainly due to a slowing of processes that follow stimulus categorization, not stimulus categorization itself, consistent with Pashler's theory. A related study by Allen Osman and Cathleen Moore showed that the lateralized readiness potential (LRP) was strongly delayed at short SOAs (Osman & Moore, 1993). As will be discussed later in this chapter, the LRP reflects response preparation, so this effect provided further evidence that the slowing of RT was mainly a result of slowed response selection.

This study is related to an important point about the design of ERP experiments (which will be discussed in more detail in chapter 4). Specifically, the P3 component itself does not arise from the process of categorizing the stimulus. Rather, stimulus categorization must occur *before*

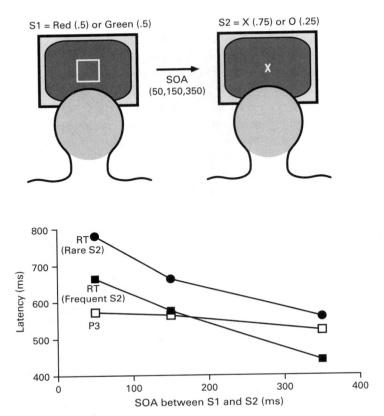

Figure 3.12
Example of how a rare-minus-frequent difference wave can be used to assess the timing of stimulus categorization in the psychological refractory period paradigm. In this study, subjects saw a red or blue box, followed after a variable stimulus onset asynchrony (SOA) by a second target (an X or an O). One letter was rare and the other was frequent. Subjects were instructed to make a rapid response indicating the color of the box and another rapid response indicating the letter identity. P3 latency and reaction time were measured for the second target (the letter) as a function of the SOA between the first target and the second target. P3 latency was measured as the 50% area latency from rare-minus-frequent difference waves. Adapted with permission from Luck (1998b). Copyright 1998 Association for Psychological Science.

the P3 can be elicited (or, more precisely, before the rare-minus-frequent difference can deviate from zero). Consequently, the latency of the P3 can be used in this manner to assess the timing of processes that must have occurred prior to the time at which the waveforms diverge for the rare versus frequent stimuli. When a component is used in this manner to assess the logically prior cognitive operations, we don't need to know what process the component itself reflects. Thus, many of the most powerful ERP experiments use a given ERP component to study the processes that logically preceded the generation of the component, not the processes that actually generated the component.

P3 and Postcategorization Processes

Although the onset of the rare-minus-frequent difference must logically follow stimulus categorization, this does not tell us whether P3 latency also depends on postcategorization processes. Several studies have provided evidence that P3 latency is sensitive *only* to the time required to perceive and categorize the stimulus and is not sensitive to the amount of time required to select and execute a response once a stimulus has been categorized (see, e.g., Kutas et al., 1977; Magliero, Bashore, Coles, & Donchin, 1984). For example, if subjects press a left-hand button when they see the stimulus LEFT and a right-hand button when they see the stimulus RIGHT, P3 latency is no faster or slower than when they are asked to make a left-hand response for RIGHT and a right-hand response for LEFT (which is known to increase the time required to perform stimulus-response mapping). In contrast, if the stimuli are perceptually degraded, then P3 latency is delayed for these stimuli.

Other researchers have disputed the conclusion that P3 latency is insensitive to manipulations of response-related processes (see review by Verleger, 1997). However, most of the research has not focused on the onset latency of rare-minus-frequent difference waves. For example, consider the study of Leuthold and Sommer (1998), who presented stimuli on the left or right side of the screen (with equal probability) and required subjects to report the side of the stimulus by making a left-hand or right-hand button-press response. In the *compatible* condition, subjects responded with the left hand for the left stimulus and with the right hand for the right stimulus; in the *incompatible* condition, this mapping was reversed. This kind of manipulation is known to primarily influence response selection processes, and it should have little or no impact on the time required to determine which side contains the stimulus. P3 latency was measured by simply finding the peak voltage between 250 and 850 ms in the ERPs elicited by the left-side and right-side stimuli. This peak latency was increased by about 20 ms in the incompatible condition relative to the compatible condition. The problem is that these ERP waveforms presumably contained many overlapping components during this broad time range, so there is no way to know whether this change in peak latency reflected the timing of the underlying P3 component. For example, the effect might have been due to motor potentials during this period, which everyone would expect to vary with stimulus-response compatibility. Moreover, the logic I have described mainly applies to the onset of the P3, not the peak. Thus, the existing evidence provides good support for the proposal that P3 latency—measured as the onset of the rare-minus-frequent difference wave—reflects the time required to categorize a stimulus and is insensitive to subsequent response-related processes.

P3 and Schizophrenia

Well over a hundred studies have examined the P3 in people with schizophrenia (see meta-analysis by Jeon & Polich, 2003). Most of these studies used simple auditory oddball tasks in which subjects silently counted the oddballs, but many other paradigms have also been used. The peak amplitude of the P3 elicited by the rare (oddball) stimuli is reliably reduced in patients relative to controls, with an average effect size (Cohen's d) of 0.89 for auditory oddball experiments. This is a large effect, and it exhibits excellent stability and reliability, making it a potential biomarker (Luck et al., 2011). Although many interesting and useful conclusions about schizophrenia have been drawn from these experiments, the fact that we do not have a well-accepted theory of the processes reflected by the P3 means that we cannot draw a precise and broadly meaningful conclusion from the simple fact that P3 amplitude is reduced in people with schizophrenia (see box 3.3 for further concerns about the P3 reduction in schizophrenia).

Language-Related ERP Components

Several ERP components are sensitive to linguistic variables (see review by Swaab, Ledoux, Camblin, & Boudewyn, 2012). The best-studied language-related component is the N400, which was first reported by Kutas and Hillyard (1980). The story of the discovery of the N400 is described at the end of the chapter, and the main results from this experiment are shown in figure 3.13.

The N400 is a negative-going wave that is usually largest over central and parietal electrode sites, with a slightly larger amplitude over the right hemisphere than over the left hemisphere. It is typically seen in response to violations of semantic expectancies. For example, if sentences are presented one word at a time on a video monitor, a large N400 will be elicited by the last word in the following sentence: "While I was visiting my home town, I had lunch with several old shirts" (see figure 3.13). Little N400 activity would be observed if the sentence had ended with *friends* rather than *shirts*. The words can be presented in naturally delivered speech rather than in discrete visual words, and the same effects are observed. An N400 can also be observed to the second word in a pair of words if the second word is semantically unrelated to the first. For example, a large N400 is elicited by the second word in "tire . . . sugar" and a small N400 elicited by the second word in "sweet . . . sugar." This may actually reflect the associative strength between the words rather than the semantic relationship per se (Rhodes & Donaldson, 2008). Some N400 activity is presumably elicited by any content word you read or hear, and relatively infrequent words like *monocle* elicit larger N400s than those elicited by relatively frequent words like *table*.

The N400 (or N400-like activity) can also be elicited by nonlinguistic stimuli, as long as they are meaningful. For example, a line drawing will elicit an N400-like component if it is incon-

Box 3.3
Is It Really the P3?

It should now be apparent to you that many different brain processes contribute to the voltage that we measure in the P3 latency range on rare trials in the oddball task, including processes that are specific to the rare stimuli and processes that are present for both the frequent and rare stimuli. As far as I can tell, researchers have not grappled much with the question of whether the reduced amplitude for rare stimuli in schizophrenia reflects a reduction in the underlying P3 component, a reduction in some other positive component, or even an increase in some negative component. One way to address this question would be to ask whether this effect interacts with target probability. For example, I described an experiment in chapter 1 in which we examined the rare-minus-frequent difference waves in schizophrenia patients and control subjects (Luck et al., 2009). Although the peak voltage in the P3 latency range was reduced on rare trials in patients compared to controls, this voltage was also reduced on frequent trials, and there was no difference in P3 between patients and controls in the rare-minus-frequent difference wave (see figure 1.4 in chapter 1). Similar results were observed in a study by Geoff Potts and his colleagues (Potts, O'Donnell, Hirayasu, & McCarley, 2002). This questions the assumption that the P3 effect in schizophrenia really reflects a bona fide reduction in the underlying P3 component (that is, the probability-sensitive P3b component).

Most oddball experiments in schizophrenia have used auditory stimuli, whereas we had used visual stimuli, so I spent some time reading the literature to see if the rare-minus-frequent difference wave is reduced in schizophrenia patients for auditory stimuli. However, I could not find a single paper that plotted rare-minus-frequent difference waves for the auditory oddball paradigm. In fact, most of the papers I read didn't even report P3 amplitude for the frequent stimuli. However, a few did, and it seems that the P3 elicited by the frequent stimulus is not reduced in amplitude as much as the P3 elicited by the rare stimuli, which implies that the rare-minus-frequent difference is smaller in patients than in controls for auditory stimuli. This still doesn't provide strong evidence that the patient deficit solely reflects a reduction in the amplitude of the underlying P3 component rather than some other component (or a combination of components). In addition, given the fact that the P3 appears to be a largely modality-independent component, the finding that the rare-minus-frequent difference wave is reduced for auditory stimuli but not for visual stimuli in schizophrenia patients suggests that the patient impairment reflects an impairment in auditory-specific processes that provide the input to the P3 rather than an impairment in the P3 generator system itself. This is just a speculation, but it seems plausible given the existing data.

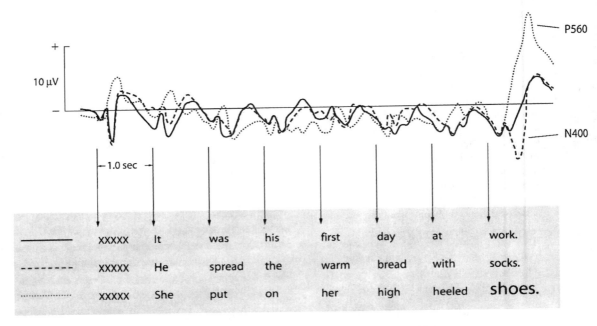

Figure 3.13
Original N400 paradigm used by Kutas and Hillyard (1980). The individual words of a sentence were presented sequentially at fixation, with one word presented per second. The last word of the sentence could be congruent with the meaning of the sentence, incongruent with the meaning of the sentence, or drawn in a larger font. An N400 was observed when the final word was incongruent, whereas as P3 (P560) was observed when the final word was drawn in a larger font. Adapted with permission from Swaab et al. (2012). Copyright 2012 Oxford University Press.

sistent with the semantic context created by a preceding sequence of words or line drawings (Holcomb & McPherson, 1994; Ganis, Kutas, & Sereno, 1996). However, it is possible that subjects in these studies name the stimuli subvocally, so it is possible that the N400 component reflects language-specific brain activity. Moreover, nonlinguistic stimuli typically elicit a more frontally distributed effect than linguistic stimuli (Willems, Özyürek, & Hagoort, 2008).

Although typically larger in right-hemisphere electrodes than left-hemisphere electrodes, the N400 appears to be generated primarily in the left temporal lobe. This apparent discrepancy can be explained by assuming that the generator dipole in the left hemisphere does not point straight upward, but instead points somewhat medially (i.e., upward and toward the right side). Studies of split-brain patients and lesion patients have shown that the N400 depends on left-hemisphere activity (Kutas, Hillyard, & Gazzaniga, 1988; Hagoort, Brown, & Swaab, 1996), and recordings from the cortical surface in neurosurgery patients have found clear evidence of N400-like activity in the left anterior medial temporal lobe (e.g., McCarthy, Nobre, Bentin, & Spencer, 1995). More recent studies also suggest that the left prefrontal cortex may also contribute to the scalp N400 (Halgren, Boujon, Clarke, Wang, & Chauvel, 2002).

Two main theories have been proposed regarding the specific process reflected by the N400. The first, proposed by Marta Kutas and her colleagues, is that the N400 component reflects the neural activity associated with finding and activating the meaning of the word. Because I'm not a psycholinguist, I'll let Marta's own words describe this theory: "Overall, the extant data suggest that N400 amplitude is a general index of the ease or difficulty of retrieving stored conceptual knowledge associated with a word (or other meaningful stimuli), which is dependent on both the stored representation itself, and the retrieval cues provided by the preceding context" (Kutas, van Petter, & Kluender, 2006, p. 669). Thus, the more work required to retrieve the knowledge associated with a word, the larger the N400 will be. The second main theory comes from Peter Hagoort, who proposed that the N400 reflects the process by which the retrieved word meaning is integrated into the preceding discourse (Hagoort, 2007; see also Friederici, Hahne, & Saddy, 2002). The more work required to perform this integration, the larger the N400 will be. Because I'm not a linguist, I do not presume to have an opinion about which theory is more likely to be correct (which may be the only time in this entire book that I do not presume to have an opinion about something).

Syntactic violations also elicit distinctive ERP components. One of these is called *P600* (see Osterhout & Holcomb, 1992, 1995). For example, the word *to* elicits a larger P600 in the sentence "The broker persuaded to sell the stock" than in the sentence "The broker hoped to sell the stock." Syntactic violations can also elicit a left frontal negativity from approximately 300 to 500 ms, which may be the same effect observed when *wh*-questions (e.g., "What is the . . .") are compared to yes–no questions (e.g., "Is the . . ."). Given the important distinction between syntax and semantics, it should not be surprising that words that are primarily syntactic elicit different ERP activity than that elicited by words with rich semantics. In particular, function words (e.g., *to*, *with*, *for*) elicit a component called *N280* at left anterior electrode sites, and this component is absent for content words (e.g., nouns and verbs). In contrast, content words elicit an N400 that is absent for function words.

ERP Components and Long-Term Memory

In addition to the NSW and CDA components that are present in working memory paradigms, several ERP components have been identified that are related to long-term memory (see the review by Wilding & Ranganath, 2012). Separate ERP components have been identified that operate during the encoding and retrieval phases of long-term memory tasks. The ERP elicited by a stimulus during memory encoding will contain many components that are unrelated to encoding. To isolate encoding-related activity, ERP studies commonly sort the single-trial EEG waveforms that were recorded during the encoding phase according to whether or not the stimulus on that trial was later remembered. One averaged ERP waveform is then constructed for trials with stimuli that were later remembered, and another averaged ERP waveform is constructed for trials with stimuli that were later forgotten. Any difference between these ERPs (seen on the remembered-minus-forgotten difference wave) is called a *Dm effect* (difference due to memory) or a *subsequent memory effect*. In most cases, the Dm effect contains a broad positivity from approximately 400 to 800 ms over centro-parietal electrode sites. However, it may also contain left anterior activity, and the details of the scalp distribution depend on whether the stimuli were words or pictures and on the instructions given to the subjects. Thus, Dm is not a single component, but instead reflects many different processes that can influence whether a stimulus is later remembered.

ERPs can also be examined at the time of memory retrieval by using a recognition task and time-locking the ERPs to probe stimuli that either match a previously studied item (*old* probes) or do not match a previously studied item (*new* probes). Two distinct effects have been observed by comparing the ERPs elicited by old versus new probes. One effect consists of a more negative voltage for new probes than for old probes from 300 to 500 ms, with a maximal voltage at midline frontal electrodes (sometimes called the *midfrontal old–new effect*, and sometimes called *FN400* because it is like a frontally distributed N400). The other consists of a more positive voltage for old probes than for new probes from 400 to 800 ms, with a maximal voltage at left parietal electrodes (called the *left-parietal old–new effect*). The left-parietal old–new effect appears to be associated with what memory researchers call *recollection*, which refers to a clear and distinct experience of the memory that is linked with a particular time and/or place (see review by Yonelinas & Parks, 2007). Some researchers have proposed that the midfrontal old–new effect is associated with *familiarity*, the more diffuse feeling that the probe has been encountered before (e.g., Curran, 2000; Rugg & Curran, 2007). However, Ken Paller and his colleagues have argued that this effect instead reflects a boost in *conceptual fluency*, which is the ease with which meaning is processed and which may be a precursor to familiarity (Paller, Voss, & Boehm, 2007; Voss, Lucas, & Paller, 2012).

Emotion-Related ERP Components

It is difficult to induce intense emotional responses over and over again in a controlled experiment because these responses tend to habituate. Consequently, ERP studies of emotion have tended to focus on how stimuli that are associated with emotions (e.g., photographs of pleasant or unpleasant scenes) are processed differently from relatively neutral stimuli (see the review by Hajcak, Wienberg, MacNamara, & Foti, 2012). It is not clear that the resulting effects reflect emotions that were elicited by the stimuli or, instead, reflect "cold" cognitive responses that are related to the emotional content of the stimuli but are not themselves emotional responses. For example, when I view the stimuli in an emotion ERP experiment and I see a dozen pictures containing snakes over a 10-min period, I don't feel the kind of fear and revulsion that I experience when I encounter an actual snake while riding my mountain bike on a trail in the woods. My attention is oriented to the picture of the snake, and I realize that it's something I'd rather not look at, but I don't feel my heart pounding (in case you can't tell, I really don't like snakes). Moreover, it takes a substantial amount of time to have a significant emotional experience, so only the relatively late portion of the ERP waveform is likely to be related to the phenomenological experience of emotion. But it's still very worthwhile to use ERPs to study emotion-related processes; you just need to make sure that you don't draw unwarranted conclusions about the nature of the processes that are evoked by a long series of photographs in a laboratory experiment.

The emotional content of a photographic image can influence many of the components that have already been described. For example, the P1, N1/N170, N2, and P3 components may all be increased for emotion-relevant stimuli compared to neutral stimuli. Two emotion-related components have been the focus of most research. First, the *early posterior negativity* is a negative potential over visual cortex in the N2 latency range that is enhanced for emotion-inducing stimuli, particularly those with a positive valence (Schupp, Junghofer, Weike, & Hamm, 2003, 2004; Weinberg & Hajcak, 2010). This component is thought to reflect the recruitment of additional perceptual processing for emotion-inducing stimuli. Second, the *late positive potential* (Cuthbert, Schupp, Bradley, Birbaumer, & Lang, 2000; Keil et al., 2002; Hajcak & Olvet, 2008) is a positive voltage that typically has the same onset time and scalp distribution as the P3 wave (i.e., onset around 300 ms and parietal maximum). It may extend for many hundreds of milliseconds and may become more centrally distributed over time. The initial portion may actually consist of an enlarged P3 component, reflecting an effect of the intrinsic task relevance of emotion-related stimuli.

Error-Related Components

In most ERP studies, trials with incorrect behavioral responses are simply thrown out. However, by comparing the ERP waveform elicited on error trials with the ERP waveform elicited on correct trials, it is possible to learn something about the cause of the error and the brain's response after detection of the error (for an extensive review, see Gehring, Liu, Orr, & Carp, 2012). For example, Gehring, Goss, Coles, Meyer, and Donchin (1993) had subjects perform a speeded response task in which they responded so fast that they occasionally made errors that were obvious right away ("Oops! I meant to press the left button!"). When the ERPs on correct trials were compared to the ERPs on error trials, a negative-going deflection was observed at frontal and central electrode sites beginning just after the time of the response. Gehring et al. called this deflection the *error-related negativity* (ERN), and it was independently discovered by Falkenstein, Hohnsbein, Joormann, and Blanke (1990), who called it N_e (see the end of the chapter for a description of Gehring's discovery of the ERN).

An example set of waveforms is shown in figure 3.14 (from Vidal, Hasbroucq, Grapperon, & Bonnet, 2000). The researchers recorded the electromyogram (EMG) from the thumb in addition to recording button-presses. This makes it possible to define trials that are fully correct (correct thumb button is pressed with no EMG from the incorrect thumb), trials that are fully incorrect (incorrect thumb button is pressed, along with EMG from that thumb), and *partial error* trials (some EMG from the incorrect thumb but without button closure). In ERN experiments, the error-related ERP activity is usually tightly time-locked to the response and more

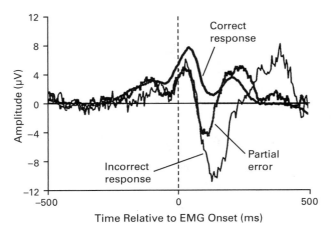

Figure 3.14
Grand average ERPs time-locked to the onset of electromyogram (EMG) activity, which show correct trials, trials with a fully incorrect response, and trials with a partial error (significant EMG without actual button closure). Adapted with permission from Vidal et al. (2000). Copyright 2000 Elsevier.

weakly time-locked to the stimulus, so the ERN is usually visualized in response-locked averages (where the response is defined either by button closure or EMG onset). The waveforms in figure 3.14 are time-locked to EMG onset. The data in this figure illustrate several common findings. First, fully incorrect trials are accompanied by a large negative-going deflection beginning near the time of the response; this is the ERN. Second, the partial error trials are accompanied by a somewhat smaller ERN. Third, a small negative-going potential can also be seen after the response on correct trials, which is termed the *correct response negativity* (CRN). A clear CRN is not always visible because of other overlapping ERPs, but many researchers believe that it is there nonetheless. Fourth, the ERN is followed by a positive deflection, peaking approximately 400 ms after the response in the study shown in figure 3.14. This is called the *error positivity* (P_e). The P_e appears to be associated with awareness of errors, whereas the ERN can occur with little or no awareness of the committed error (Endrass, Reuter, & Kathmann, 2007).

Some complicated technical issues arise in quantifying the ERN. A difference wave approach can be used in which the ERN is quantified as the difference in amplitude between error trials and correct trials. This has the advantage of eliminating any brain activity that is identical on error and correct trials, isolating error-related processes. However, this is not a complete solution. One problem is that the difference may partly reflect a later onset of the P3 wave on error trials than on correct trials rather than a distinct error process. A second problem is that activity prior to the response may differ between error trials and correct trials, distorting the baseline. Both of these problems are related to the fact that the ERN is a correlational effect rather than an experimental effect. That is, the researcher does not control which trials have errors and which trials don't, and some of the differences in the ERPs between correct trials and error trials may reflect factors that are correlated with errors but not causally related to them. A third problem with difference waves is that a difference across conditions or groups in the strength of the process that generates the ERN may be equally present on correct trials and on error trials. For example, people with obsessive-compulsive disorder appear to have larger ERNs on both correct trials and error trials, perhaps reflecting an aberrant sense of error even when a correct response is made (Gehring, Himle, & Nisenson, 2000). Thus, quantifying ERN amplitude from error-minus-correct difference waves may underestimate the size of increased ERN in these individuals and makes it impossible to test the hypothesis that they also exhibit an enlarged ERN on correct trials. Although difference waves are problematic for quantifying the ERN, it can be even more problematic to measure the ERN from the correct and incorrect trials separately, because that virtually guarantees that the results will be distorted by the many overlapping ERP components (for an extensive discussion, see Gehring et al., 2012).

Most investigators believe that the ERN reflects the activity of a system that monitors responses, is sensitive to conflict between intended and actual responses, or generates emotional reactions depending on responses. Indeed, these processes may be closely interrelated (Yeung, 2004). It is often assumed that the ERN is generated in the dorsal portion of the anterior cingulate cortex (dACC) because fMRI and single-unit studies show that error-related activity is present in this region and because dipole source modeling studies have found that the scalp distribution of the ERN is consistent with a dACC generator location. However, dipole modeling cannot easily distinguish between a single deep dipole and a distributed set of superficial dipoles, and intracranial recordings have found ERN-like responses in multiple cortical areas (Brazdil et al., 2002). Thus, as we have seen with several other ERP components, it is likely that multiple neural sources contribute to the scalp ERN (see discussion in Gehring et al., 2012).

Response-Related ERP Components

If subjects are instructed to make a series of occasional manual responses, with no eliciting stimulus, the responses are preceded by a slow negative shift at frontal and central electrode sites that begins up to 1 s before the actual response. This is called the *bereitschaftspotential* (BP) or *readiness potential* (RP), and it was independently discovered by Kornhuber and Deecke (1965) and Vaughan, Costa, and Ritter (1968). The scalp topography of the readiness potential depends on which effectors will be used to make the response, with differences between the left and right sides of the body and differences depending on which effector is used within a given side (see review by Brunia et al., 2012). The BP/RP can also be observed in tasks that require subjects to make responses to stimuli.

The lateralized portion of the readiness potential is called the *lateralized readiness potential* (LRP), and the LRP has been widely used in cognitive studies (see the review by Smulders & Miller, 2012). As discussed in chapter 2, the LRP is particularly useful because it can be easily isolated from other ERP components. That is, because it is lateralized with respect to the hand making the response, whereas other components are not lateralized in this manner, it is easy to

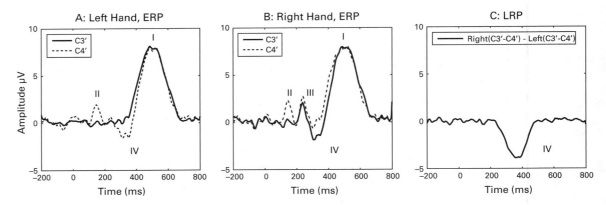

Figure 3.15
Simulated data from a lateralized readiness potential (LRP) experiment in which a stimulus is presented that triggers either a left-hand response or a right-hand response. Adapted with permission from Smulders and Miller (2012). Copyright 2012 Oxford University Press.

tell when a given experimental manipulation has affected the time or amplitude of the LRP. In contrast, it is difficult to be certain that a given experimental manipulation has influenced the P3 component rather than some other overlapping component, and this is one of the main reasons why it has been so difficult to determine what cognitive process is reflected by the P3 component.

Figure 3.15 shows how the LRP is isolated by means of a contralateral-minus-ipsilateral difference wave in an imaginary experiment in which subjects make either a left-hand or right-hand response depending on what stimulus is presented. Panels A and B show the waveforms on left-response and right-response trials, recorded from the C3′ and C4′ electrode sites (which are just lateral and anterior to the C3 and C4 sites, overlying left and right motor cortex, respectively). There is a large peak (labeled *peak I*) that is the same size irrespective of electrode side or response side (like the P3 wave). There is also a peak (labeled *peak II*) that is larger over the right hemisphere than over the left hemisphere for both left-hand and right-hand responses. Another peak (labeled *peak III*) is larger for right-hand responses than for left-hand responses over both hemispheres. Finally, there is a peak (labeled *peak IV*) that is more negative over the right hemisphere for left-hand responses than for right-hand responses (panel A) and more negative over the left hemisphere for right-hand responses than for left-hand responses (panel B). The LRP is this fourth peak; a greater negativity over the contralateral hemisphere than over the ipsilateral hemisphere (relative to the response hand). The first step in isolating the LRP from the other types of peaks is to compute contralateral and ipsilateral waveforms that combine the two hemispheres. That is, the contralateral waveform is the average of the right hemisphere activity for left-hand responses and the left hemisphere activity for right-hand responses; the ipsilateral waveform is the average of the left hemisphere activity for left-hand responses and the right hemisphere activity for right-hand responses. The second step is to compute the difference between the contralateral and ipsilateral waveforms. As shown in panel C of figure 3.15, this difference isolates the LRP (peak IV) from the nonspecific (peak I), hemisphere-specific (peak II), and hand-specific (peak III) components. Beautiful! Note, however, that there are two slightly different variations on this approach, one of which doubles the amplitude relative to the other (see Smulders & Miller, 2012).

The LRP is generated, at least in part, in motor cortex (Coles, 1989; Miller, Riehle, & Requin, 1992). The most interesting consequence of this is that the LRP preceding a foot movement is opposite in polarity to the LRP preceding a hand movement, reflecting the fact that the motor cortex representation of the hand is on the lateral surface of the brain, whereas the representation of the foot is on the opposed mesial surface. The LRP appears to reflect some aspect of response preparation: Responses are faster when the LRP is larger at the moment of stimulus onset (Gratton et al., 1988).

The RP and LRP may be present for hundreds of milliseconds before the response, but other components can be observed that are more tightly synchronized to the response. The early view was that a positive-going deflection is superimposed on the RP beginning 80–90 ms before the response, followed by a negative-going deflection during the response and another positive-going deflection after the response. However, subsequent research identified several additional movement-related components (see, e.g., Shibasaki, 1982; Nagamine et al., 1994).

Steady-State ERPs

In a typical ERP experiment, stimuli are presented at a relatively slow rate so that the brain has largely completed processing one stimulus before the next stimulus is presented. The waveform elicited in this situation is a *transient* response that occurs once and then ends. However, if many identical stimuli are presented sequentially at a fast, regular rate (e.g., 8 stimuli per second), the system will stop producing a *transient* response to each individual stimulus and enter into a *steady state*, in which the system resonates at the stimulus rate. Typically, steady-state responses will look like two summed sine waves, one at the stimulation frequency and one at twice the stimulation frequency.

Examples of transient and steady-state responses are shown in figure 3.16. The upper-left portion of the figure shows the response obtained when the on–off cycle of a visual stimulus repeats at 2 Hz (two on–off cycles per second). This is slow enough that each stimulus onset elicits a transient neural response with multiple distinct peaks. When the stimulation rate is increased to 6 cycles per second (upper-right panel), it is still possible to see some distinct peaks, but the overall waveform now appears to repeat continuously, with no clear beginning or end. As the stimulation rate is increased to 12 and then 20 cycles per second (bottom two panels), the response is predominantly a sine wave at the stimulation frequency (plus a small, hard-to-see oscillation at twice the stimulation frequency).

This steady-state response can be summarized by four numbers; the amplitude (size) and phase (temporal shift) of each of the two sine waves. This is a lot simpler than a transient response containing many different components. In addition, it is possible to collect hundreds of trials in

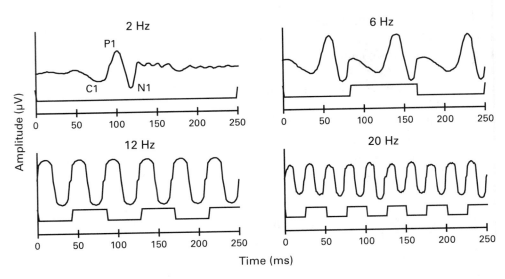

Figure 3.16
Examples of visual responses to repeating stimuli at a rate of 2, 6, 12, or 20 cycles per second.

a very short period of time owing to the fast stimulation rate. As a result, steady-state ERPs are commonly used in the diagnosis of sensory disorders. They have also been used to study the effects of attention on visual processing (Morgan, Hansen, & Hillyard, 1996; Di Russo, Teder-Sälejärvi, & Hillyard, 2003).

Steady-state ERPs have a significant shortcoming, however, which is that they do not provide precise temporal information. For example, if stimuli are presented every 150 ms, the voltage measured at 130 ms after the onset of one stimulus consists of the sum of the response to the current stimulus at 130 ms, the response to the previous stimulus at 280 ms, the response to the stimulus before that at 430 ms, and so forth. Nonetheless, steady-state ERPs are very useful when high temporal resolution is not needed.

The General Problem of Determining the Process Indexed by a Given ERP Component

I have now discussed a large number of interesting and distinctive ERP components, each reflecting a different aspect of perception, attention, memory, language, emotion, or cognitive control. For most of these components, I have tried to make a link between the component and the specific psychological or neural process that the component reflects. However, I will be the first to admit that most of these links are somewhat tenuous. In particular, the lack of a good theory of the functional significance of the P3 wave seems like a notable failure. If you are interested in reading more about the general problem of linking an ERP component with a specific neural or psychological process, see the online chapter 3 supplement.

The Discovery of N2pc, CDA, N400, and ERN

In this section, I'd like to share some stories of how various ERP components were discovered. Each of these stories has some lessons that you might find useful, and I hope they will inspire *you* to go out and discover a new ERP component!

The Discovery of N2pc (and Advice about Listening to Your Mentor)

Shortly after I started graduate school in 1986, we had a lab meeting where we read a *Scientific American* article that Anne Treisman had written about feature integration theory (Treisman, 1986). Most of the evidence for this theory—and much of the behavioral research of the time—centered on visual search tasks, in which subjects report whether a target item is present in an array of distractors. I showed quite a bit of interest in the paper, and at the end of the lab meeting, Steve Hillyard suggested that I "ought to ERP that task." Very little ERP research had been conducted using visual search tasks, so I agreed that this would be a good idea. It became my first real ERP project in the Hillyard lab.

I programmed the stimuli for a search task, and Steve came to the lab to see it. He made a variety of minor suggestions, and then asked, "Do you have different event codes to indicate whether the target is on the left or right side?" The stimulus arrays (like those in figure 3.8) had an equal number of objects on each side, and it seemed obvious to me that we wouldn't get different ERPs depending on which side contained the target. So I thought to myself, "That's a dumb idea." But, because I was a first-year graduate student (and because I grew up in a very polite Lutheran household in Wisconsin), my response was, "No, I didn't program it that way. But I will add the separate event codes." And I did.

Several months later, after I had recorded all the data, I figured I should separate the waveforms according to whether the target appeared on the left side or on the right side, and I immediately saw the N2pc component. I didn't know what it meant, so when we published the study, we merely mentioned this effect in a footnote (Luck & Hillyard, 1990). Hajo Heinze and I conducted a pair of related studies shortly after that, and we both saw the N2pc component again (Heinze, Luck, Mangun, & Hillyard, 1990; Luck et al., 1990). To figure out what this new component meant, I conducted about a dozen follow-up experiments, and I published eight of these experiments in a pair of papers (Luck & Hillyard, 1994a, 1994b). These two papers came out in 1994, 8 years after I first recorded the N2pc. That's how long it took for me to feel confident that I understood what the N2pc meant. I've published 18 N2pc papers since then, including one in *Nature* (Woodman & Luck, 1999), and Google Scholar reports more than 2000 papers that include the term *N2pc*. I think it's fair to say that the N2pc component has been very, very good to me.

The moral of this story, of course, is that graduate students should listen when their mentors make suggestions like including separate event codes for left- and right-side targets. However, I should point out that graduate students should sometimes ignore their mentors when their mentors tell them *not* to do something. For example, Vince Clark came up with a new approach to studying the effects of stimulus location on C1 polarity, but Steve Hillyard told him it wouldn't work and that he shouldn't try it. I suggested to Vince that he collect some pilot data: if the experiment turned out well, he could show the data to Steve; if it didn't turn out well, Steve would never know. Vince collected the pilot data, and when he showed the data to Steve, Steve realized that it would work, and it turned into a beautiful paper (Clark et al., 1995).

This happened with one of my own graduate students, Weiwei Zhang, as well. One day, Weiwei came to me with an idea for a new way to measure the precision of visual working memory. I told him that there were at least four reasons why it would never work and that he shouldn't waste his time with it. He did the experiment anyway (without telling me), and he came back to me a month later with beautiful data. I was convinced by the data, and that experiment became the beginning of a study that was eventually published in *Nature* (Zhang & Luck, 2008). The conclusion is that you should listen when your mentor suggests adding something to your experiments, but you should feel free to close your ears when your mentor tells you something won't work, especially when you believe in your heart that it will.

The Discovery of the CDA

After completing his dissertation in my lab at the University of Iowa, Ed Vogel went to UCSD to do a postdoc with Steve Hillyard. Ed's dissertation included some behavioral experiments that compared the effects of attention on perceptual encoding and on working memory encoding. In the Hillyard lab, Ed ran an ERP version of these experiments to see whether the P1 was influenced by attention under these conditions. He initially examined the first few hundred milliseconds after the stimulus to look at the P1 wave. The P1 effects were weak, but he saw a big difference between the contralateral and ipsilateral sites toward the end of this time window. As Ed tells the story:

After being underwhelmed by the P1 data, I went back and re-averaged to include the whole retention interval to take a peek at that late effect. At the time it felt like blasphemy to even look at ERP activity that late into the trial. I remember being skeptical about anything that was later than the P3. But there it was. Plain as day.

Ed showed the data to Steve Hillyard, but Steve is a dedicated attention researcher and wasn't interested in pursuing a working memory effect. As Ed told me:

I decided to pocket the effect and chase it down once I had my own lab. Which I did about 6 months later in Oregon. They were the absolute first experiments I ran once my lab was operational.

The first experiment on how CDA was modulated by set size was done in spring 2003. I was going to the Cognitive Neuroscience Society meeting in New York City, and I had my new master's student (Maro Machizawa) collect the data while I was away. Because it was a new experiment and a new student flying solo, I told him to just run one or two subjects while I was gone and that we'd look over the data together when I returned. That way, if there was a problem with the data (e.g., lots of artifacts, missing event codes, etc.) it would be a minimal loss. When I got back to Oregon, I was absolutely irate to find out that he'd run 12 subjects and hadn't looked at any of the data. So for all we knew, the data were noisy or missing event codes and would be unusable. Fortunately the data were clean and showed the first of many demonstrations of the set size modulation and asymptote of the CDA.

I'd like to point out that Ed was right about not running a large number of subjects before checking to make sure everything was okay. There are so many things that can go wrong with an ERP experiment that you need to do a reasonably complete set of analyses on the first few subjects to make sure that you have all the right event codes, that there is nothing weird about the data, and so forth. I have seen many experiments that had to be re-run because of a problem that could have been detected by a closer look at the data from the first few subjects. In this particular experiment, Ed and Maro got lucky and everything turned out okay. In fact, this became the first experiment in a paper that was published in *Nature* (Vogel & Machizawa, 2004). But you should still do a full set of analyses (including behavioral analyses) after the first few subjects to make sure there aren't any problems.

The Discovery of the N400 Component

I asked Marta Kutas to tell me about the discovery of the N400, and here is the story she told me:

The discovery of the N400 was a result of a "failed" P3 experiment that I ran in 1978 when I was a postdoc in Steve Hillyard's lab. We were interested in language, but almost all language ERP experiments up to that point had been oddball experiments with single words, in which participants were asked to make a binary decision about each word (e.g., male versus female). Our goal was to extend the oddball paradigm to sentence materials so that we could use P3 latency to investigate the role of context on word recognition.

Steve and I designed an initial experiment using an oddball paradigm with simple sentences—75% of the sentences were ordinary and meaningful and ended in a relatively predictable way (e.g., "He shaved off his mustache and beard."), but a random 25% of them were oddballs and ended with an interpretable but surprising word (e.g., "He shaved off his mustache and eyebrows."). The standard stimuli were further subdivided into idioms or proverbs (e.g., "A bird in the hand is worth two in the bush.") and facts ("The capital of California is Sacramento.") for which only one final word was acceptable, as well as more open-ended (less contextually constraining) sentences for which multiple endings were possible (e.g., "He returned the book to the library.").

In the long run, the plan was to examine how P3 latency changed as a function of variations in contextual constraint. However, we gave up on that plan because the ERPs elicited by the final words of the oddball sentences were not quite what we expected. Rather than a large P3, oddball words elicited a small negativity followed by a positivity. We thought the negativity might have been an N2, but it was later than usual; the positivity could have been a P3, but we weren't sure. Thus, we went back to the experimental drawing board. After some head scratching, we decided that we should shake the system as much as we could by presenting sentence-final words that were not just improbable and unexpected but semantically anomalous (e.g., "He shaved off his mustache and city." or "I take my coffee with cream and dog."). This was before computers were readily available for presenting stimuli, so it took quite a bit of work to change the experiment. We typed the each word in the middle of a frame the size of a slide, photocopied the words onto clear plastic sheets, cut the sheets into small rectangles, mounted them in slide holders, and then projected them using a slide projector.

The results of this new experiment were much clearer—the anomalous word elicited a large centro-parietal negativity peaking around 400 ms. This negativity (rather than the positivity, whether or not it was a P3) seemed to be where the action was. The rest is N400 history.

An important lesson of this story is that when you find an unexpected result in a complicated experiment, you should conduct a much simpler follow-up experiment so that you can see this new result more clearly. If Marta had just published her original experiment without running the now-classic follow-up experiment, it may have been years before anyone realized that a distinct and important new ERP had been discovered. As Bob Galambos would say, "you've got to get yourself a phenomenon" (see box 4.2 in chapter 4).

The Discovery of the ERN

The ERN was independently discovered by Michael Falkenstein and by Bill Gehring. I asked Bill how he happened to discover the ERN, and he told me that he first saw it in the context of a categorization experiment that he ran when he was in graduate school (Gehring, Coles, Meyer, & Donchin, 1995). In this experiment, two words were simultaneously presented, one above the other, and subjects were instructed to squeeze a response device with one hand if the upper word was a member of the category indicated by the lower word (e.g., *robin* above *bird*) and to squeeze with the other hand if not (e.g., *spoon* above *bird*). Although the experiment had been designed to look at the LRP, Bill found some interesting differences in N400 between these trial types. He wanted to know if they were related to task performance, so he did a variety of comparisons between correct trials and error trials. He initially did this for stimulus-locked averages (figure 3.17), but there were not any impressive differences between correct and incorrect trials. In retrospect, the lack of a large ERN in these waveforms probably reflected a large amount of RT variability, causing any activity related to the correctness of the response to be smeared out over a long time period.

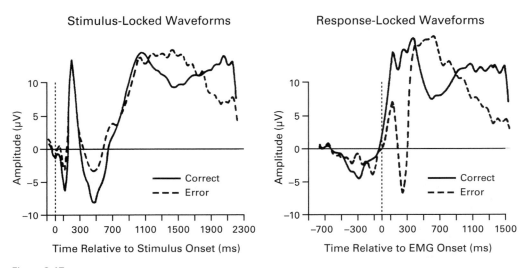

Figure 3.17
Stimulus-locked and EMG-locked ERPs from the experiment in which Bill Gehring first observed the error-related negativity (ERN). Courtesy of Bill Gehring.

Bill might have given up on the ERN at that point except for a conversation he had with Marta Kutas. Bill was a graduate student in the famous Cognitive Psychophysiology Laboratory (CPL) at the University of Illinois, where many classic ERP studies have been conducted over the decades. Marta Kutas had done her Ph.D. work at the CPL many years before, and she had coffee with Bill one day when she was back at the CPL for a visit. As Bill told me:

She suggested looking at response-locked data: that N400/performance relationships might come out better if we looked at N400 in the response-locked waveforms. So that's where the ERN popped up…. I remember vividly when I first saw the ERN: our Harris computer green CRT displays plotted the waveforms out one pixel at a time, so you could see it draw the waveform in bright green from left to right. It sort of looked liked an oscilloscope plotting out the waveform in green against a black background. First the correct waveform emerged, and then the error trials were plotted out, and at the time of the response a huge peak happened—it looked just like a big blip on an oscilloscope. (For a while in the lab we nicknamed the ERN the "blunder blip.") I'm sure I had a big P3 at just that moment. The ERN was so huge, in fact, that we were worried it was some kind of artifact, so I spent the better part of the next couple of years re-analyzing other data sets to see if the ERN was also in those data…. One of the reasons it took us so long to publish the data was because we had decided to look at all of the data we could to make sure we were dealing with something real.

The response-locked waveforms are shown in figure 3.17, and you can see that the ERN was quite impressive in these waveforms.

I love this story because a graduate student made an important discovery after following a suggestion from a more senior researcher, and I also love it because the results were not published for several years because the graduate student wanted to make sure it was real. This is just what happened with the discovery of the N2pc.

Suggestions for Further Reading

Donchin, E., & Coles, M. G. H. (1988). Is the P300 component a manifestation of context updating? *Behavioral & Brain Sciences, 11*, 357–374.

Folstein, J. R., & Van Petten, C. (2008). Influence of cognitive control and mismatch on the N2 component of the ERP: A review. *Psychophysiology, 45*, 152–170.

Gehring, W. J., Goss, B., Coles, M. G. H., Meyer, D. E., & Donchin, E. (1993). A neural system for error-detection and compensation. *Psychological Science, 4*, 385–390.

Gratton, G., Coles, M. G. H., Sirevaag, E. J., Eriksen, C. W., & Donchin, E. (1988). Pre- and post-stimulus activation of response channels: A psychophysiological analysis. *Journal of Experimental Psychology: Human Perception and Performance, 14*, 331–344.

Kutas, M., & Hillyard, S. A. (1980). Reading senseless sentences: Brain potentials reflect semantic incongruity. *Science, 207*, 203–205.

Luck, S. J., & Hillyard, S. A. (1994). Spatial filtering during visual search: Evidence from human electrophysiology. *Journal of Experimental Psychology: Human Perception and Performance, 20*, 1000–1014.

Luck, S. J., & Kappenman, E. S. (Eds.). (2012). *The Oxford Handbook of Event-Related Potential Components*. New York: Oxford University Press.

Näätänen, R., & Picton, T. (1987). The N1 wave of the human electric and magnetic response to sound: A review and an analysis of the component structure. *Psychophysiology, 24*, 375–425.

Verleger, R. (1988). Event-related potentials and cognition: A critique of the context updating hypothesis and an alternative interpretation of P3. *Behavioral & Brain Sciences, 11*, 343–427.

4 The Design of ERP Experiments

Overview

Chapter 2 described the very difficult problems that arise in trying to isolate individual ERP components. If it has been a while since you read that chapter, you might want to remind yourself by looking at figures 2.5 and 2.6. These problems can be summarized in a single paragraph:

Peaks and components are not the same thing, and this makes it difficult to isolate an underlying component from the observed ERP waveform. For example, a change in one component may influence the amplitudes and latencies of multiple peaks. Similarly, an effect that occurs in the time range that is usually associated with a given component may actually reflect a change in a different component that is also active during that time period (e.g., an effect during the time range of the N2 component may actually reflect a change in P3 amplitude during this period). Moreover, a change in component amplitude may lead to a change in peak latency, and a change in component timing may lead to a change in peak amplitude. The averaging process may also lead to incorrect conclusions because differences in latency variability will produce differences in peak amplitude. Similarly, the onset and offset of an effect in an averaged waveform will reflect the earliest single-trial onset and latest single-trial offset times rather than the average single-trial onset and offset times. Finally, it is difficult to be sure that an effect in one experiment reflects the same component that was observed in previous experiments.

In addition, chapter 3 described several examples of ERP effects that we ordinarily consider to reflect a single component but that actually reflect multiple distinct brain areas that are engaged in related (but probably not identical) processes.

Together, these problems often make it difficult to isolate highly specific psychological or neural processes in ERP experiments. This is a very real challenge because many ERP experiments make predictions about the effects of an experimental manipulation on a given component, and the conclusions of these experiments are valid only if the observed effects really reflect changes in that component. For example, the N400 component is widely regarded to be a sensitive index of the degree of mismatch between a word and a previously established semantic context, and it would be nice to use this component to determine which of two sets of words is perceived as being more incongruous. However, the N400 and P3 components have similar timing and scalp distributions, so it can be difficult to tell the difference between an increased N400 and a reduced P3. If one group of words produces a smaller negativity than another, this could mean that the first group of words is perceived as less incongruous (leading to a smaller N400) or is more resource-demanding (leading to a larger P3). This sort of difficulty is present in a very large proportion of ERP experiments.

But don't get depressed! You should first realize that it is always challenging to study the human mind and brain, and many of the difficulties involved in ERP research are also present in fMRI research (see the online chapter 3 supplement for a discussion of the challenges involved in linking physiological measures with the underlying processes). And the temporal resolution of the ERP technique solves many of the problems that complicate fMRI studies. Moreover, the fact that ERPs have made many important contributions to our understanding of the human mind and brain demonstrates that these problems are not insurmountable.

This chapter focuses on experimental design, with the goal of helping you figure out how to design experiments in which ERPs—despite their limitations—can provide definitive answers to important questions about the mind and brain. I will begin by describing eight time-tested strategies you can use when you design your experiments to solve or avoid the problems involved in isolating ERP components. The design phase is definitely the best time to address these potential problems; once your data have been collected, the right design will make your life much easier. After I cover those eight strategies, I will describe a variety of other experimental design problems that commonly arise in ERP experiments and provide some good solutions to these problems. An online supplement to this chapter provides several examples of experiments that successfully overcame the challenges associated with ERPs and had a real impact on our understanding of the mind and brain.

Designing experiments is my favorite part of being a scientist. The design process is where theory meets experiment, and it is a point where the inevitable imperfections of real data have not yet marred the beauty of the scientific ideas. The excellent book on the physics of EEG by Paul Nuñez and Ramesh Srinivasan notes that engineers and physicists—compared to neuroscientists and psychologists—tend to spend more of their time thinking about what experiments are worth conducting and less time implementing actual experiments (Nunez & Srinivasan, 2006). I'm not sure I agree 100% with their view of what we should be doing, but I certainly believe that people should spend more time (and intellectual effort) designing their experiments before they start collecting data (see box 4.1 for additional thoughts about experimental design).

Strategies for Avoiding Ambiguities in Interpreting ERP Components

The following eight strategies have proved to be very useful for designing experiments that minimize or avoid the problem of identifying specific ERP components. Note that they're not listed in any particular order, and you do not need to implement all strategies in all experiments. In fact, if you follow some of these strategies, others may not be necessary. Concrete examples

Box 4.1
The Craft of Experimental Design

Apple Computer Inc. is one of the world's leaders in industrial design, and one of the most influential executives at Apple is Jonathan Ive, Vice President of Design. Ive was the very first winner of the Designer of the Year Award from the Museum of Design, and he gave a very interesting interview about Apple's design process to the museum (see http://designmuseum.org/design/jonathan-ive/). In this interview, he said the following: "Perhaps the decisive factor is fanatical care beyond the obvious stuff: the obsessive attention to details that are often overlooked." I love this quote! It captures the difference between a good experimental design and a great experimental design.

I'm a fan of the Craftsman style of furniture and architecture. As shown in the following photograph, Craftsman designs take utilitarian elements such as joints and braces and turn them into aesthetic elements. By analogy, a clever counterbalancing scheme in an ERP experiment may be a thing of beauty. In addition, Craftsman designs often include elements that are hidden inside the object and cannot be seen, but these elements add to the strength and durability of the object. Similarly, if you devote "obsessive attention to details that are often overlooked," such as the temperature of the recording chamber and the quality of the signal coming from every electrode site, your results will be stronger and more durable. Craftsman designs also feature carefully selected wood that has only a few simple coats of clear finish rather than multiple layers of paint and fabric, allowing the beauty of the underlying design to be seen. I find that the most convincing ERP experiments similarly feature flawless data with only a few simple layers of processing, allowing the beauty of the underlying design to be seen.

of the application of these strategies in previous research are provided in the online supplement to this chapter.

Strategy 1: Focus on a Single Component

My first experimental design strategy is to focus a given experiment on only one or perhaps two ERP components, trying to keep all other components from varying across conditions. This reflects the tension between the conceptual and operational definitions of the term *ERP component* that were discussed in chapter 2. Manny Donchin's operational definition was that an ERP component is "a source of controlled, observable variability" (Donchin, Ritter, & McCallum, 1978, p. 354). This means that any activity that differs across a given set of conditions is equivalent to a single component. However, my conceptual definition states that an ERP component is "generated in a given neuroanatomical module when a specific computational operation is performed." If multiple processes that reflect different neuroanatomical modules and different computational operations get lumped together because they covary across the conditions of your experiment, then you will have a mess. But if you use very precise manipulations that cause only a single computational operation in a single neuroanatomical module to vary across conditions, then you will be able to isolate a single ERP component according to both the operational and conceptual definitions of *component*.

The most complicated, uninterpretable, and downright ugly results often come from experiments in which someone simply takes a previously used behavioral paradigm and runs it while recording the EEG. However, a "fishing expedition" of this sort can be very useful when you are beginning a new program of research with a task that no one has ever used with ERPs before. If you try this, you will probably find that many components vary across the conditions of the experiment and that you cannot draw any strong conclusions from the results. But this experiment may give you great ideas for more focused experiments, so it may be very worthwhile. Just make sure you treat that first experiment as a pilot study or experiment 1 in a multiple-experiment paper, and don't "pollute" the literature with complicated, uninterpretable results. For some advice about starting a new program of research, see box 4.2.

It is sometimes possible to use a factorial experimental design in which one factor is used to isolate one component and a different factor is used to isolate a different component. For example, the schizophrenia study described in chapter 1 (see figure 1.4) used a rare/frequent manipulation to isolate the P3 wave, and this was factorially crossed with a left-hand/right-hand manipulation to isolate the LRP (Luck et al., 2009). Emily Kappenman and I took this even further in a proof-of-concept study in which we used four different factors to isolate four different components. We call this the MONSTER paradigm (for *Manipulation of Orthogonal Neural Systems Together in Electrophysiological Recordings*) (Kappenman & Luck, 2011). Although this approach appears to be inconsistent with the strategy of focusing on a single component, it is actually an extension of this strategy because it is like running a sequence of four different experiments to isolate the four components.

Box 4.2
Getting Yourself a Phenomenon

My graduate school mentor was Steve Hillyard, who inherited his lab from his own graduate school mentor, Bob Galambos (shown in the photograph that follows). Dr. G (as we often called him) was still quite active after he retired. He often came to our weekly lab meetings, and I had the opportunity to work on an experiment with him. He was an amazing scientist who made fundamental contributions to neuroscience. For example, when he was a graduate student, he and fellow graduate student Donald Griffin provided the first convincing evidence that bats use echolocation to navigate. He was also the first person to recognize that glia are not just passive support cells (and this recognition essentially cost him his job at the time). You can read the details of his interesting life in his autobiography (Galambos, 1996) and in his *New York Times* obituary (http://www.nytimes.com/2010/07/16/science/16galambos.html).

Bob was always a font of wisdom. My favorite quote from him is this: "You've got to get yourself a phenomenon" (he pronounced *phenomenon* in a slightly funny way, like "pheeeenahmenahn"). This short statement basically means that you need to start a program of research with a robust experimental effect that you can reliably measure. Once you've figured out the instrumentation, experimental design, and analytic strategy that enables you to measure the effect reliably, then you can start using it to answer interesting scientific questions. You can't really answer any interesting questions about the mind or brain unless you have a "phenomenon" that provides an index of the process of interest. And unless you can figure out how to record this phenomenon in a robust and reliable manner, you will have a hard time making real progress. So, you need to find a nice phenomenon (like a new ERP component) and figure out the best ways to see that phenomenon clearly and reliably. Then you will be ready to do some real science! For examples, see the descriptions of how several ERP components were discovered at the end of chapter 3.

Strategy 2: Focus on Large Components

When possible, it is helpful to study large components such as P3 and N400. When the component of interest is very large compared to the other components, it will dominate the observed ERP waveform, and measurements of this component will be relatively insensitive to distortions from the other components. As an example, take a look at the large P3 and the small N2 in the schizophrenia study described in chapter 1 (see figure 1.4).

It is not always possible to focus on large components because sometimes a smaller component provides an index of the process that you're trying to study. However, you may be able to figure out a clever and nonobvious way to use the P3 or N400 component to answer the question you are asking.

Strategy 3: Hijack Useful Components from Other Domains

If you look at ERP experiments that have had a broad impact in cognitive psychology or cognitive neuroscience, you will find that many of them use a given ERP component that is not obviously related to the topic of the experiment. For example, the attentional blink experiment described in the online supplement to this chapter used the language-related N400 component to examine the role of attention in perceptual versus postperceptual processing (Luck, Vogel, & Shapiro, 1996; see also Vogel, Woodman, & Luck, 2005). The N400 has also been used to determine the stage of processing at which a specific variety of visual masking operates (Reiss & Hoffman, 2006). Similarly, although the LRP is related to motor preparation, it has been used to address the nature of perception without awareness (Dehaene et al., 1998) and syntax processing in language (van Turennout, Hagoort, & Brown, 1998). One of my former graduate students, Adam Niese, refers to this as *hijacking* an ERP component.

Strategy 4: Use Well-Studied Experimental Manipulations

It is usually helpful to examine a well-characterized ERP component under conditions that are as similar as possible to conditions in which that component has previously been studied. For example, when Marta Kutas first started recording ERPs in language paradigms, she focused on the P3 wave and varied factors such as "surprise value" that had previously been shown to influence the P3 wave in predictable ways. Of course, when she used semantic mismatch to elicit surprise, she didn't observe the expected P3 wave but instead discovered the N400 component. However, the fact that her experiments were so closely related to previous P3 experiments made it easy to determine that the effect she observed was a new negative-going component and not a reduction in the amplitude of the P3 wave (as discussed in the final section of chapter 3).

In my own research, I have almost always included a manipulation of target probability in experiments that look at P3 (Luck, 1998b; Vogel, Luck, & Shapiro, 1998; Luck et al., 2009) and a manipulation of semantic/associative relatedness in experiments that look at N400 (Luck et al., 1996; Vogel et al., 1998; Vogel et al., 2005). Virtually everyone uses a manipulation of stimulus location to look at N2pc and CDA and a manipulation of response hand to look at the LRP, because these manipulations are intrinsic to the definition of these components.

Strategy 5: Use Difference Waves

This is probably the most important and widely applicable of all the experimental design strategies. The use of difference waves was discussed extensively in chapters 2 and 3 (see figures 2.5, 2.7, and 3.11), but here is an example that will serve as a reminder.

Imagine that you are interested in assessing the N400 for two different noun types, *count nouns* (words that refer to discrete items, such as *cup*) and *mass nouns* (words that refer to entities that are not divisible into discrete items, such as *water*). The simple approach would be to present one word at a time, with count nouns and mass nouns randomly intermixed, and have subjects do some simple semantic task (e.g., judge the pleasantness of each word). This would yield two ERP waveforms, one for count nouns and one for mass nouns. However, it would be difficult to know if any differences observed between the count noun and mass noun waveforms were due to differences in N400 amplitude or due to differences in some other ERP component.

To isolate the N400, the experiment could be redesigned so that each trial contained a sequence of two words, a context word and a target word, with a count noun target word on some trials and a mass noun target word on others. In addition, the context and target words would sometimes be semantically related and sometimes be semantically unrelated. You would then have four trial types:

Count noun, related to context word (e.g., "plate ... cup")
Mass noun, related to context word (e.g., "rain ... water")
Count noun, unrelated to context word (e.g., "sock ... cup")
Mass noun, unrelated to context word (e.g., "garbage ... water")

The N400 could then be isolated by constructing difference waves in which the ERP waveform elicited by a given word when it was preceded by a semantically related context word is subtracted from the ERP waveform elicited by that same word when preceded by a semantically unrelated context word. Separate difference waves would be constructed for count nouns and mass nouns (unrelated minus related count nouns and unrelated minus related mass nouns). Each of these difference waves should be dominated by a large N400 component, with little or no contribution from other components (because most other components aren't sensitive to semantic mismatch). You could then see if the N400 was larger in the count noun difference wave or in the mass noun difference wave (a real application of this general approach is described in the online supplement to this chapter).

Although this approach is quite powerful, it has some limitations. First, difference waves constructed in this manner may contain more than one ERP component. For example, there may be more than one ERP component that is sensitive to the degree of semantic mismatch, so an unrelated-minus-related difference wave might consist of two or three components rather than just one. However, this is still a vast improvement over the raw ERP waveforms, which will probably contain at least ten different components.

A second limitation of this approach is that it is sensitive to the *interaction* between the variable of interest (e.g., count nouns versus mass nouns) and the factor that is varied to create the difference waves (e.g., semantically related versus unrelated word pairs). Imagine, for example, that the N400 amplitude is 1 μV larger for count nouns than for mass nouns, regardless of the degree of semantic mismatch. If the N400 is 2 μV for related mass nouns and 12 μV for unrelated mass nouns, then it would be 3 μV for related count nouns and 13 μV for unrelated count nouns (i.e., 1 μV bigger than the values for the mass nouns). If we then made unrelated-minus-related difference waves, this difference would be 10 μV for the mass nouns (12 μV minus 10 μV) and would also be 10 μV for the count nouns (13 μV minus 3 μV). Fortunately, when two factors influence the same ERP component, they are likely to interact multiplicatively. Imagine, for example, that N400 amplitude is 50% greater for count nouns than for mass nouns. If the N400 is 2 μV for related mass nouns and 12 μV for unrelated mass nouns, then it would be 3 μV for related count nouns and 18 μV for unrelated count nouns (i.e., 50% bigger for the count nouns than for the mass nouns). An unrelated-minus-related difference wave would then be 10 μV for the mass nouns and 15 μV for the count nouns, so now we would be able to see the difference between count nouns and mass nouns in the related-minus-unrelated difference waves.

Of course, the interactions could take a more complex form that would lead to unexpected results. For example, count nouns could elicit a larger N400 than mass nouns when the words are unrelated to the context word, but they might elicit a smaller N400 when the words are related to the context word. Thus, although difference waves can be very helpful in isolating specific ERP components, care is necessary when interpreting the results.

Strategy 6: Focus on Components That Are Easy to Isolate

To use strategy 4 and strategy 5, it is helpful to focus on those ERP components that are relatively easy to isolate by means of well-studied manipulations and difference waves. Not just any manipulation or any difference wave will do, because you want the difference wave to eliminate all components except for the one that tells you something about the process you are trying to study. For example, you could have an "easy" condition and a "difficult" condition and use these to make difficult-minus-easy difference waves, but the resulting difference waves would likely contain many different components reflecting the many different processes that might differ between difficult and easy conditions. This would not usually be very helpful.

Some components are easier to isolate than others, especially if you plan to use one manipulation to isolate the component and then factorially combine this manipulation with another manipulation designed to ask how this component varies across conditions. The best examples are the components that are defined by a contralateral-minus-ipsilateral difference wave, including the lateralized readiness potential (LRP), the N2pc component, and the contralateral delay activity (CDA; see chapter 3). For example, the LRP is defined by the difference in amplitude between the hemisphere contralateral to a response and the hemisphere ipsilateral to the response hand (reviewed by Smulders & Miller, 2012). The contra-minus-ipsi difference wave is sensitive to motor-related activity because of the contralateral organization of the motor system, but it subtracts away all other brain responses that are not contralaterally organized. It also subtracts away any brain activity prior to the time at which the brain has determined whether a left-hand response or a right-hand response should be made for the current stimulus. Consequently, the LRP can be used to determine with high levels of certainty that the brain has begun to prepare a given response at a given moment in time (see review by Smulders & Miller, 2012). The LRP has been used in many high-impact studies, showing that the brain sometimes prepares an incorrect response even when the correct response is ultimately emitted (Gratton, Coles, Sirevaag, Eriksen, & Donchin, 1988), that partial results from one stage of processing are transmitted to the next stage (Miller & Hackley, 1992, described in the online supplement to this chapter), and that subliminal stimuli are processed all the way to response selection stages (Dehaene et al., 1998).

Similarly, the N2pc component is isolated by a contra-minus-ipsi difference wave relative to the location of an attended stimulus in a bilateral visual stimulus array (see figure 3.8 in chapter 3 and the review by Luck, 2012b). Because the overall array is bilateral, the initial sensory response is bilateral, as are the ERPs corresponding to higher-level postperceptual processes. The only processes that remain in the contra-minus-ipsi subtraction are those that are both influenced by the allocation of attention to the target and generated in contralaterally organized areas of visual cortex. Moreover, this contra-minus-ipsi difference can easily be combined with other experimental manipulations or group comparisons. For example, one can manipulate whether the attended item is associated with large or small rewards (Kiss, Driver, & Eimer, 2009) or is surrounded by nearby distractors (Luck, Girelli, McDermott, & Ford, 1997), or one can ask whether the contra-minus-ipsi difference varies as a function of age (Lorenzo-Lopez, Amenedo, & Cadaveira, 2008) or psychiatric diagnosis (Luck et al., 2006). This makes it possible to ask very precise questions about the operation of attention (see, e.g., Woodman & Luck, 1999, 2003b; Eimer & Kiss, 2008; Lien, Ruthruff, Goodin, & Remington, 2008).

Strategy 7: Use a Component to Study the Processes That Precede It

Strategy 7 is based on the idea—which was described previously in the section on P3 latency in chapter 3—that the occurrence of a difference between conditions logically entails that certain processes must have already occurred. For example, the P3 wave is larger for stimuli that belong to a rare category than for stimuli that belong to a frequent category, and the difference in amplitude cannot begin until the brain has begun to determine the category to which a stimulus belongs. Consequently, the presence of a difference between the rare and frequent categories indicates that the brain has determined whether the stimulus belongs to the rare or frequent category, and the brain must have begun to categorize the stimulus by the time the difference has exceeded 0 μV (assuming that the experiment is appropriately designed). The P3 does not itself reflect the categorization process; instead, categorization is a necessary precondition for the occurrence of a P3 probability effect.

A made-up example is shown in figure 4.1. The goal of this experiment is to measure the amount of time required to identify a digit, determine whether it is odd or even, and add it to another digit. In the experiment, subjects view a sequence of digits at the center of the monitor, with one digit every 1500 ± 100 ms (the reasons for this timing will be explained later in the

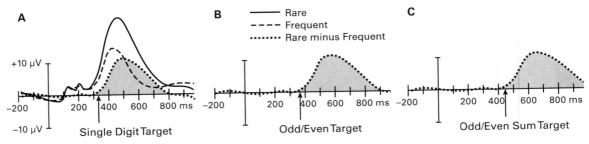

Figure 4.1
Simulated data from an imaginary oddball experiment in which the digits 0–9 are presented in the center of the screen in random order. A target category is defined at the beginning of each trial block, and the subject presses one button for target stimuli and a different button for nontarget stimuli (with the same hand). Ten percent of the stimuli are targets, and 90% are nontargets. In the single-digit condition (A), a specific digit is designated as the target, and the other nine digits serve as nontargets. ERPs elicited by stimuli in the rare (target) and frequent (nontarget) categories are shown, along with the rare-minus-frequent difference wave. In the odd/even condition (B), the target is defined as any odd digit, and the nontarget category consists of any even digit (or vice versa). Again, 10% of the stimuli are targets. Only the difference wave is shown for this condition. In the odd/even sum condition (C), the target is defined as a digit that, when added to the previous digit, is an odd number, and the nontarget is defined as a digit that, when added to the previous digit, is an even number (or vice versa). Again, 10% of the stimuli are targets. Only the difference wave is shown for this condition. For all three conditions, an arrow marks the onset latency of the difference wave, which is earliest in the single-digit condition, later in the odd/even condition, and latest in the odd/even sum condition.

chapter). Each of the 10 digits occurs in an unpredictable order. At the beginning of each trial block, a target is designated for that block. Subjects press one button when the current stimulus is a target and a different button when it is a nontarget. The target occurs on 10% of trials, and the remaining 90% are nontargets. In the *single-digit* condition, the target is defined as a specific digit (e.g., the number 3). Figure 4.1A shows the ERPs elicited by the rare stimuli (targets), the frequent stimuli (nontargets), and the rare-minus-frequent difference wave. Logically, this difference wave cannot exceed 0 μV until the brain has begun to determine whether the current stimulus is the target digit or not. Thus, the onset of the difference wave gives us an upper bound on the amount of time required for the brain to classify a number. It's an upper bound rather than the exact time because the brain might have made this classification earlier, but with no effect on the observable ERP waveform. The arrow in figure 4.1A shows the onset time of the difference, defined as the time at which it exceeds 1 μV (see chapter 9 for a discussion of methods for quantifying onset latencies).

This experiment also includes an *odd/even* condition to determine how much additional time is required to determine whether a digit is odd or even. In this condition, the target is defined as any odd digit and the nontarget is any even digit (or vice versa, counterbalanced across trial blocks). Again, the target category occurs on 10% of trials and the nontarget category occurs on 90% of trials. It presumably takes longer for the brain to determine whether a digit is odd or even than it takes to identify a specific digit, so it will take longer for the brain to determine if a given stimulus belongs to the rare category or the frequent category in this condition than in the single-digit condition. Thus, the onset of rare-minus-frequent difference should be later in the odd/even condition than in the single-digit condition (see the arrow in figure 4.1B). Moreover, the difference in onset latency between the odd/even condition and the single-digit condition tells us how much more time is required to make an odd/even decision than is required to identify a specific digit.

The experiment also includes an *odd/even sum* condition in which the subject must determine whether the sum of the current digit and the previous digit is odd or even. Again, the sum would be odd on 10% of trials and even on 90% of trials (or vice versa). This task should take even longer than determining whether the current digit is odd or even. Thus, the onset of rare-minus-frequent difference should be even later in this condition (see the arrow in figure 4.1C) than in the odd/even condition. And the difference in onset latencies provides an estimate of the additional time required to combine two digits.

You may be asking why we included a rare/frequent manipulation in this experiment. After all, we are really interested in things like odd versus even, and combining this with rare versus frequent may seem like an unnecessary complication. However, it is in fact necessary to include the rare versus frequent manipulation. If we used 50% odd and 50% even stimuli in the odd/even condition, the ERP waveform would be virtually identical for odd stimuli and for even stimuli, and the difference wave would be a flat line. This would make it impossible to assess the time required to make the odd/even categorization. Moreover, if we compared the waveforms across the three conditions directly, without first making difference waves, then we would not be able to isolate a specific process, and we would not be able to determine the time at which the target/nontarget discrimination was made. This is a little "trick" that is often useful in ERP experiments: You take a manipulation of interest (e.g., odd versus even) and combine it with another manipulation (e.g., rare versus frequent) that will allow the manipulation of interest to generate differential ERP activity. However, you must make sure that you counterbalance the combinations so that you don't have a confound. In our odd/even condition, for example, we wouldn't want odd to be rare and even to be frequent in all trial blocks because this would confound the odd/even and rare/frequent manipulations. Instead, we would have even be rare and odd be frequent in half of the trial blocks for the odd/even condition.

Keep in mind that the P3 wave does not itself reflect the *process* of determining whether a digit is a particular number, whether it is an odd or even number, or whether the sum of two digits is odd or even. Instead, it reflects the *consequences* of making the relevant categorization. In other words, these processes must logically occur before the brain can determine whether a given stimulus falls into the rare or frequent category in this experiment. This is what I mean when I recommend using a component to assess the processes that must precede it.

Example 3 in chapter 1 describes an experiment that used this logic to study the time course of perception and categorization in schizophrenia. Subjects performed a task in which they determined whether a given stimulus was a letter or a digit; one category was rare and the other was frequent. The timing of the P3 wave, measured from the rare-minus-frequent difference wave, was virtually identical in patients and control subjects (see figure 1.4 in chapter 1 and Luck et al., 2009). This demonstrates that the patients were able to perceive and categorize simple alphanumeric stimulus just as rapidly as the control subjects, even though behavioral RTs were substantially longer in the patients.

In these examples, ERPs are being used to measure the time course of processing, which is what ERPs do best. However, this general approach can also be used to determine whether a specific process happened at all. As described in the online supplement to this chapter, for example, the N400 component can be used to determine whether a word has been identified. Specifically, if the N400 is larger for a word that mismatches a semantic context than for a word that matches the context, the word must have been identified. This logic was used to demonstrate that words can be identified even when they cannot be consciously reported in the *attentional blink* paradigm (Luck et al., 1996; Vogel et al., 1998). The N400 does not itself reflect word identification, but word identification is a necessary precondition for seeing a difference between semantically matching and semantically mismatching words.

Strategy 8: Component-Independent Experimental Designs

Many of the previous strategies focused on ways to isolate specific ERP components. In many cases, an even better strategy is to completely sidestep the issue of identifying a specific component. For example, Thorpe, Fize, and Marlot (1996) conducted an experiment that asked how quickly the visual system can differentiate between different abstract classes of objects. To answer this question, they presented subjects with two sets of photographs—pictures that contained animals and pictures that did not. They found that the ERPs elicited by these two classes of pictures were identical until approximately 150 ms, at which point the waveforms diverged. From this result, it is possible to infer that the brain can detect the presence of an animal in a picture by 150 ms (but note that the onset latency represents the trials and subjects with the earliest onsets and not necessarily the average onset time, and this is an upper bound on the initial point at which the brain detected the animal).

This experimental effect occurred in the time range of the N1 component, but it did not matter at all whether the effect consisted of a change in the amplitude of that particular component; the conclusions depended solely on the time at which the effect occurred. This was a component-independent design because the conclusions did not depend on which component was influenced by the experimental manipulation (see also the first experiment described in the online supplement to this chapter, which determined the latency at which attention influences the ERP response to a stimulus).

The use of a component to assess the processes that logically precede it (strategy 7) usually leads to a component-independent experimental design. The fact that the rare-minus-frequent difference waves were nearly identical for people with schizophrenia and healthy control subjects (see figure 1.4 in chapter 1) tells us that these two groups were able to perceive and categorize the stimuli at the same rate, even if we don't assume that the difference waves consisted of a modulation of the P3 wave. Similarly, the conclusions of the N400 experiment described in the online supplement do not actually depend on whether the effects consisted of an N400 or some other component. This is the essence of a component-independent design.

Additional examples of prior research using component-independent designs are provided in the online supplement to this chapter. For other examples of high-impact studies using this strategy (resulting in papers published in *Science* and *Nature*), see van Turennout et al. (1998) and Dehaene et al. (1998).

Common Design Problems and Solutions

Although isolating specific components is typically the most difficult aspect of ERP research, there are several other challenges that you are likely to encounter when designing your experiments. In this section, I will describe several of the most common confounds and misinterpretations in ERP research. For each problem, I will also describe one or more solutions that you can use when you design your own experiments.

Online chapter 15 will formalize this set of common problems into a list of things that you should look for when you are about to submit a paper for publication or when you are reviewing a manuscript that someone else has submitted. I'm hoping that this checklist will help expunge these common errors from the literature. See box 4.3 for a discussion of some big-picture issues in experimental design.

To make these challenges and solutions concrete, I will describe a *Gedankenexperiment* (thought experiment) that is a conglomeration of many of the bad ERP experiments I have encountered over the years (including some of my own experiments that I wish I had designed differently). This Gedankenexperiment is designed to examine the effects of task difficulty on P3 amplitude. As shown at the top of figure 4.2, the target is the letter X, and the nontarget is the letter O. The stimuli are presented at the center of the video display with a duration of 500 ms, and one letter is presented every 1000 ms (SOA = 1000 ms; ISI = 500 ms). X is presented on 20% of trials, and O is presented on the other 80%. The letter X never occurs twice in succession because the P3 is reduced for the second of two consecutive targets. Subjects press a button with the index finger of the right hand whenever an X is detected, making no response for O. In the *bright condition*, the stimuli are bright and therefore easy to discriminate; in the *dim condition*, the stimuli are very dim and therefore difficult to discriminate. The bright and dim conditions are tested in separate blocks of trials (in counterbalanced order). At the end of the experiment, typical artifact rejection and averaging procedures are conducted, and P3 amplitude is quantified as the peak voltage at the Pz electrode site, separately for the rare and frequent trials in the dim and bright conditions.

When I teach about experimental design in the ERP Boot Camp, I describe this same Gedankenexperiment, and then I ask the participants to tell me the problems they see with the design and to propose solutions for these problems. As a group, they come up with almost all of the problems that I had in mind when I designed the experiment, and they generate some interesting solutions. I would recommend that before you read further, you make a list of all the design problems that you see with this experiment and potential solutions for these problems. You can then compare your list with mine.

Box 4.3
Confounds and Side Effects

Anyone who has taken a basic course on experimental design knows that the most fundamental principle of experimentation is to make sure that a given experimental effect has only a single possible cause. This is usually discussed in terms of avoiding *confounds*. A confound occurs when more than a single factor differs between conditions. For example, if the rare stimuli in an oddball paradigm are red and the frequent stimuli are blue, then these stimuli differ in two ways (i.e., rare versus frequent is confounded with red versus blue). A true confound can almost always be detected by a careful reading of the methods section of a journal article.

A related but subtler problem occurs when the experimenter varies only one factor, but this factor has *side effects* that are ultimately responsible for the effect of interest. To take a simple example, imagine you observe that heating a beaker of water causes a decrease in the mass of the water. This might lead you to the erroneous conclusion that hot water has a lower mass than cool water, even though the actual explanation is that some of the heated water turned to steam, which escaped through the top of the beaker. To reach the correct conclusion, it is necessary to seal the beaker so that water does not escape. Similarly, imagine that the P3 wave is smaller when you increase the temperature inside the recording chamber. You might conclude that warmer brains produce smaller P3 waves, but this might instead be a side effect of the fact that heating the chamber causes the subjects to become sleepier. You only varied one factor explicitly (the temperature of the chamber), but this change had multiple consequences (warmer brains, sleepier subjects, and, as described in chapter 5, more skin potentials). Side effects are more difficult to detect than confounds because they are secondary consequences of the manipulation of a single factor. Some imagination on your part may be required to realize that an experimental manipulation had a problematic side effect.

People who conduct experimental research sometimes have a condescending attitude toward people who conduct correlational research, because correlations can always be explained by some unknown third factor. But experimental effects can always be explained by some unforeseen side effect. That is, even though experiments have the advantage that one can be certain that *something* related to the experimental manipulation caused the effect, the exact cause is never known with certainty. So we should keep in mind that science is always prone to alternative explanations for both experimental and correlational research.

Sensory Confounds

There are some obvious sensory confounds in this Gedankenexperiment. First, the target is the letter X and the nontarget is the letter O, so the target and nontarget stimuli differ in terms of both shape and probability. This sort of confound is present in the vast majority of oddball ERP experiments. You might be asking why people design experiments with an obvious confound and why reviewers allow these experiments to be published. If you ask the experimenter about this kind of confound, you'll almost certainly get a response like this: "I can't imagine how that small sensory difference could produce a difference in the ERP at 400 ms, where the P3 is being measured" (see box 4.4 for a general discussion of this type of logic). This is probably true for the P3 wave, and this kind of sensory confound is therefore relatively benign for experiments focusing on late components. However, sensory confounds may produce significant effects as

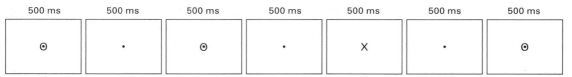

80% O's, 20% X's (no X repetitions), Bright or Dim (in separate blocks), Press button for X's

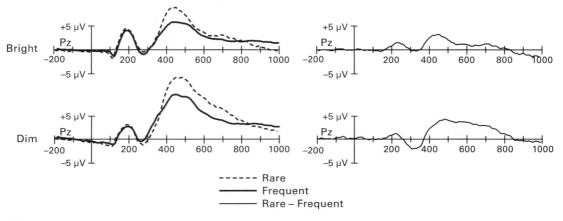

------ Rare
——— Frequent
——— Rare – Frequent

Figure 4.2
Experimental design (top) and simulated data (bottom) from the Gedankenexperiment. ERPs are overlaid for the rare and frequent stimulus categories, separately for bright stimuli and dim stimuli (left column). Rare-minus-frequent difference waves are also shown for the bright and dim stimuli (right column).

Box 4.4
Ignorance and Lack of Imagination

When someone says, "I can't imagine how that little confound could explain my results," this is a case of a general logical fallacy that philosophers call the *argument from ignorance*. In fact, it's a special case that is called (with a touch of humor) the *argument from lack of imagination*. The fact that someone can't imagine how a confound could produce a particular effect might just mean that the person doesn't have a very good imagination! I myself have occasionally used the "I can't imagine how ..." type of reasoning and then found that I was suffering from a lack of imagination (see, e.g., box 4.5). But now that I realize that this is not a compelling form of argument, I usually catch myself before I say it.

late as 200–300 ms. Moreover, a benign confound is a little bit like a benign tumor; wouldn't you rather not have it? It's usually a trivial matter to design an experiment to avoid confounds of this nature, so I would recommend not marring your beautiful experimental design with an ugly little confound.

The obvious solution to this problem is to counterbalance the Xs and Os. That is, you could have rare Xs and frequent Os in half of the trial blocks and then switch to rare Os and frequent Xs in the other half. If you are working with a cognitively impaired population, this might cause some confusion, so you could instead counterbalance across subjects.

However, this experiment contains a subtler sensory confound that cannot be solved by simple counterbalancing. Specifically, a difference in the probability of occurrence between two stimuli creates differences in sensory adaptation, which will in turn create differences in the sensory response to the two stimuli. The basic idea is that when the visual system encounters a particular stimulus many times, the neurons that code that stimulus will produce smaller and smaller responses. This is known as *stimulus-specific adaptation* or *refractoriness*. The fact that O occurs more frequently than X in this Gedankenexperiment means that the neurons that code O will be more adapted than the neurons that code X, and this may lead to a smaller sensory response for the nontarget O stimuli than for the target X stimuli. This will be true even if you switch the stimuli, making O rare and X frequent in half of the trial blocks. In each block, the neurons coding the frequent stimulus will become more adapted than the neurons coding the rare stimulus, leading to a smaller sensory response for the frequent stimulus. Box 4.5 provides an example of how this sort of adaptation created a significant and replicable but bogus effect in some of my own experiments.

How might we solve this adaptation problem? The solution to this and virtually all sensory confounds is to follow the following precept:

The Hillyard principle To avoid sensory confounds, you must compare ERPs elicited by *exactly* the same physical stimuli, varying only the psychological conditions.

I call this *the Hillyard principle* because Steve Hillyard made his mark on the field by carefully designing his experiments to rule out confounds that had plagued other experiments (and also because this principle was continually drilled into my head when I was a grad student in the Hillyard lab). The key to this principle is that you should be able to conduct your experiments by presenting exactly the same stimulus sequences, using instructions to create the different experimental conditions. Imagine, for example, that we try to solve the sensory confound in our Gedankenexperiment by simply counterbalancing whether X or O is the rare stimulus. We can't do this simply by using the same stimulus sequences and varying whether we tell the subjects to respond to the X or to the O, because this does not change which physical stimulus is rare. Note that it is not enough to equate the ERP-eliciting stimulus across conditions; the whole sequence of stimuli must be equated to avoid all possible sensory confounds.

Here is how we could change the experiment to follow the Hillyard principle. Instead of using X and O as the stimuli, we could use the letters A, B, C, D, and E. Each of these letters would

Box 4.5
Example of an Adaptation Confound

Many years ago, I conducted a series of experiments in which I examined the ERPs elicited by visual search arrays consisting of seven randomly positioned "distractor" bars of one orientation and one randomly positioned "pop-out" bar of a different orientation (see the illustration in this box). In several experiments, I noticed that the P1 wave was slightly but significantly larger over the hemisphere contralateral to the pop-out item relative to the ipsilateral hemisphere (this tiny effect is highlighted with a circle in the figure). I thought this might reflect an automatic capture of attention by the pop-out item, although this didn't fit very well with what we knew about the time course of attention. My officemate, Marty Woldorff (who is now a professor at Duke), suggested that this effect might actually reflect an adaptation effect. Specifically, the location of the pop-out bar on one trial typically contained an opposite-orientation distractor bar on the previous trial, whereas the location of a given distractor bar on one trial typically contained a same-orientation distractor bar on the previous trial. Thus, the neurons that coded the distractor bars had usually responded to a bar of the same orientation in a similar location in the previous trial, whereas the neurons that coded the pop-out bar had not usually responded to a bar of that orientation at that location in the previous trial. Consequently, the neurons coding the pop-out bar may have been less adapted, generating a larger response, and this response would have been especially visible over the hemisphere contralateral to the pop-out bar.

At first, I refused to believe that this kind of adaptation could impact the P1 wave (in fact, Marty tells me that my initial response involved a phrase that is not suitable for an academic book). I couldn't imagine how this small adaptation effect could have a significant impact, especially because the screen was blank for 750 ms between trials. However, Marty kept bugging me about it, and eventually I designed an experiment to prove he was wrong. But it turned out that he was absolutely right (see experiment 4 of Luck & Hillyard, 1994b). I guess Marty's imagination was better than mine.

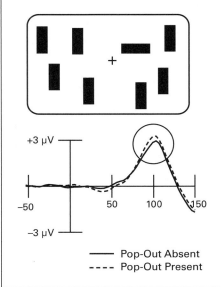

occur on 20% of trials. We would then have five different trial blocks, with a different letter designated as the target in each block (e.g., in one block, we would say that D is the target and all of the other letters were nontargets). We could then present exactly the same sequence of stimuli in each trial block, with the target category (e.g., D) occurring on 20% of trials and the nontarget category (e.g., A, B, C, and E) occurring on the remaining 80%. This would solve the sensory adaptation problem, because the probability of any given physical stimulus is 20%, whether it is the target stimulus or a nontarget stimulus (for an even fancier example of the experimental "backflips" that are sometimes necessary to avoid sensory confounds and satisfy the Hillyard principle, see Sawaki & Luck, 2011).

There is one small proviso to the Hillyard principle: The stimulus sequences for the different conditions should be *equivalent in principle*, but not actually the same sequences. You wouldn't want to create a situation in which people could potentially learn (whether implicitly or explicitly) the stimulus sequences by repeating them in different conditions. Nonetheless, when you generate the sequences, you should be able to use a given sequence in any condition just by changing the instructions.

Our Gedankenexperiment also contains another obvious sensory confound; namely, that the stimuli in the dim and bright conditions are physically different. The goal of the dim/bright manipulation is to change the difficulty of the task, but the dim and bright stimuli will elicit different ERP waveforms irrespective of any differences in task difficulty. For example, they would elicit different ERPs even in a passive viewing task. There are two general approaches to solving this problem. The first is to use the same stimuli for both the easy and difficult tasks and have subjects discriminate different aspects of the stimuli for the different tasks. For example, subjects could discriminate between X and O in the easy condition, and they could make a subtle size discrimination for the same stimuli in the difficult condition. To implement this, you would have four stimuli: a larger X, a smaller X, a larger O, and a smaller O. All four stimuli would be presented in every trial block, but subjects would make an X/O discrimination (ignoring size) in the easy condition and a large/small discrimination (ignoring shape) in the difficult condition. There are many different stimuli and tasks that you could use with this approach, but in all cases you would have the same physical stimuli in both the easy and difficult conditions, and difficulty would be manipulated by means of the task instructions that you give at the beginning of each trial block.

The other general approach is to use difference waves to factor out the sensory confound. To make this clear, let's imagine that we first dealt with the X/O sensory confound by using five different letters (A–E), each occurring on 20% of trials, and we instructed subjects that one of these letters was the target for a given trial block. Again, we would have bright stimuli in some trial blocks for the easy condition and dim stimuli in other trial blocks for the difficult condition. To eliminate the brightness confound, we would compute rare-minus-frequent difference waves separately for the dim and bright stimuli. First we would average across each bright letter when that letter was the target, creating a bright-rare waveform (see figure 4.2). We would also average across each bright letter when that letter was the nontarget, creating a bright-frequent waveform. Then we would make a difference wave between these two waveforms, giving us a rare-minus-frequent difference wave for the bright stimuli. Because the bright-rare and bright-frequent waveforms were created from exactly the same stimuli, differing only in the task instruction, the difference between these waveforms no longer contains any pure sensory activity. You can see this in the difference waves shown in figure 4.2, in which the difference does not begin to deviate from 0 μV until approximately 175 ms, whereas the "parent" wave-forms have a negative-going dip at 125 ms followed by a large positive wave at around 200 ms. These initial sensory responses are eliminated in the difference wave, leaving only brain activity that reflects the task-induced differential processing of the rare and frequent stimulus categories. This difference wave can then be compared with the rare-minus-frequent difference wave for the dim stimuli. Any differences between these difference waves cannot be attributed to pure sensory differences between the bright and dim stimuli and instead reflect the interaction between brightness and the task (which is what we are interested in studying in this experiment).

It is not always feasible to implement the Hillyard principle, especially when you are using naturally occurring stimulus classes. For example, a language experiment might be designed to examine the ERPs elicited by closed-class words (*the*, *for*, *with*, etc.) and open-class words (nouns, verbs, etc.), and these are by definition different stimuli. However, it is almost always possible to include a control condition that can demonstrate that the differences observed in the main condition are not a result of the sensory confound. For example, you could present closed-class and open-class words in two conditions, a main condition in which subjects are reading sentences containing these words and a control condition in which subjects are monitoring for a word that is presented in a different color. If the differences between open- and closed-class words disappear in the control condition, then you know that they are not pure sensory effects. But what if the effects are automatic and are therefore present even during the control condition? A simple approach would be to flip the words upside down and show that this causes the effects to disappear in the control condition. This would rule out low-level sensory differences as the cause of the ERP effects. Alternatively, if you are really ambitious and want to impress everyone with your experimental design abilities, you could test two groups of subjects who speak differ-ent languages, presenting open- and closed-class words of both languages to both groups of subjects. Any differences in the ERPs elicited by the open- and closed-class words that are linguistic rather than sensory in nature should be present only for subjects who speak the lan-guage in which the words are presented, and this would give you a beautiful double dissociation. It would be difficult and time-consuming for you to conduct the experiment this way, but if an experiment is worth doing, isn't it worth doing well?

I often violated the Hillyard principle early in my research career (before I started teaching other people about this principle, which has made me much more careful). However, every time I violated the Hillyard principle, I later regretted it. And I often ended up running a new experi-ment to rule out sensory confounds, so I would have saved a lot of time by designing the experi-ment properly to begin with.

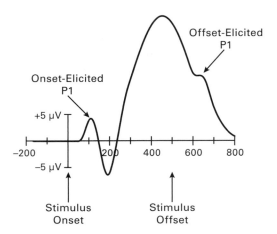

Figure 4.3
Simulated effect of a 500-ms stimulus duration. A P1 wave is triggered by the onset of the stimulus, peaking approximately 100 ms after stimulus onset, and another P1 wave is triggered by the offset of this stimulus, peaking approximately 600 ms after stimulus onset (100 ms after stimulus offset). The offset-elicited P1 adds a "bump" to the P3 wave, which might be misinterpreted as a part of the P3 wave rather than being a sensory response.

Stimulus Duration Problems

The stimuli in the Gedankenexperiment are shown for a duration of 500 ms. This duration does not create a confound, exactly, but the offset of the stimuli will generate a sensory response during the time period of the P3 wave, making the ERPs look a little weird. As shown in figure 4.3, the offset of the stimulus leads to a P1 wave approximately 100 ms after stimulus offset (600 ms after stimulus onset), which is added onto the onset-elicited P3 wave. This makes the waveforms look strange, and it could complicate the interpretation of the results because the time period of the P3 wave contains sensory activity as well as postperceptual activity.

In most cases, the offset response will be negligible if stimulus duration is short, but it becomes progressively larger as the duration becomes longer. Thus, you should choose a stimulus duration that is either so short that no significant offset activity is present or so long that the offset is after the period of time that you will be examining in your ERPs. For example, you might use a duration of 750 ms if you are interested in the ERP components that occur within the first 500 ms after stimulus onset.

How short is short enough? In the visual modality, the retina integrates photons over a period of approximately 100 ms, and a stimulus with a duration of less than ~100 ms is perceptually equivalent to a 100-ms stimulus of lower brightness (I am simplifying a bit here; if you want to know more, find a textbook that describes *Bloch's law*). Consequently, there isn't usually a reason to use a duration of less than 100 ms—you might as well just use a dimmer stimulus and a 100-ms duration. A stimulus of ~100 ms or less doesn't have a distinct onset and offset, but is just perceived as a flash. However, once the duration of the stimulus exceeds ~100 ms, increases in duration are perceived as increases in duration rather than as increases in brightness. At this point, you will begin to perceive distinct onsets and offsets, and the ERP waveform will contain an offset response as well as an onset response. Consequently, I typically use a duration of 100 ms for visual stimuli (unless I want to use a very long duration). However, the offset response is very small for durations that are only a little more than 100 ms, and I sometimes use a duration of 200 ms if I want to give the subject a little more time to perceive the stimuli (especially when working with subjects who may have diminished sensory or cognitive abilities).

For simple auditory stimuli (e.g., sine wave tones), a duration of 50–100 ms is usually appropriate. If a sine wave (or other repeating wave) begins and ends suddenly, a clicking sound will be audible at the onset and offset. This clicking sound is great for producing very early components, but it can be distracting and can interfere with the perception of the tone's pitch (especially for short stimuli). To avoid this, the amplitude of the sound wave should ramp up over a period of 5–20 ms at the beginning of the sound and ramp down over a period of 5–20 ms at the offset of the sound (these are termed the *rise time* and *fall time* of the stimuli).

Motor Confounds

Our Gedankenexperiment also contains an obvious motor confound, because subjects make a motor response to the targets and not to the nontargets. Consequently, any ERP differences between the targets and nontargets could be contaminated by motor-related ERP activity. This is a very common confound in oddball experiments, but not as common in more sophisticated designs. In oddball experiments, one common solution is to have subjects silently count the oddballs. This doesn't really solve the problem, because the act of silently counting the targets involves additional brain activity that is not present for the standards. Also, it is difficult to assess how well the subject is performing the task (box 4.6). The best solution is usually to require the subject to press one button for the target stimuli and a different button for the standard stimuli.

Overlap Confounds

One of the most common problems that I encounter when reading ERP studies arises when the ERP elicited by one stimulus is still ongoing when the next stimulus is presented. If the overlapping activity differs across conditions, then an effect that is attributed to the processing of the current stimulus may actually be the result of continued processing of the preceding stimulus. In our Gedankenexperiment, this problem arises because the stimulus sequences were constrained so that the target letter never occurred twice in succession. Consequently, the target letter was always preceded by a nontarget letter, whereas nontarget letters could be preceded by either targets or nontargets. This is a common practice because the P3 to the second of two targets tends to be reduced in amplitude. Using nonrandom sequences like this is usually a bad idea, however, because the response to a target is commonly very long-lasting and can extend past the next stimulus and influence the waveform recorded for the next stimulus.

Figure 4.4 illustrates how overlap could contaminate our Gedankenexperiment (for a detailed discussion of overlap, see Woldorff, 1993). Panel A shows what the ERPs for the targets and

Box 4.6
The Problem with Counting

When I was a college student and just starting to learn about ERPs, I happened to be sitting on an airplane next to an engineer from Nicolet Instruments, a manufacturer of clinical EEG/ERP systems. The engineer told me that they were just finishing a new portable ERP system, and he asked me if I'd like to earn a little money by serving as a test subject. I was interested in both ERPs and money, so I agreed.

One of the tasks they were testing was an auditory oddball task. I was asked to sit in a dimly lit room, stare at a point on the wall, and count high-pitched target tones that were occasionally embedded among frequent low-pitched tones. I did this for about 10 minutes, but it was incredibly boring and seemed to take hours. I found it surprisingly difficult to keep track of how many high-pitched tones I had heard. I rehearsed the current count in my head until I heard the next target: "33, 33, 33, 33, 33, 33, 33, 33, 33, 33, 34, 34, 34, 34, 34, 35, 35, 35, 35 …." But as I was doing this, my mind was wandering a bit, and at some point I couldn't remember whether I was at 37 or 47. The typical procedure in counting tasks is to ask the subject how many targets he or she counted at the end of the trial block, and this experience made it clear to me that someone could detect every target and still report a very wrong number at the end. It's also possible that someone could miss one target and also misperceive one of the nontargets as a target and yet report the correct number at the end. Because of this experience, I've never been fond of counting tasks.

---

standards would look like without any overlap. Panel B shows the overlap that would occur with a relatively short period between trials (an SOA of approximately 700 ms). As you can see from the figure, the tail end of the P3 from the previous trial is present during the baseline period of the current trial. When the current trial is a target, the previous trial was always a nontarget, and so the overlap is simply the tail end of the ERP elicited by a nontarget (this is labeled *Overlap Y* in the figure). When the current trial is a standard, the previous trial was 80% likely to be a standard and 20% likely to be a target, and so the overlap is a weighted sum of the tail end of the standard-elicited ERP and the tail end of the target-elicited ERP (this is labeled *Overlap Z* in the figure).

In this example, the overlap from the previous trial is over by 100 ms after the onset of the current stimulus, so you might think that it wouldn't have much effect on the P3 wave on the current trial. However, the overlap is still a major problem because of *baseline correction*. For reasons that will be described in chapter 8, it is almost always necessary to use the average voltage during the prestimulus interval as a baseline, which is subtracted from the entire waveform. If the baseline is contaminated by overlap from the previous trial, the baseline correction procedure will "push" the whole waveform downward (if the contamination is from a positive component like P3) or upward (if the contamination is from a negative component). The overlap in our Gedankenexperiment is a positive voltage (from the P3 wave), so it causes the waveform to be pushed downward (which you can see in panel C of figure 4.4). The overlap is greater when the current stimulus is a standard than when it is a target (because the preceding stimulus

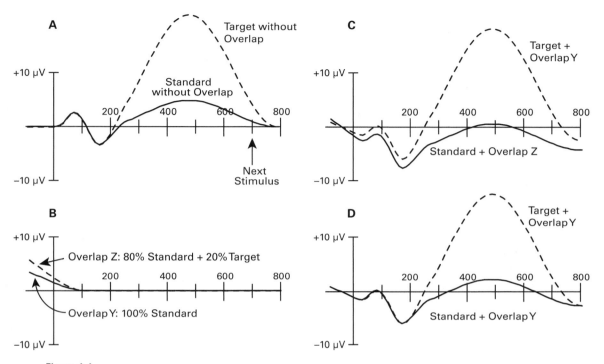

Figure 4.4
Example of differential overlap in the Gedankenexperiment, in which a target stimulus was always preceded by a standard stimulus but a standard stimulus was preceded by a target on 20% of trials and by a standard on 80% of trials. (A) ERPs for target and standard trials without any overlap. (B) Overlap from the last portion of the previous trial. The overlap prior to a target (labeled *Overlap Y*) consists of the last portion of a standard trial. The overlap prior to a standard (labeled *Overlap Z*) consists of the last portion of a standard trial mixed with the last portion of a target trial (80% standard and 20% target). (C) ERP waveforms with the overlap added in. Because the waveforms are baseline corrected, the overlap ends up distorting the waveforms (by pushing them downward). (D) Waveforms that would occur if the ERP waveform for both standards and targets included only trials for which the previous stimulus was a standard. Note that the overlap in this example assumes a very short interval between stimuli and that the subsequent trial would also produce overlapping activity that is not shown here.

was sometimes a target when the current stimulus is a standard), and so the waveform is pushed down farther by the overlap when the current stimulus is a standard than when it is a target. You can see this by comparing the P3 elicited by the frequent stimulus in the presence of overlap (panel C), which has a peak amplitude of approximately 0 μV, with the P3 elicited by the frequent stimulus in the absence of overlap (panel A), which has an amplitude of approximately 5 μV. In contrast, the P3 peak amplitude for the rare stimulus is only slightly smaller in the presence of overlap (panel C) than in the absence of overlap (panel A).

Differential overlap often leads to this sort of pattern, in which an artifactual difference between conditions arises very early (e.g., within 50–200 ms of stimulus onset) and then persists for a very long time (see chapter 6 for a discussion of the importance of looking at the baseline period). If you see this pattern of waveforms in a given study, you should think carefully about whether there might be differential overlap.

Note that differential preparation can produce similar effects. That is, if the onset time of a stimulus is more predictable in one condition than in another, the baseline in these two conditions will be differentially contaminated by preparatory activity (such as the CNV component). The baseline correction procedure will then cause this baseline difference to push the ERP waveform upward or downward.

The overlap problem will be described in more detail in chapter 8 and online chapter 11, but here I will discuss some common solutions. First, however, I would like to stress the fact that overlap per se is not usually a problem. In fact, some overlap (or preparatory activity) is present in almost every ERP experiment. The problem arises only when the overlap or preparatory activity differs between conditions (with one exception that will be described in chapter 6). In our Gedankenexperiment, for example, the problem arises because the waveform elicited by a target is always overlapped by the ERP from a standard, whereas the waveform elicited by a standard is sometimes overlapped by the ERP from a target.

One easy way to solve this problem would be to exclude trials that were preceded by a target during averaging. That way, both the target and standard ERPs would be from trials preceded by a standard, and the overlap would be the same for both. This is illustrated in figure 4.4D. As you can see in the figure, overlap still distorts the data (compare panel D with panel A). However, this overlap is equivalent for the rare and frequent ERPs and therefore has no impact on the difference between the rare and frequent ERPs.

Another solution would be to change the design so that the stimulus sequences were random, eliminating the constraint that a rare stimulus could never be preceded by a rare stimulus. This might make the P3 slightly smaller, but not by very much, because only 20% of targets would be preceded by another target in a fully randomized sequence (assuming that the targets occurred on 20% of trials). In my experience, adding nonrandom sequential constraints usually creates more problems than it solves, so random is usually best. For example, if we constrain the sequence so that a target is never preceded by a target, then this also means that a target is never followed by a target; consequently, subjects could (in principle) just close their eyes and make a guess with 100% accuracy that the next stimulus will be a standard. Thus, my general advice is to use fully random sequences unless there is a very compelling reason to add a sequential constraint.

Arousal Confounds

In our Gedankenexperiment, the stimuli are bright in some trial blocks to make the task easy, and the stimuli are dim in other blocks to make the task difficult. Whenever conditions differ across trial blocks in a way that changes difficulty, changes in arousal may occur as a side effect because subjects are likely to be more aroused in a difficult condition than in an easy condition. Arousal can increase the amplitude of some ERP components (e.g., the P1 wave, as described in chapter 3), and any differences in the ERPs across conditions may therefore be a side effect of differences in arousal. This could also change the preparatory activity before each stimulus, producing different baselines for the bright and dim conditions.

The best way to avoid arousal confounds is usually to vary the conditions unpredictably within each trial block rather than having different trial blocks for the different conditions. In our Gedankenexperiment, for example, it would be trivial to randomly intermix the dim and bright stimuli. However, there are some experiments in which it is necessary to test different conditions in different trial blocks. In these cases, it is sometimes possible to ensure that behavioral accuracy is identical across conditions, which typically equates the arousal level (see, e.g., Leonard, Lopez-Calderon, Kreither, & Luck, 2013).

Confounds Related to Noise and the Number of Trials

In ERP research, the term *noise* refers to random variations in the ERP waveform that are unrelated to the brain activity that you are trying to record (e.g., electrical activity from external devices, skin potentials, etc.). In most cases, noise simply adds random variability to our measurements, reducing the probability that real effects in the data will be statistically significant (i.e., reducing *statistical power*). More noise usually means that we need more trials per subject or more subjects per experiment to achieve statistical significance. In some cases, however, noise can *bias* the data in a particular direction, artificially creating the appearance of an effect. Measurements of peak amplitude are particularly problematic in this regard. All else being equal, the peak amplitude will tend to be larger on average in noisier waveforms than in cleaner waveforms. However, the mean amplitude over a given time range (e.g., 400–600 ms) is not biased in this manner. A fuller discussion of this issue is provided in the online supplement to chapter 9.

In our Gedankenexperiment, this issue arises if we try to compare the peak amplitude for the targets with the peak amplitude for the standards, because fewer trials contribute to the target waveforms than to the standard waveforms, thus making the target waveforms noisier than the standard waveforms. This will tend to bias the peak amplitude to be greater for rare than for frequent stimuli. There are two common ways to solve this problem. First, you can measure mean amplitude rather than peak amplitude (which has a number of additional advantages, as will be discussed in chapter 9). Second, you can create an average of a subset of the standard trials so that the same number of trials contributes to the target and standard waveforms (which should equate the noise levels for these two waveforms). Although this is sometimes the best approach, it decreases the signal-to-noise ratio and therefore decreases your power to find significant effects. I find that some people fail to worry about this problem at all and end up using biased measurements, whereas other people worry too much and end up needlessly sacrificing statistical power when they could have solved the problem simply by measuring mean amplitude instead of peak amplitude.

Tips for Avoiding Confounds

The following list distills the confounds that I just described into six tips for designing ERP experiments:

Tip 1 Whenever possible, avoid physical stimulus confounds by using the same physical stimuli across different psychological conditions (i.e., follow the Hillyard principle). This includes "context" confounds, such as differences in sequential order. Difference waves can sometimes be used to subtract away the sensory response, making it possible to compare conditions with physically different stimuli.

Tip 2 When physical stimulus confounds cannot be avoided, conduct control experiments to assess their plausibility. Don't assume that a small physical stimulus difference cannot explain an ERP effect, especially when the latency of the effect is less than 300 ms.

Tip 3 Although it is often valid to compare averaged ERPs that are based on different numbers of trials, be careful in such situations and avoid using peak-based measures.

Tip 4 Avoid comparing conditions that differ in the presence or timing of motor responses. This can be achieved by requiring responses for all trial types or by comparing subsets of trials with equivalent responses.

Tip 5 To prevent confounds related to differences in arousal and preparatory activity, experimental conditions should be varied within trial blocks rather than between trial blocks. When conditions must be varied between blocks, arousal confounds can be prevented by equating task difficulty across conditions.

Tip 6 Think carefully about stimulus timing so that you don't contaminate your data with offset responses or overlapping activity from the previous trial.

Advice about Timing: Duration, SOA, ISI, and ITI

Almost every experimental design requires you to make decisions about the timing of the stimuli, including durations, SOAs, ISIs, and ITIs. These decisions are often based on the specific goals of an experiment, but there are some general principles that apply to most experiments.

As described earlier, you will usually want to choose a stimulus duration that avoids offset responses, either by choosing a duration that is so short that it doesn't produce a substantial offset response (e.g., 100–200 ms for visual stimuli) or is so long that the offset response occurs after the ERP components of interest (e.g., 1000 ms). In behavioral experiments, it is common for the stimulus to offset when the subject makes a behavioral response to the stimulus. This can sometimes be a good approach in ERP experiments, but it could lead to problems if you decide to look at the postresponse period in response-locked averages (because the sensory offset response will be visible in the response-locked waveforms).

My typical approach for experiments with simple visual stimuli is to use a duration of 100 ms for college student subjects and 200 ms for subjects with poorer perceptual or cognitive abilities. For simple auditory tones, I would typically use a duration of 50–100 ms, including 5-ms rise and fall times. For experiments with more complex auditory or visual stimuli, I typically use a duration of 750–1000 ms.

Determining the optimal amount of time between trials requires balancing several factors. On the one hand, you want to keep the amount of time between trials as short as possible to get the maximal number of trials in your experimental session, thereby maximizing the signal-to-noise ratio of your average ERP waveforms. On the other hand, several factors favor a slower rate of stimulus presentation. First, sensory components tend to get smaller as the SOA and ISI decrease, and this reduction in signal might outweigh the reduction in noise that you would get by averaging more trials together. Second, if the subject is required to make a response on each trial, it becomes tiring to do the task if the interval between successive trials is very short. Third, a short SOA will tend to increase the amount of overlapping ERP activity (which may or may not be a problem, as described earlier). However, using a very long interval between stimuli to minimize overlap may lead to a different problem; namely, anticipatory brain activity prior to stimulus onset (especially if stimulus onset time is relatively predictable).

My typical approach is to use an SOA of 1500 ms for experiments in which each trial consists of a single stimulus presentation and the subject must respond on each trial (e.g., an oddball experiment). I typically reduce this to 1000 ms if the subject responds on only a small proportion of trials, and I might increase it by a few hundred milliseconds for more difficult tasks or for groups of subjects who respond slowly. Of course, there are times when the conceptual goals of the experiment require a different set of timing parameters, but I find that this kind of timing is optimal for most relatively simple ERP experiments.

I almost always add a temporal jitter of at least ±100 ms to the SOA, which avoids the possibility that alpha oscillations will become time-locked to the stimuli and also helps filter out overlapping activity from the previous trial. In a typical oddball experiment, for example, I would use an SOA of 1400–1600 ms. When someone specifies a range like this, it almost always means that the range is broken up into very small increments (e.g., increments of a single screen refresh cycle), and each possible increment in the range is equally likely. For example, with a typical refresh rate of 60 Hz, the time between two stimuli is always a multiple of 16.67 ms, and a range of 1400–1600 ms means that the SOA is equally likely to be 1400.00 ms, 1416.67 ms, 1433.33 ms, . . . 1600 ms. This is called a *rectangular distribution* of SOAs (see chapter 11 for a definition of probability distributions).

Examples from the Literature

Concrete examples can be helpful in clarifying the general principles described in this chapter. To read three of my favorite examples, see the online supplement to chapter 4.

Suggestions for Further Reading

Dehaene, S., Naccache, L., Le Clec'H, G., Koechlin, E., Mueller, M., Dehaene-Lambertz, G., van de Moortele, P. F., & Le Bihan, D. (1998). Imaging unconscious semantic priming. *Nature, 395*, 597–600.

Gratton, G., Coles, M. G. H., Sirevaag, E. J., Eriksen, C. W., & Donchin, E. (1988). Pre- and post-stimulus activation of response channels: A psychophysiological analysis. *Journal of Experimental Psychology: Human Perception and Performance, 14*, 331–344.

Handy, T. C., Solotani, M., & Mangun, G. R. (2001). Perceptual load and visuocortical processing: Event-related potentials reveal sensory-level selection. *Psychological Science, 12*, 213–218.

Hillyard, S. A., & Münte, T. F. (1984). Selective attention to color and location: An analysis with event-related brain potentials. *Perception and Psychophysics, 36,* 185–198.

Hillyard, S. A., Hink, R. F., Schwent, V. L., & Picton, T. W. (1973). Electrical signs of selective attention in the human brain. *Science, 182,* 177–179.

Miller, J., & Hackley, S. A. (1992). Electrophysiological evidence for temporal overlap among contingent mental processes. *Journal of Experimental Psychology: General, 121,* 195–209.

Paller, K. A. (1990). Recall and stem-completion priming have different electrophysiological correlates and are modified differentially by directed forgetting. *Journal of Experimental Psychology: Learning, Memory and Cognition, 16,* 1021–1032.

Van Petten, C., & Kutas, M. (1987). Ambiguous words in context: An event-related potential analysis of the time course of meaning activation. *Journal of Memory & Language, 26,* 188–208.

van Turennout, M., Hagoort, P., & Brown, C. M. (1998). Brain activity during speaking: From syntax to phonology in 40 milliseconds. *Science, 280,* 572–574.

Vogel, E. K., Luck, S. J., & Shapiro, K. L. (1998). Electrophysiological evidence for a postperceptual locus of suppression during the attentional blink. *Journal of Experimental Psychology: Human Perception and Performance, 24,* 1656–1674.

Winkler, I., Kishnerenko, E., Horvath, J., Ceponiene, R., Fellman, V., Huotilainen, M., Naatanen, R, & Sussman, E. (2003). Newborn infants can organize the auditory world. *Proceedings of the National Academy of Sciences, 100,* 11812–11815.

Woldorff, M., & Hillyard, S. A. (1991). Modulation of early auditory processing during selective listening to rapidly presented tones. *Electroencephalography and Clinical Neurophysiology, 79,* 170–191.

5 Basic Principles of ERP Recording

Overview

This chapter describes how electrodes, amplifiers, filters, and analog-to-digital converters work together to record the EEG. The main goal of this chapter is to help you record clean data so that your results are valid and statistically significant.

The first section of this chapter will describe why clean data are so important. We will then discuss how EEG amplifiers work and why you need both a ground electrode and a reference electrode (in most systems). This includes a discussion of how to choose a reference electrode location and how to re-reference your data offline. We will then discuss how electrodes work and how the impedance of the electrode–skin connection can have an enormous impact on the quality of your data and the probability that your effects will be statistically significant. We will then finish the chapter by discussing how the continuous EEG signal is amplified and then converted into a set of discrete voltage samples that can be stored in your computer and why filters must be applied to your data prior to this conversion.

This chapter provides information that will be relevant if you are deciding what kind of EEG recording system to purchase for your laboratory. However, I have not tested every system on the market, and I have not done extensive side-by-side comparisons of different systems, so I will not recommend specific systems. Nonetheless, I will provide some useful information that you can use when evaluating different systems. Box 5.1 provides financial disclosures that you may want to read when you consider my advice.

The Importance of Clean Data

Before I get into the details of how to record the EEG, I want to convince you that it is worthwhile to spend some significant time and effort making sure that you are recording the cleanest data possible. This is the reason in a nutshell: The cleaner your data are, the fewer trials you will need per subject and the fewer subjects you will need to obtain clear and reliable results in a given experiment. If you don't need as many trials and subjects in each experiment, you can run more experiments per year that yield publishable results. By recording cleaner data, you can

Box 5.1
Money

Whenever someone publishes a book or journal article that includes information or data that you might use in making decisions about spending money, you should ask whether the author has any financial incentives that might lead to intentional or unintentional biases. I am therefore disclosing my (meager) financial interests that are related to ERPs so you can decide if I might be biased.

I receive honoraria for providing Mini ERP Boot Camp workshops at universities, conferences, and industry sites. I receive royalties from publishers for books, such as this one (although this is no way to get rich!).

My laboratory uses EEG recording systems manufactured by BioSemi but has not received any free or discounted equipment or any other financial considerations from BioSemi or from any other manufacturers. The UC-Davis ERP Boot Camp was given electrode caps and a small amount of money by Cortech Solutions, the U.S. distributor for BioSemi. The ERP Boot Camp has also received financial support from several other vendors of ERP recording and analysis systems, including Brain Products GmbH, EasyCap GmbH, and Advanced Neuro Technologies, and software has been provided for the ERP Boot Camp by Compumedics Neuroscan, Megis GmbH, and Brain Products GmbH. The cash donations from these companies have been used to support Boot Camp activities (e.g., group meals). I receive no personal income from any manufacturers or distributors of ERP-related products. I don't think I have any financial incentives that might bias my advice, but now you can decide for yourself.

publish more (and better) papers. This will also increase the likelihood that you will get a great postdoc/faculty position, large grants, awards, fame, and fortune. If you're not interested in publishing a lot of high-quality journal articles, then go ahead and record noisy data. But if you would like to have a successful career and make important contributions to science, then you will want to record clean data. This may seem obvious, but I often find that people spend much more time on applying fancy analyses to their data than on making sure the data they are analyzing is clean.

The background EEG obscures the ERPs on individual trials, and the ERPs are isolated from the EEG noise by means of averaging across multiple trials. As you average together more and more trials, the amount of residual EEG noise in the averages will become progressively smaller, and it is therefore crucial to include a sufficient number of trials in your ERP averages. However, increasing the number of trials eventually has diminishing returns because the effect of averaging on noise is not linearly proportional to the number of trials; instead, the noise decreases as a function of the square root of the number of trials in the average. As a result, you won't be able to cut the noise in half by doubling the number of trials. In fact, doubling the number of trials decreases the noise only about 30%, and you have to quadruple the number of trials to reduce the noise by 50%. This will be discussed in more detail in chapter 8.

It should be obvious that you can quadruple the number of trials only so many times before your experiments will become absurdly long, so increasing the number of trials is only one part of the solution. The other part is to reduce the noise before it is picked up by the electrodes.

Much of the noise in an ERP recording arises not from the EEG, but from other biological signals such as skin potentials and from nonbiological electrical noise sources in the environment. It is possible to reduce these sources of noise directly. In fact, if you spend a few days tracking down and eliminating these sources of noise, the resulting improvement in your averaged ERPs could be equivalent to doubling the number of trials for each subject. This initial effort will be well rewarded in every experiment you conduct.

In addition to tracking down noise sources and eliminating them before they contaminate your recordings, it is possible to reduce noise by the use of data-processing techniques such as filtering. As will be discussed in chapter 7, these techniques are essential in ERP recordings. However, it is important not to depend too much on postprocessing techniques to "clean up" a set of noisy ERP data because these techniques are effective only under limited conditions and because they almost always distort the data in significant ways. This leads us to an important principle that I call *Hansen's axiom*:

Hansen's axiom There is no substitute for good data.

The name of this principle derives from Jon Hansen, who was the technical guru in Steve Hillyard's lab when I was a graduate student at UCSD. As Jon put it in the documentation for a set of artifact rejection procedures:

There is no substitute for good data. It is folly to believe that artifact rejection is going to transform bad data into good data; it can reject occasional artifactual trials allowing good data to be better. There is no way that artifact rejection can compensate for a subject who consistently blinks in response to particular events of interest or who emits continuous high-amplitude alpha activity. In other words, data that are consistently noisy or have systematic artifacts are not likely to be much improved by artifact rejection. (J. C. Hansen, unpublished software documentation)

Jon made this point in the context of artifact rejection, but it applies broadly to all postprocessing procedures that are designed to clean up the data, ranging from averaging to filtering to independent component analysis. Some postprocessing procedures are essential, but they cannot turn bad data into good data. You will save a great deal of time in the long run by eliminating electrical noise at the source, by encouraging subjects to minimize bioelectric artifacts, and by designing experiments in a way that maximizes the size of the effects relative to the amount of noise.

Online chapter 16 describes a practical approach for finding and eliminating sources of electrical noise in your laboratory. However, you need to understand how the noise is picked up by your recording system before you can effectively eliminate it. A key goal of this chapter is to provide you with this understanding.

Active, Reference, and Ground

Before I can discuss noise in detail, I need to describe one of the most basic aspects of ERP recording; namely, the use of three electrodes (active, reference, and ground) to record the signal from a single scalp site. These three electrodes are combined to provide a single *channel* of EEG. The reference electrode plays a particularly important role that is not always appreciated by ERP researchers. I will provide a fairly detailed description of how the reference electrode works, because this is an absolutely fundamental aspect of EEG/ERP recordings. If you don't fully understand referencing, you won't understand the signal that you are recording. So bear with me as I describe the details of active, reference, and ground electrodes. Note that all of the following information is relevant even if your system does referencing in software rather than during the recording.

Voltage as a Potential between Two Sites

As described in chapter 2, voltage is the potential for current to move from one place to another (if you did not read the section "Basic Electrical Concepts" at the beginning of chapter 2, this would be a good time to go back and read that section). As a result, there is no such thing as a voltage at a single electrode.

Consider, for example, a typical 120-V household electrical outlet, which has two main terminals. The voltage measurement of 120 V represents the potential for current to move between the two terminals, and it doesn't make sense to talk about the voltage at one terminal in isolation. For example, you could touch one terminal without being shocked (assuming you weren't touching any other conductors), but if you touch both terminals at the same time, you will allow the outlet's potential to be realized as a strong current that passes from one terminal through your body into the other terminal. Similarly, you can never record the voltage at a single scalp electrode. Rather, the EEG is always recorded as a potential for current to pass from one electrode (called the *active* electrode) to some other specific place. This "other specific place" is usually a *ground* electrode (see box 5.2 if you'd like to know why this is called the "ground").

Although voltage is always recorded between two sites and there is no such thing as a voltage at a single electrode site, we can use the term *absolute voltage* to refer to the potential between a given active site and the average of the rest of the head (because physics dictates that the average of the EEG activity across the entire surface of the head must be 0 μV). This allows us to use some very simple math to describe how the active and ground electrodes work. We will use A to refer to the absolute voltage at the active electrode (i.e., the potential between the active electrode and the average of the scalp), and we will use G to refer to the absolute voltage at the ground electrode (i.e., the potential between the ground electrode and the average of the scalp). The potential between the active electrode and the ground electrode is simply the difference between these two absolute electrodes. In other words, the voltage recorded between an active site and a ground site is simply A – G. Note that neural activity will be present in both A and G, not just in A, and that anything that is equivalent in A and G will be subtracted away.

There is a practical problem that arises in the design of an EEG amplifier: To create a working amplifier, the ground electrode must be connected to a *ground circuit*, which is necessarily connected to other parts of the amplifier that generate electrical noise. This noise—which is an inevitable fact of electrical circuits—is present in the G signal but not in the A signal. Conse-

Box 5.2
Ground and Earth

You've probably heard the term *ground* used in the context of electrical systems, but most people don't know that this term originated from the actual ground (i.e., the dirt that covers much of the planet). Specifically, in household electrical systems, a metal stake is driven deep into the ground beneath the house and serves as an important reference point for electrical devices. To distinguish between the term *ground* as it is used in electrical circuits and the stake driven into the ground under a house or other building, I will use the term *earth* to refer to the stake that is driven into the physical ground.

The earth is used in electrical systems, in part, to avoid the buildup of large static electrical charges between electrical devices inside a building and the rest of the world. If, for example, lightning strikes your house, the large current from the lightning will be discharged into the earth. The "ground pin" in a standard electrical outlet is therefore connected to the earth. The term *ground* is now used more generally to refer to a common reference point for all voltages in a system, whether or not this reference point is actually connected to a stake in the dirt, and this is how I will use this term. The term *common* is also sometimes used as a synonym for *ground*.

If we measured the electrical potential between an electrode on a subject's scalp and a stake driven into the earth, the voltage would reflect any surplus of electrical charges that had built up in the subject (assuming the subject was not touching a conductor that was connected to earth), and this static electricity would obscure any neural signals. We could put an electrode somewhere on the subject's body that was connected to earth, and this would cause any static electricity in the subject to be discharged into the earth, eliminating static differences and making it easier to measure changes in neural signals over time. However, it is dangerous to directly connect a subject to earth, because the subject might receive a shock if touched by an improperly grounded electrical device (like a button box used to collect behavioral responses). Thus, EEG amplifiers create a virtual ground that is used as a reference point but is not directly connected to earth. The virtual ground of the EEG amplifier is connected to a ground electrode located somewhere on the subject, and you record the voltage between each active electrode and this ground electrode.

quently, the measured potential between A and G will contain this noise along with electrical activity picked up from the scalp.

The Reference Electrode

As illustrated in figure 5.1A, EEG recording systems typically solve the problem of noise in the ground circuit by using *differential amplifiers*. With a differential amplifier, a *reference* electrode (R) is used along with the active (A) and ground (G) electrodes. The amplifier records the potential between A and G (which is equal to A − G) and the potential between R and G (which is equal to R − G). The output of the amplifier is the difference between these two voltages ([A − G] − [R − G]). This is equivalent to A − R (because the Gs cancel out), so any noise in G will be removed because it is the same for A − G and R − G. In other words, electrical noise from the amplifier's ground circuit will be the same for the A − G and R − G voltages and will therefore be eliminated by the (A − G) − (R − G) subtraction.

A

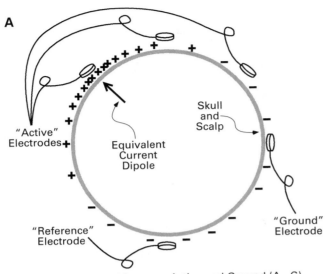

"Active"
Electrodes

Equivalent
Current
Dipole

Skull
and
Scalp

"Ground"
Electrode

"Reference"
Electrode

Voltage is measured between Active and Ground (A - G)
Voltage is measured between Reference and Ground (R - G)
Output is difference between these voltages
[A - G] - [R - G] = A - R
It's as if the ground does not exist
Any noise (or signals) in common to A and R will be eliminated

B

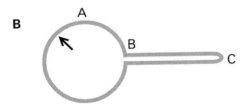

Figure 5.1
Active and reference electrodes. (A) Example of an equivalent current dipole (arrow) inside a spherical head, with the resulting surface voltages ("+" and "–" for positive and negative) on the surface. The recorded voltage will be the difference between the absolute voltage at the active and reference electrodes. (B) Example of the use of a distant reference source. If the active electrode is at point A, it will not matter whether point B or point C is used as the reference, because the absolute voltage at point C will be approximately the same as at point B despite its distance.

The output of a differential amplifier is equivalent to the electrical potential between the active and reference electrodes (A – R), as if the ground electrode does not exist (although a good electrical connection for the ground electrode is necessary for the amplifier to work correctly). Why, then, do we need a ground electrode? As mentioned in the previous paragraph, when voltage is initially measured between two electrodes, one of the two electrodes must be connected to the amplifier's ground circuit, and this circuit picks up noise from the amplifier circuitry. There is no way to record the voltage directly between A and R without connecting R to the ground circuit (in which case it would simply be the ground electrode). Thus, most systems use a differential amplifier to cancel out the noise in the ground circuit.

The electrical potential between the subject's body and the amplifier's ground circuit is called the *common mode* voltage (because it is in common with the active and reference electrodes). To achieve a clean EEG recording, this common mode voltage must be completely subtracted away. Although the subtraction seems like a perfect solution when it is written out as an equation (as in figure 5.1), the subtraction will not be perfect because the equation is implemented in physical electronic circuitry. For example, if the A – G signal is amplified a little more strongly than the R – G signal, the subtraction will not be perfect, and some of the noise in the ground circuit will remain in the amplifier's output. The ability of an amplifier to subtract away the common mode voltage accurately is called *common mode rejection*, and it is usually measured in decibels (dB; see the glossary). A good EEG amplifier will have a common mode rejection of at least 70 dB. As we will discuss in a little bit, electrode impedances can impact the common mode rejection.

Note that some systems work a little differently, and one of these (the BioSemi ActiveTwo EEG system) is described in box 5.3.

Box 5.3
The BioSemi ActiveTwo System

The BioSemi ActiveTwo system works a little bit differently from most EEG recording systems. First, instead of a single ground electrode, it includes a *common mode sense* (CMS) electrode and a *driven right leg* (DRL) electrode. CMS is much like a traditional ground electrode; the system records the potential between each active electrode and CMS. The DRL electrode is part of a feedback circuit that drives the potential of the subject very close to the potential of the amplifier's ground circuit, resulting in a very low common mode voltage. That is, it actually injects a small current into the head so that the average voltage is close to the potential of the ground circuit. This improves the effective common mode rejection of the system because there is less common mode voltage to be subtracted away (for the technically inclined reader, details can be found in Metting van Rijn, Peper, & Grimbergen, 1990). The second unusual thing about this system is that, because of optimized electronics, the BioSemi system does not need to subtract the reference signal from the active signal prior to digitizing the signal. That is, the system provides a *single-ended* recording rather than a *differential* recording (see the glossary). The subtraction of the reference ([A – G] – [R – G]) is conducted in software after the recording is complete. The end result is conceptually equivalent to a differential recording, but with lower noise and more flexibility.

The original idea behind the terms *active electrode* and *reference electrode* was that the active electrode was assumed to be near the active neural tissue, and the reference electrode was assumed to be at some distant site that doesn't pick up any brain activity, such as the earlobe. As the neural signals near the active electrode changed, it was assumed that this would influence the voltage at the active site but not at the reference site. This would work well if the earlobe contained no neurally generated activity. The problem is that neural activity is conducted to the earlobe (and to every other place on the head or body). Consequently, the voltage recorded between an active site and a so-called reference site will reflect neural activity at both of the sites. This leads to perhaps the most important thing you should remember from this chapter, which I call the *no-Switzerland principle* (because Switzerland is famous for being politically neutral):

The no-Switzerland principle There is no electrically neutral site on the head or body. An ERP waveform therefore reflects the difference in voltage between two sites that both contain neural activity.

Consider, for example, the tip of the nose. This seems like it ought to be a neutral site because it is separated from the brain by a bunch of non-neural tissue, but it's not. To make the example more extreme, imagine a head with an extremely long nose, like that of Pinocchio. Electrical activity from a dipole inside the head flows to the entire surface of the head, including the base of the nose (see figure 2.2C in chapter 2). Pinocchio's long and skinny nose is much like a wire, so neural activity generated in the brain flows readily from the base of the nose to the tip (see figure 5.1B). Consequently, the absolute voltage at the tip of the nose will be nearly identical to the absolute voltage at the base of the nose. It doesn't really matter, therefore, whether the reference electrode is placed at the tip of the nose or where the nose joins the head; either place will pick up neural activity, just like any other spot on the surface of the head. This is not to say that the tip of the nose is an inappropriate site for a reference electrode—it can work quite well. Rather, my point here is that there is no such thing as an electrically neutral reference site, so you must always keep in mind that an ERP waveform reflects contributions from both the active site and the reference site.

The choice of a reference electrode can be somewhat difficult, and some key factors will be described in the next two sections. Fortunately, choosing the location of the ground electrode is trivial: Because the signals at the ground electrode are subtracted out by the referencing process, you can put the ground electrode anywhere on the head that is convenient. The location of the ground electrode will not ordinarily influence your recordings.

Re-referencing Your Data Offline

The whole reference issue is a bit of a pain, but one nice thing is that you can easily change the reference offline, after the data have been recorded. And you can do this many times to see what your data look like with different references (which I highly recommend you do). An implication of this is that it doesn't really matter what reference site was used during recording, because you can always re-reference offline. Some systems allow you to select any reference you like during recording, whereas other systems require that you use a particular location (e.g., Cz). Because you can easily re-reference offline, it's not a problem if your system forces you to use a location that is not optimal.

Re-referencing your data offline is very simple. Recall that hardware referencing is accomplished by taking the potential between A and G and subtracting the potential between R and G (i.e., $[A - G] - [R - G]$). The result is the potential between A and R ($A - R$). You can do the same type of thing in software by literally subtracting one channel from another. For example, if you've recorded the voltage at electrodes A and B using a differential amplifier, with R as the reference, you can compute the voltage between A and B by simply subtracting channel B from channel A ($A - B$). In other words, the output of your amplifier for channel A is equivalent to $A - R$, and the output of your amplifier for channel B is equivalent to $B - R$, so the potential between these channels ($A - B$) is equivalent to $(A - R) - (B - R)$.

Now let's take a more complicated example, where you want to re-reference using the average of the left and right mastoids (the bony protrusions behind the ears) as the new reference. In other words, you want to look at the potential between an active site, A, and the average of the left mastoid (Lm) and the right mastoid (Rm), which is ($[Lm + Rm] \div 2$). This is the same thing as $A - ([Lm + Rm] \div 2)$. Imagine that you've recorded from site A using a differential amplifier, with the Lm as the reference (which is $A - Lm$). You've also recorded the signal at the Rm electrode with Lm as the reference (which is $Rm - Lm$). After the data have been recorded, you can re-reference to the average of the two mastoids using the formula $a' = a - (r/2)$, where a' is the re-referenced waveform for site A (i.e., the waveform using the average of the two mastoids as the reference), a is the original waveform for channel A (recorded with an Lm reference), and r is the original waveform for the Rm channel (recorded with an Lm reference).

Here's how this formula works. Because we originally used Lm as the reference, the output of the amplifier for channel A is equal to $A - Lm$. Similarly, the voltage recorded from the Rm channel with an Lm reference is equal to $Rm - Lm$. We need to recombine these two recorded signals into something that equals $A - ([Lm + Rm] \div 2)$. This is achieved with some simple algebra:

$a = A - Lm$ The voltage recorded at site A is the absolute voltage at A minus the absolute voltage at Lm.

$r = Rm - Lm$ The voltage recorded at Rm is the absolute voltage at Rm minus the absolute voltage at Lm.

$a' = A - ([Lm + Rm] \div 2)$ This is what we are trying to compute (the potential between A and the average of Lm and Rm).

$a' = A - (Lm \div 2) - (Rm \div 2)$ This is just an algebraic reorganization of the preceding equation.

$a' = A - (Lm - [Lm \div 2]) - (Rm \div 2)$ This works because $Lm \div 2 = Lm - (Lm \div 2)$.

a′ = (A − Lm) − ([Rm − Lm] ÷ 2)   This is just an algebraic reorganization of the preceding equation.

a′ = a − (r ÷ 2)   Here we've taken the previous equation and replaced (A − Lm) with a and replaced (Rm − Lm) with r.

In other words, you can compute the voltage corresponding to an average mastoids reference for a given site simply by subtracting one-half of the voltage recorded from the other mastoid. The same thing can be done with earlobe reference electrodes.

If you have a system that does not use a reference during recording (e.g., BioSemi ActiveTwo), this is even easier. Imagine, for example, that you recorded signals from A, Lm, and Rm (each recorded relative to the ground electrode). To reference channel A to the average of the two mastoids, you would just use a′ = A − ([Lm + Rm] ÷ 2).

One convenient aspect of re-referencing is that you can re-reference your data as many times as you want. You can go back and forth between different reference schemes, with no loss of information. I highly recommend that you look at your data with multiple different references.

In most ERP experiments, the EEG electrodes are recorded with respect to a single common reference (which may be a single electrode or the average of two or more electrodes). These are sometimes called *monopolar recordings*, but this term is something of a misnomer because monopolar recordings reflect contributions from both the active and reference sites, not the absolute voltage at the active site. The term *bipolar recording* is typically used when each channel uses a different reference. In the clinical evaluation of epilepsy, for example, each electrode is typically referenced to an adjacent electrode. In cognitive and affective neuroscience experiments, bipolar recordings are often used to measure the electrooculogram (EOG), which is the electrical potential caused by blinks and eye movements. To measure horizontal eye movements, for example, the active electrode is usually placed adjacent to one eye, and the reference electrode is placed adjacent to the other eye (see chapter 6 for more details). However, all the scalp electrodes use the same reference site in most cases.

The Average Reference

It is possible to re-reference your data to the average of all of your scalp sites, which is often called the *average reference*. This is relatively easy to do, and it has become very common, but it has some conceptual complexities that I will describe after I explain the basics of how it works.

To understand how the average reference works, imagine that the head was not connected to the rest of the body so that you could place electrodes around the entire head (including the bottom, which is normally inaccessible because of the neck). By using the average voltage across all of the electrodes as the reference, you could obtain the absolute voltage at each electrode site. The mathematics of this would be trivial: The absolute voltage at a given site can be obtained by simply subtracting the average of all of the sites from each individual site, assuming that all sites were recorded with the same reference electrode. Although this would be ideal, it isn't practical for the simple reason that the neck and face get in the way of putting electrodes over the entire surface of the head. Nonetheless, many researchers use the average of whatever scalp electrodes they happen to record as the reference.

The following list of equations provides the logic behind re-referencing to the average of all the electrodes. In this example, we have recorded from three active electrodes using Lm as the reference. In the equations, A1, A2, and A3 refer to the absolute voltage at these three active sites, and a1, a2, and a3 refer to the voltage actually recorded at these sites with the Lm reference. If we simply subtract the average of the recorded voltages (denoted *avg[a1, a2, a3]*) from each of the recorded voltages, this is equivalent to converting each recording into the difference in absolute voltage between each site and the average of the absolute voltage at each of the active sites. For example, the re-referenced version of a1 is equal to the potential between A1 and avg(A1, A2, A3). Note that the original reference (Lm) no longer contributes to the reference. As an exercise, you can figure out how to re-reference the data so that the original reference is one of the sites that contributes to the average reference.

$$a1 = (A1 - Lm)$$

$$a2 = (A2 - Lm)$$

$$a3 = (A3 - Lm)$$

$$\begin{aligned} avg(a1, a2, a3) &= (a1 + a2 + a3) \div 3 \\ &= [(A1 - Lm) + (A2 - Lm) + (A2 - Lm)] \div 3 \\ &= [(A1 + A2 + A3) - 3Lm] \div 3 \\ &= avg(A1, A2, A3) - Lm \end{aligned}$$

$$\begin{aligned} a1 - avg(a1, a2, a3) &= (A1 - Lm) - (avg[A1, A2, A3] - Lm) \\ &= A1 - avg(A1, A2, A3) \end{aligned}$$

$$a2 - avg(a1, a2, a3) = A2 - avg(A1, A2, A3)$$

$$a3 - avg(a1, a2, a3) = A3 - avg(A1, A2, A3)$$

The average reference is quite appealing. First, it is convenient: You simply record from a large set of scalp electrodes and then re-reference to the average of all sites offline. Second, it is not biased toward a particular hemisphere (assuming that you use a symmetrical set of electrode locations). Third, because it reflects the average across a large number of sites, it will tend to minimize noise. Fourth, as long as the set of electrodes covers a very large proportion of the scalp, it is unlikely that the average reference will subtract away most of the voltage for a given component (as can happen for the average mastoids reference when the component is large near the mastoids).

However, the use of the average reference has several side effects that are not obvious (for additional discussion, see Dien, 1998). These side effects are not necessarily disadvantages, but they can lead to unexpected results and misinterpretations if they are not fully understood. To explain these side effects, let's take a closer look at what really happens when the average reference is used. Figure 5.2 illustrates the effect of taking recordings that were obtained using an Lm reference and re-referencing to the average of all sites. These are artificial data, designed to be similar to what one might expect for a broadly distributed positive component, such as the P3 wave. For the sake of simplicity, we are assuming that data were recorded only from 11 sites along the midline. Figure 5.2A shows the absolute voltage (the potential between each electrode and the average of the entire surface of the head, including the bottom side). Figure 5.2B shows what the data would look like in the original recordings, referenced to Lm, if we assume that the absolute voltage at Lm was zero (which could happen if the band of zero voltage for the dipole just happened to pass under the Lm electrode). In this unusual situation, the scalp distribution of the recorded voltage is simply equal to the absolute voltage, because 0 μV (the voltage at Lm) is subtracted from each active site by the initial referencing process. Figure 5.2C shows the scalp distribution after we re-reference to the average reference by subtracting the average of all the original values (3.27 μV) from each of the original values. Note that re-referencing all of the electrodes to a new value simply pushes the distribution of values upward or downward, without changing the relative differences in voltage between electrodes. Panels E and F of figure 5.2 show the same thing, but under the more realistic assumption that the absolute voltage at Lm is not zero. Instead, we are assuming that Lm was 2 μV (which is just an arbitrary value). With this assumption, all the values are pushed downward by 2 μV for the data recorded using the Lm reference, but we end up with the same values as before when we use the average reference.

This example shows the first problem with the average reference; namely, that researchers often assume that it provides a good estimate of the absolute voltage at each site. Although this will sometimes be true (in cases where the positive and negative sides of the dipole are captured equally by the set of recording electrodes), it will often be false. If your electrodes cover only a portion of the head, then you are missing much of the voltage that is needed to compute the average of the entire surface of the head. Moreover, even if you put electrodes everywhere you can, you will still be missing the bottom portion of the head (unless you figure out a way to insert electrodes through the neck to reach the bottom half of the head). Thus, even under the best of circumstances, the average reference yields an imperfect approximation of the absolute voltage. Dien (1998) provided examples in which the average reference led to a good approximation of the absolute voltage, but there is no guarantee that this will be true for a given ERP component. For example, it is clearly not true for the example shown in figure 5.2. Of course, the use of the mastoids, earlobes, or any other site also fails to yield the absolute voltage in most cases, but at least these references don't lead anyone to believe that they are recording an absolute voltage. Thus, the no-Switzerland principle arises even when the average reference is used.

A second side effect of the average reference is that the ERP waveforms and scalp distributions you will get when you use the average reference will change depending on what electrode sites you happened to record from. An extreme version of this problem is illustrated in figure 5.3, which shows waveforms from the Fz, Cz, and Pz electrode sites recorded with a left mastoid

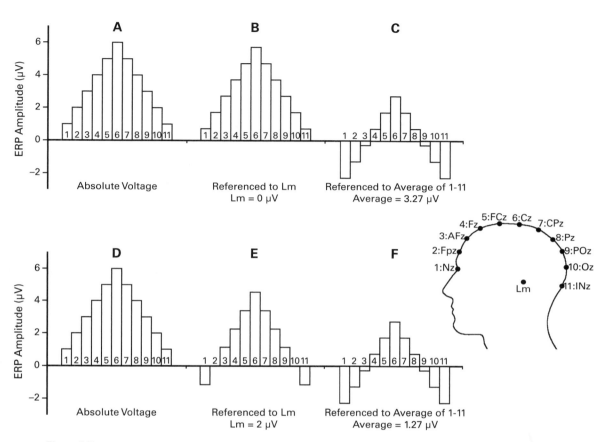

Figure 5.2

Examples of how the reference configuration can influence the measured scalp distribution. In this made-up example, the amplitude is measured at a particular time point from 11 scalp electrodes, distributed along the midline between the nasion and the inion (see cartoon head on the right), using the left mastoid (Lm) as the reference. Panels A–C assume that the absolute voltage at Lm was 0 μV (a very unlikely occurrence). Panels D–E assume that the absolute voltage at Lm was 2 μV (a purely arbitrary but more realistic value used for the sake of this example). Panels A and D show the true absolute voltage (i.e., the difference in potential between each scalp site and the average of the entire surface of the head, including the part that is ordinarily obscured by the neck). This absolute voltage is purely theoretical; there is no way to actually measure it in a living person. Panels B and E show the voltage that would actually be recorded using the Lm reference. Panels C and F show the voltage that would be obtained if the recorded voltages in panels B and E, respectively, were re-referenced to the average voltage at sites 1–11 (eliminating any contribution from Lm). Even though the Lm value differs between panels B and E, the values in panels C and F are identical. Note that all six scalp distributions are exactly the same except for a vertical shift upward or downward. However, the actual value at some electrodes is positive for some reference configurations and negative for others. Thus, different references may make the scalp distributions look quite different, even though they differ only in terms of a vertical shift.

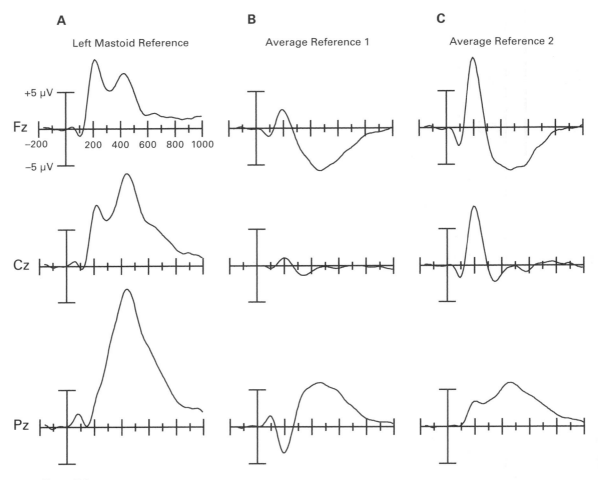

Figure 5.3
Effects of the average reference on ERP waveforms. (A) Voltage recorded at each of three sites (Fz, Cz, and Pz) using a left mastoid reference. (B) Waveforms obtained when the average of the three sites in column A is used as the reference. (C) Waveforms obtained when the average of these three sites and five additional occipital and temporal electrode sites was used as the reference.

reference (column A), the same data using the average of these three sites as the reference (column B), and the same data using the average of these three sites plus five occipital and temporal sites as the reference (column C). Even though the same Fz, Cz, and Pz signals were present in all three cases, the resulting waveforms look very different. For example, the large and broadly distributed P3 that can be seen with a left mastoid reference (column A) is converted by the use of an average reference into a medium-sized positive value at the Pz site, a flat line at the Cz site, and a medium-sized negative value at the Fz site (column B). Moreover, the shapes of the ERP waveforms at each site are very different for the average reference data shown in columns B and C, because the electrodes contributing to the average reference are different in these two columns.

To make this clear, imagine that you record ERPs from Fz, Cz, Pz, and 30 other electrodes, and I record ERPs in the same paradigm from Fz, Cz, Pz, and 30 other electrodes, but I don't use the same 30 other electrodes that you use. And imagine that we both publish papers showing the waveforms at Fz, Cz, and Pz, using an average reference. Your waveforms could look very different from mine if those "other electrodes" were quite different in my experiment than in your experiment. However, to someone casually reading the methods sections of our papers, it would sound as if our data should look the same. This is obviously a recipe for confusion. Many ERP researchers who use the average reference appear to be unaware of this problem, as I have seen many papers in which the researchers used the average reference but did not specify all of the electrodes that were used in the recordings (and that therefore contributed to the average reference and influenced the waveforms). That is, the researchers indicated which active electrode sites were used in the statistical analysis, but they said something vague like "recordings were obtained from a total of 33 scalp electrodes, using the average reference, but the analyses were limited to Fz, Cz, and Pz." This suggests that the researchers were unaware that the other, unspecified electrodes had a large effect on the data they analyzed from Fz, Cz, and Pz (because the unspecified electrodes contributed to the average reference).

A third side effect of the average reference is that the voltage will always sum to zero across all of the electrode sites at every point in time (which is a simple consequence of the math). This can be seen in figure 5.2, where several of the electrodes have a negative voltage after the data were re-referenced to the average reference. This can also be seen in the average reference waveforms in figure 5.3B, where the short-duration, positive-going peak at around 400 ms at the Fz electrode site in column A becomes a long-duration, negative-going peak in column B. This occurred because of the large P3 wave at Pz in the column A waveforms; to achieve a summed voltage of zero, a large negative voltage had to be added onto the Fz site in the average reference waveforms. To be fair, it is important to note that the dipolar nature of ERP components means that every component is actually positive over some parts of the head and negative over other parts, summing to zero over the entirety of the head (although you may not have electrodes over enough of the head to see both the positive and negative sides). The transition point between the positive and negative portions is influenced by the reference voltage, whether the absolute reference or some other reference is used (see, e.g., the artificial negative voltages at electrode

1 and electrode 11 in the left mastoid–referenced data in figure 5.2E). However, this distortion can be extreme when the average reference is used, especially when the electrodes cover less than half the head.

Many researchers do not appear to realize that the voltage will necessarily be positive at some electrode sites and negative at others when the average reference is used. For example, I have reviewed manuscripts in which an average reference was used and the authors made a great deal out of the finding that an experimental effect was reversed in polarity at some sites relative to others. But this is necessarily the case when the average across sites is used as the reference.

The fourth and biggest side effect of the average reference is that it makes it difficult to compare waveforms and scalp distributions across studies (as noted by Dien, 1998). This is truly a disadvantage. As I mentioned earlier, even when two researchers show data from the same active electrodes and say that they used the average reference, their waveforms and scalp distributions may look quite different if the entire set of electrodes was not the same across studies. Thus, unless everyone uses the same set of electrodes to form the average reference, the use of the average reference may lead to confusion and misinterpretations, with researchers thinking that the pattern of results differs across experiments when in fact only the set of recording electrodes differs. This problem does not arise when people use, for example, the mastoids as the reference, making it easy to compare results across experiments. The consistent use of exactly the same reference configuration across experiments is therefore quite beneficial to the field. Emerson famously wrote that "a foolish consistency is the hobgoblin of little minds," but this particular consistency is not usually foolish.

Choosing a Reference Site

Now that we have discussed several different options for the reference, you probably want to know what you should use for the reference in your own experiments. There is no simple answer to this question, and you need to consider the following five factors.

First, given that no site is truly electrically neutral, you might as well choose a site that is convenient and comfortable. The tip of the nose, for example, is a somewhat distracting place for an electrode.

Second, you will want to avoid a reference site that is biased toward one hemisphere. For example, if you use the left earlobe as the reference, then people might be concerned that you have introduced an artificial asymmetry between the left and right hemispheres. This is not usually a real problem, but reviewers might hassle you about it, and it's always better to avoid this kind of hassle. To avoid the potential for a hemispheric bias, you could use a single electrode somewhere on the midline (e.g., Cz). Another option is to place two electrodes in mirror-image locations over the left and right hemispheres and then combine them as the reference. For example, sometimes people place electrodes on both the left mastoid and the right mastoid, physically connect the wires from these electrodes, and then use the combined signal as the reference (the same thing can be done with earlobe electrodes). This is called the *linked mastoids* (or linked earlobes) reference, and it is not biased toward either hemisphere. However, physically linking the wires from these two electrodes creates a zero-resistance electrical bridge between the hemispheres, which may distort the distribution of voltage over the scalp and reduce hemispheric differences. This also invalidates the use of source localization techniques. Thus, I would advise against using linked mastoids (or linked earlobes) as the reference. It is better to re-reference the data offline to the average of the separately recorded left and right mastoids or earlobes, as described in the previous section, creating an *average mastoids* or *average earlobes* reference (Nunez, 1981). Note that some researchers use the term *linked mastoids* or *linked earlobes* to refer to the average of the mastoids or earlobes, but to avoid any ambiguities about what was actually done, the term *linked* should be used only when the two sides have been physically linked.

Third, you should avoid using a reference that introduces a lot of noise into your data. For example, a reference electrode near the temporalis muscle (a large muscle on the side of the head that is used during eating and talking to control the jaw) will pick up a lot of muscle-related activity, and this activity will then be present in all channels that use this reference (box 5.4). Averaging tends to reduce noise, so using the average of many electrode sites as the reference may lead to cleaner data.

Fourth, it is usually a good idea to avoid using a reference that is near the place on the scalp where the effect of interest will be largest. For example, the absolute voltage for the N170 component is largest over the lateral posterior scalp, near the mastoids. A large N170 is therefore present in the absolute voltage recorded at the mastoids, and this large N170 will be subtracted from all your electrodes if you use the mastoids as the reference. This will make the N170 appear to be relatively small in the region of the scalp where the absolute voltage is actually largest, and it will create a large positive voltage at distant sites (because subtracting a negative voltage from the reference electrodes creates a positive voltage). This may cause confusion about the scalp distribution of the N170, so the mastoid is not usually used as the reference in N170 experiments. Similarly, Cz is the default reference in systems made by Electrical Geodesics, and

Box 5.4
Where's That Noise Coming From?

Whether your system subtracts the reference in hardware or in software, any noise in the reference electrode will be present in all electrodes that use this reference (but will be upside down because it is being subtracted). Thus, if you see some noise in all the channels that share a particular reference site, then the noise is almost certainly coming from the reference site. This is a simple consequence of the fact that the referenced data is equal to the absolute voltage at the active site (A) minus the absolute voltage at the reference site (R). In other words, the referenced voltage is A − R. For example, a reference electrode on the mastoid may pick up the electrocardiogram (EKG) voltage from the heart, and an upside-down version of this EKG will then be present in all electrodes that use this reference. One implication of this is that you should avoid using as reference a site that is picking up a lot of noise. Another implication is that you can figure out where the noise is coming from by asking whether it is present in all channels that use a particular reference.

researchers who use these systems almost always re-reference offline to a different site, especially if they are looking at components like P3 and N400 that are very large at Cz.

The last and perhaps most important factor is that you want your ERP waveforms and scalp distributions to be comparable with those published by other researchers. That is, because an ERP waveform for a given active site will look different depending on the choice of the reference site, it is usually a good idea to use a site that is commonly used by other investigators in your area of research. Otherwise, you and others may incorrectly conclude that your data are inconsistent with previously published data.

Now that we've spelled out all the relevant factors, we can finally discuss what the "best" reference site actually is. Unfortunately, there is no single best site, and the best site according to one factor might not be the best according to another factor. My lab usually uses the average of the mastoids as the reference because it is convenient, unbiased, and widely used. The earlobes and mastoids are close enough to each other that the resulting ERP waveforms should look about the same no matter which is used. I prefer to use the mastoid rather than the earlobe because I find that an earclip electrode becomes uncomfortable after about an hour and because I find it easier to obtain a good electrical connection from the mastoid (because the skin is not so tough).

A downside of the mastoids is that they are near the neck muscles, and they therefore tend to pick up muscle-related artifacts. They are also more likely to pick up electrocardiogram (EKG) artifacts, although this is not usually a significant problem in practice (see chapter 6). As I mentioned earlier, the mastoids can also be problematic for the N170 and other components that are largest at lateral posterior electrode sites. However, for a large proportion of studies, the mastoids are the best option. A similar alternative is to use the average of the P9 and P10 electrodes, which are near the mastoids but don't pick up quite as much muscle noise from the neck.

My lab rarely uses the average reference. The main reason is that the average reference makes it difficult to compare waveforms across different laboratories that record from different electrode sites. Even within my lab, we use different electrode configurations for different experiments, and using the average reference would make it difficult to compare across experiments. In addition, I like to avoid encouraging the incorrect view that the average reference provides a means of recording the absolute voltage from each site. However, there are certainly situations in which the average reference is acceptable or even preferable.

Because there are many issues involved in choosing the reference, I will not make a single simple recommendation for your research. After all, I don't know what you are studying! However, I've created the following list, which provides several different pieces of advice that you can use to decide what reference is best for your research:

• Look at your data with multiple different references. This will keep the no-Switzerland principle active in your mind.
• Concentrate on the pattern of differences in voltage among electrodes, not on the specific voltage at each site.

• In many cases, you will want to use the average mastoids or average earlobes as the reference simply because this is the most common practice in your area of research.

• You will probably want to use something other than the mastoids or earlobes if some other reference is standard in your area of research or if the component of interest is largest near the mastoids and earlobes (as in the case of N170).

• If the data look noisy with a mastoid or earlobe reference, you may gain some statistical power by using the average reference.

• If you wish to use the average reference, you should use a large number of evenly spaced electrodes that cover more than 50% of the surface of the head, and you absolutely must report all electrodes that contributed to the average reference in the methods sections of your papers.

Current Density

It is possible to completely avoid the reference problem by transforming your data from voltage into *current density* (sometimes called *scalp current density* [SCD] or *current source density* [CSD]). Current density is the flow of current out of the scalp at each point on the scalp. Because current is the flow of charges past a single point, rather than the potential for charges to flow between two points, current is naturally defined at a single point and does not require a reference. Conveniently, it is possible to convert the distribution of voltage across the scalp into an estimate of the distribution of current. This is done by taking the second derivative of the distribution of voltage over the scalp (for a detailed description, see Pernier, Perrin, & Bertrand, 1988). This is often called the *surface Laplacian* (after the mathematician Pierre-Simon Laplace).

To understand what it means to take the second derivative of the voltage distribution, start by imagining that you have recorded data from thousands of tightly packed electrodes, so you have a nearly continuous measure of the voltage distribution over the scalp at each moment in time. Now imagine that you replace the voltage at each electrode with the difference in voltage between that electrode and the average of its nearest neighbors (at each moment in time). The result would be the first derivative of the voltage distribution. Now imagine that you repeat this process again, taking the difference between each newly computed difference value and the average of the difference values from the nearest neighbors. This would be the second derivative. Derivatives are ordinarily calculated from continuous data, but you could get a good approximation of the true derivative if you had thousands of tightly packed electrodes. In a real experiment, however, you might have only 32 electrodes, and taking differences between adjacent electrodes would not give you a very good estimate of the true derivative. The typical approach is therefore to use an interpolation algorithm that provides an estimate of the continuous voltage distribution and then use this continuous distribution to compute the second derivative (Perrin, Pernier, Bertrand, & Echallier, 1989). The accuracy of the interpolation will, of course, depend on how many electrodes you're using. In practice, a reasonably good estimate of current density can be obtained with 32 electrodes. However, the interpolation becomes less accurate near the edge of electrode array, so you shouldn't place much faith in the current density estimates at the outermost electrodes.

In addition to eliminating the reference issue, current density has the advantage of "sharpening" the scalp distributions of ERP components. In other words, current density is more focally distributed than voltage. This may help you to separately measure two different electrodes that overlap in time, because the current density transformation may make them more spatially distinct. You should also know that current density minimizes activity from dipoles that are deep in the brain and preferentially emphasizes superficial sources (this is because the current from a deep source dissipates widely over the entire scalp and is therefore not very dense). This can make broadly distributed components like P3 difficult to see. However, this can sometimes be a benefit because it reduces the number of overlapping components in your data. You should also keep in mind that current density is an estimated quantity that is one step removed from the actual data, and it is therefore usually a good idea to examine both the voltage waveforms and the current density waveforms. Current density can be a very valuable tool, and it is probably used less than it should be.

Electrodes and Impedance

Electrodes

Now that we've discussed the nature of the voltages that are present at the scalp, let's discuss the electrodes that are used to pick up the voltages and deliver them to an amplifier. Basically, a scalp electrode is just a way of attaching a wire to the skin. You could just tape a wire to the skin, but that would not create a very stable electrical connection. The main consideration in selecting an electrode is to create a stable connection that does not vary rapidly over time (e.g., if the subject moves slightly).

Electrode Composition   In most cases, an electrode is a metal disk or pellet that does not directly contact the scalp but instead makes an electrical connection to the scalp via a conductive gel or saline. The choice of metal is fairly important, because some metals quickly become corroded and lose their conductance. In addition, the circuit formed by the skin, the electrode gel, and the electrode can function as a capacitor that attenuates the transmission of low frequencies (i.e., slow voltage changes).

The most common choice of electrode material is silver covered with a thin coating of silver chloride (these are typically called *Ag/AgCl* electrodes). These electrodes have many nice electrical properties, and contemporary manufacturing methods lead to good reliability. In the 1980s, many investigators started using electrode caps made by Electro-Cap International, which feature tin electrodes. In theory, tin electrodes will tend to attenuate low frequencies more than Ag/AgCl electrodes (Picton, Lins, & Scherg, 1995), but Polich and Lawson (1985) found essentially no difference between these two electrode types when common ERP paradigms were tested, even for slow potentials such as the CNV and sustained changes in eye position. This may reflect the fact that the filtering caused by the electrodes is no more severe than the typical filter settings of an EEG amplifier (Picton et al., 1995). Either tin or Ag/AgCl should be adequate for most purposes, but Ag/AgCl may be preferable when very slow potentials are being recorded.

Electrode Placement Figure 5.4 illustrates the most common system for defining and naming electrode positions (Jasper, 1958; American Encephalographic Society, 1994a). This system is called the *International 10–20 System* because the original version placed electrodes at 10% and 20% points along lines of latitude and longitude (5% points are now sometimes used). The first step in this system is to define an equator, which passes through the nasion (the depression between the eyes at the top of the nose, labeled Nz), the inion (the bump at the back of the head, labeled Iz), and the left and right pre-auricular points (depressions just anterior to the ears). A longitudinal midline is then drawn between Iz and Nz, and this line is then divided into 10 equal sections (defining the 10% points along the line). The equator is similarly broken up at the 10%

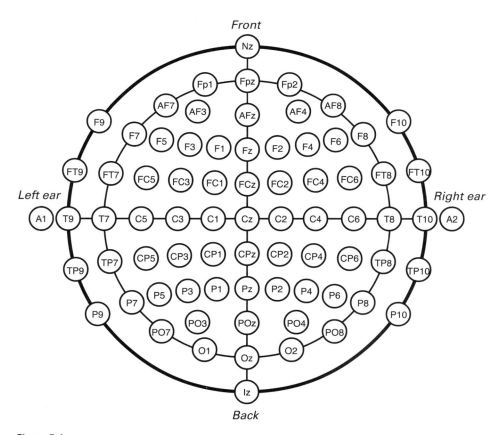

Figure 5.4
International 10/20 System for electrode placement. Note that earlier versions of this system used labels of T5 and T6 in place of P7 and P8 and used T3 and T4 in place of T7 and T8.

and 20% points, allowing electrodes such as F7, F8, P7, and P8 to be defined. Additional sites are then placed at equal distances along arcs between the midline sites and these equatorial sites.

Each electrode name begins with one or two letters to indicate the general region of the electrode (Fp = frontal pole; F = frontal; C = central; P = parietal; O = occipital; and T = temporal). Each electrode name ends with a number or letter indicating distance from the midline, with odd numbers in the left hemisphere and even numbers in the right hemisphere. Larger numbers indicate greater distances from the midline, with locations on the midline labeled with a "z" for zero (because the number 0 looks too much like the letter O).

Other systems are sometimes used, especially when very large numbers of electrodes are present. The most common alternative is a geodesic arrangement, which provides the same distance between any two adjacent electrodes. You can use any system you like, but you will need to be able to describe how your electrode sites are related to the 10/20 system.

How Many Electrodes Do You Need?   You might think that it is always better to record from more electrodes and therefore obtain more information. However, it is very difficult to ensure that you are obtaining clean data when you record from more than 30–40 electrodes. It's just too much information to monitor. My general advice is to record from between 16 and 32 active electrode sites in most experiments. You may want to go up to 64 electrodes on rare occasions, but I see very little value in going beyond 64 except in rare cases (e.g., when you are doing very serious source localization, including structural MRI scans from each subject). A more detailed discussion of this issue can be found in the online supplement to chapter 5.

Safety Precautions
EEG recordings carry a theoretical possibility of disease transmission between subjects via contaminated electrodes and between the subject and the experimenter during the electrode application and removal process. To my knowledge, there have been no reported cases of serious disease transmission in research studies using scalp electrodes. Risks are slightly greater in clinical practice (because of the greater possibility of people who have infectious diseases). A report from the American Encephalographic Society noted that "transmission of infection during routine EEG recording procedures is almost unheard of except under special conditions, such as depth electrode placement in patients with spongiform encephalopathy" (American Encephalographic Society, 1994b, p. 128). This reflects, in part, the fact that most EEG/ERP laboratories have implemented procedures to reduce the risk of disease transmission.

For many serious infectious diseases, such as hepatitis B, the risk comes from pathogens in the blood and other fluids. In most kinds of ERP research, risks of disease transmission can be nearly eliminated by thoroughly disinfecting the electrodes after each subject and by ensuring that the experimenter wears gloves when touching the subject (or touching the electrodes before they have been disinfected). There are many different ways of disinfecting electrodes, some of which may harm the electrodes. I therefore recommend that you follow the manufacturer's instructions for disinfecting your electrodes. If you are working with high-risk groups (i.e., individuals who are likely to have infectious diseases or who may have compromised immune systems), you should take more extensive precautions. For a more extensive discussion and more detailed recommendations, see Sullivan and Altman (2008).

Impedance, Common Mode Rejection, and Skin Potentials

It is obviously important to have a good electrical connection between the electrode and the scalp. As described in chapter 2, the quality of the connection is usually quantified by the *imped-ance* of the electrode–scalp connection (the *electrode impedance*). Impedance is related to resistance, which was defined in chapter 2. Resistance is the tendency of a material to impede the flow of a constant current, whereas impedance is the tendency to impede the flow of an alternating current (a current that fluctuates rapidly over time). The EEG is an alternating current, so it makes more sense to measure impedance than resistance in EEG recordings. Impedance is actually a combination of resistance, capacitance, and inductance, and the properties of the skin, the electrode gel, and the electrode can influence all three of these quantities. Thus, to assess the extent to which current can flow between the scalp and the recording electrode, it is important to measure the impedance rather than the resistance. Impedance is typically denoted by the letter Z and is measured in units of ohms (Ω) or thousands of ohms (kilohms; kΩ). As will be described in detail later in this chapter, the scalp is covered by a layer of dead skin cells and an oil called sebum, and these can dramatically increase electrode impedance, which can reduce the quality of the EEG recordings. You might think that a higher electrode impedance would make the recorded EEG voltages smaller, but this is not a problem with modern EEG amplifiers. Higher impedance may increase the noise in the data however, which is a problem.

Many EEG recording systems require the experimenter to lower the electrode impedance, which is typically achieved by cleaning and abrading the skin under each individual electrode. However, many newer EEG systems can tolerate high electrode impedances (these are often called *high-impedance systems*). I used low-impedance systems in my own lab for many years, and now I use a high-impedance system. Note that low-impedance systems require low electrode impedances to work properly, whereas high-impedance systems have special circuitry that allows them to work with either low or high electrode impedances.

High-impedance systems have three significant advantages. First, they reduce the amount of time required to prepare each subject (by eliminating the time involved in abrading the skin under each electrode, which can be considerable if you are recording from a lot of electrodes). Second, they reduce the likelihood of disease transmission between subjects because they elimi-nate abrasion of the skin (which may lead to a small amount of blood). Of course, disease transmission can be minimized by thoroughly disinfecting the electrodes between subjects, but it is even safer to avoid abrading the scalp at all. Third, because abrasion can be uncomfortable, high-impedance recordings may be more pleasant for the subject.

However, high electrode impedances can potentially lead to two problems. The first problem is poor common mode rejection. Recall from earlier in this chapter that common mode rejection is the ability of a differential amplifier to accurately subtract the noise that is present in the amplifier's ground circuit. To do this subtraction accurately, the active and reference signals ([A – G] and [R – G]) must be treated exactly equivalently. If the impedances differ among the active, ground, and reference electrodes, they may not be treated equivalently in the subtraction process, and some of the common mode noise may remain in the referenced data. Differences in impedance typically become more profound as the impedance increases. Thus, as the electrode impedance increases, the ability to reject common mode noise decreases. However, it is possible to design EEG amplifiers that are less prone to this problem (by increasing the *input impedance* of the amplifiers). Thus, if you have an amplifier that is designed to work with high electrode impedances, the effect of increased electrode impedance on common mode noise rejection may not have much impact on the quality of your data. In my lab's high-impedance system, we see very little common mode noise (but we also spend a lot of time making sure that we've eliminated sources of electrical noise).

The second problem associated with high electrode impedance is that this may increase skin potential artifacts. Unfortunately, high-impedance EEG recording systems do nothing to address this problem. Skin potentials are a very significant factor in ERP studies. As described in the online supplement to this chapter, the skin potentials in high-impedance recordings can add so much noise that you might need to double or triple the number of trials needed to get a statistically significant result. Consequently, it is worthwhile to understand how skin potentials are generated, which is described in detail in the next section. A later section will provide concrete advice about how to deal with skin potentials and whether you should use low- or high-impedance recordings.

Skin, Impedance, and Skin Potentials   There is a tonic voltage between the inside and the outside of the skin, and the magnitude of this voltage changes as the impedance changes. Thus, if the electrode impedance goes up and down over time, this will lead to voltages that go up and down in your EEG recordings. These voltage changes are called *skin potentials*, and they are a major source of noise in ERP experiments. Understanding these skin potentials requires that you learn a little about the skin.

The skin is a very complex organ, and figure 5.5A provides a simplified diagram of some of the major components. The skin is divided into a thick lower layer called the *dermis* and a thin top layer called the *epidermis*. Skin cells die and are replaced at a rapid rate, and the top portion of the epidermis is a layer of dead skin cells. These dead skin cells flake off at a rapid rate (thousands of cells per minute, which is more than a little disgusting!). But before they flake off, they provide a very important layer of protection between the living skin below and the harsh, dangerous world around us. These dead skin cells are poor conductors of electricity, and they are a big part of the high impedance between the electrode and the living dermis (which itself has a very low impedance). A second major factor is that the outside of the skin is covered with a thin layer of an oil called *sebum* that is secreted by *sebaceous glands* in the dermis. Like the layer of dead skin cells, this oil serves to protect us from the surrounding environment but is a poor conductor of electricity.

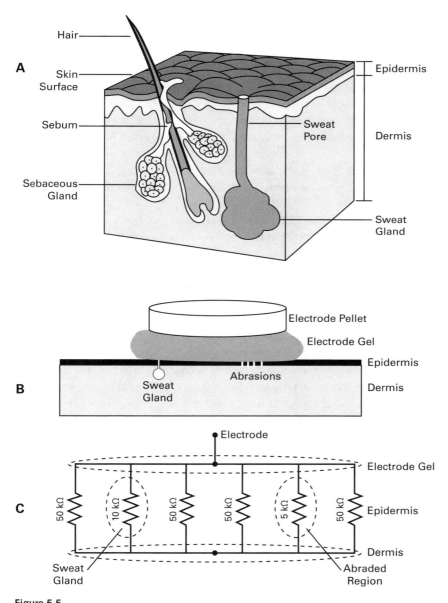

Figure 5.5
(A) Major components of the skin. Adapted from a public-domain image provided by the National Institute of Arthritis and Musculoskeletal and Skin Diseases (www.niams.nih.gov/Health_Info/Acne/default.asp). (B) Cartoon of the skin and electrode after abrasion of the skin under the electrode. (C) Simplified electrical diagram of the skin's resistance. The "squiggly line" symbols represent resistors, and each resistor reflects the resistance or impedance of a small patch of the skin under the electrode.

The final major factor in determining the impedance between the outside of the skin and the electrode is sweat, which is generated in sweat glands and secreted through sweat pores on the surface of the skin (figure 5.5A). Sweat contains salt and is therefore highly conductive. When the sweat glands start filling with sweat, they become a means of conducting electricity between the outside of the skin and the dermis. This does not require that the sweat is actually pouring out onto the skin—the presence of sweat in the tube leading up to the sweat pore is sufficient to reduce the impedance of the skin at the location of that sweat pore. The amount of sweat in the sweat glands reflects both environmental factors (e.g., temperature and humidity) and psychological factors (e.g., stress). As these factors change, the level of sweat changes within the sweat glands, and this leads to changes in the impedance of the skin. These changes in impedance lead to changes in the standing voltage between the dermis and the outside of the skin, thus producing skin potentials.

To illustrate how sweat influences your EEG recordings, figure 5.5B shows an electrode pellet making an electrical connection with the epidermis via a conductive electrode gel. The epidermis ordinarily has a high impedance (due to oils and dead skin cells) that impedes the flow of brain-generated voltages through the dermis to the electrode gel and then to the electrode. The impedance may decrease gradually over time as the dead skin cells become hydrated by the electrode gel, because hydration makes the dead skin cells better conductors. This will lead to a very gradual shift in the voltage recorded by the electrode over a period of many minutes. In addition, the impedance will also decrease if a sweat gland in the skin under the electrode gel starts filling with sweat, providing a path for electricity to flow more readily across the epidermis. This will cause the recorded voltage to shift over a period of several seconds. In addition, the impedance will increase if the skin starts to become less hydrated or if the level of sweat in the sweat glands starts to decrease (e.g., if the subject begins to relax), and this will also cause changes in the skin potential. Thus, as the impedance shifts up and down over time, this creates a voltage that shifts up and down over time in the recording electrode. The very gradual shifts caused by hydration of the skin are not ordinarily a problem, but the faster changes caused by the sweat glands can add random variance to the EEG and reduce your ability to find statistically significant effects.

You can think of the electrode, electrode gel, epidermis, and dermis as a simple electrical circuit consisting of several resistors in parallel. As illustrated in figure 5.5C, each tiny patch of epidermis serves as a resistor. The impedances of the electrode, electrode gel, and dermis are negligible, so they are just represented as wires and dots. The overall impedance between the electrode and the dermis is mainly determined by the impedance of the little patches of epidermis between the electrode gel and the dermis. Because electricity follows the path of least resistance (and least impedance), decreasing the impedance of one little patch allows current to flow more easily from the dermis to the electrode pellet, changing the overall impedance for that electrode. More formally, the impedances of the little patches combine according to a basic principle of electricity: When multiple resistors are arranged in parallel, the overall resistance (e.g., the resistance between the electrode gel and the dermis in the figure) is less than the smallest of the individual resistances. This is just a simple consequence of the fact that electricity tends to follow the path of least resistance. Thus, if each patch of epidermis has an impedance of about 50 kΩ, but a sweat gland starts filling with sweat and its impedance decreases to 10 kΩ, the overall resistance between the electrode gel and the epidermis will be a little less than 10 kΩ. This is why tiny sweat glands can have such a large impact on the overall impedance for a given electrode.

We can use the same basic principle of electricity to minimize the effect of the sweat glands on impedance. Specifically, if we abrade the skin under the electrode (by making tiny scratches), this will disrupt some of the dead skin cells and oils on the surface of the skin, providing a low-impedance path for electricity to travel through the epidermis. If the abraded region has a lower impedance than a filled sweat gland, then the overall impedance between the electrode gel and the dermis will be determined almost entirely by the impedance of the abraded region, and changes in the impedance of the sweat gland will not have much impact on the overall impedance. Box 5.5 describes some methods for reducing impedance.

Advantages and Disadvantages of High-Impedance Systems   The online supplement to this chapter describes a study that Emily Kappenman conducted in my lab to determine whether skin potentials are actually a significant problem in high-impedance recordings (Kappenman & Luck, 2010). She ran an oddball paradigm and recorded the EEG with a high-impedance system, but she abraded half of the electrodes to reduce the impedance for those electrodes. She found that the high-impedance (unabraded) electrodes contained much more low-frequency noise than the low-impedance (abraded) electrodes, which reduced the statistical power by a large amount in an analysis of P3 amplitude. Thus, many more trials or subjects may be needed to obtain statistically significant effects if you use high electrode impedances than if you abrade the skin to provide low electrode impedances. However, impedance had much less impact on statistical power in analysis of the N1 wave (see the online supplement to this chapter for details).

It would be tempting to conclude from Emily's study that high-impedance systems are a bad idea and that everyone should rely on old-fashioned low-impedance systems. However, this would be a mistake. In fact, after we collected the data and saw the initial results, I bought a second high-impedance system that was identical to the first, and my lab has conducted and published many ERP studies using these high-impedance systems over the past several years. However, it is important for you to carefully consider the pros and cons of high-impedance systems, and if you use a high-impedance system, you need to take steps to minimize the problem of skin potential artifacts.

A very important thing to keep in mind is that different EEG recording systems typically vary along many dimensions that may influence data quality, and a high-impedance system from manufacturer A is not ordinarily inferior to a low-impedance system from manufacturer B. In fact, it is not the recording system that is the issue, but the actual electrode impedances in your subjects (which is a consequence of how you prepare the subject, not the recording system itself). All else being equal, including the actual electrode impedances, recordings from

Box 5.5
Reducing Electrode Impedances

There are several ways to reduce the electrode impedance. Some can be done before the electrodes are applied. For example, you can rub the skin with alcohol to reduce oils (or ask your subjects to wash their hair immediately before coming to the lab). You can also rub an abrasive paste or gel on the skin using a cotton-tipped swab. A product called Nuprep (made by Weaver and Company) is widely used for this purpose. Some labs ask subjects to comb or brush their hair vigorously when they arrive at the lab. These are good things to do even if you have a high-impedance system.

To get the impedance low enough for a traditional low-impedance system, it is almost always necessary to abrade the skin individually at each electrode site, but you don't know where the scalp electrodes will be until you've placed the electrode cap on the subject's head. Each electrode has a hole for squirting electrode gel into the space between the electrode and the skin, and you can lower something through this hole to abrade the skin. The goal of this is to disrupt some of the dead skin cells and expose the living skin cells below. One common method is to use a blunt, sterile needle to squirt the gel into the hole and then rub the tip of this blunt needle against the skin. I don't like this method because a blunt needle isn't very good for displacing the top layer of dead skin cells, and you have to rub pretty hard to get the impedance down to an acceptable level. When I've had this done to me, I found it quite unpleasant.

A better alternative is to put a drop of Nuprep on the end of a thin wooden dowel (e.g., the wooden end of a cotton-tipped swab), lower it into the electrode, and twirl it on the skin.

My favorite technique uses a sharp, sterile hypodermic needle. You lower it into the electrode hole, sweeping it back and forth as you are lowering it so that you don't poke directly into the skin. When you feel some resistance to the back-and-forth sweeping motion, you know that you're starting to touch the skin. You can also have the subject tell you when the needle is contacting the skin. Once the tip of the needle has contacted the skin, you just drag it lightly across the skin a few times. Remember, your goal is to disrupt the dead skin cells, not to puncture this skin. In my experience, this does a great job of reducing the impedance with minimal discomfort. However, it requires considerable practice to do this without occasionally poking the subject. You can practice on yourself first, then try a few friends or labmates before you start working with real subjects. This approach is a little scary (and also increases the risk of disease transmission), and you need a biohazard waste disposal container, so you might just want to use the wooden dowel with Nuprep.

Whatever you do, the method should be clearly stated in your ethics approval form and on your consent form.

a high-impedance system should be as good as or better than recordings from a low-impedance system. Although I haven't done a formal assessment, I think the data quality from my current high-impedance recordings is nearly as good as the data quality from my previous low-impedance recordings (from a traditional low-impedance system). In addition, my informal observations of a variety of EEG recording systems suggests that some high-impedance systems yield much higher data quality than others. I won't "name names," because I haven't formally compared different systems, but you should certainly compare systems carefully before purchasing one (for a brief direct comparison, see Kayser, Tenke, & Bruder, 2003).

When deciding what kind of system to use, it is important to remember that abrading the skin carries a slight risk of disease transmission, and the reduction in this risk is a benefit of high-impedance recordings. This was a factor in my decision to buy a high-impedance system.

High-impedance recordings also mean that you will save time when preparing the subject, because you won't need to abrade the skin under each electrode. However, it is not clear that this will save you any time in the long run, because you may need many more subjects or trials per subject to obtain statistical significance in high-impedance recordings. Moreover, as discussed in the online supplement to this chapter, you may not really need to record from a large number of electrodes. Thus, it's not clear that high-impedance recordings really save you any time.

An exception to this arises in recordings from infants and small children, who cannot tolerate a long period of electrode preparation. It is very difficult to record from these subjects if you need to abrade the skin under each electrode. Moreover, parents are not happy when their children have a red spot or a scab at the electrode locations at the end of the recording session. Consequently, the advantages of high-impedance recordings are very important in studies with infants and small children. However, not all high-impedance systems are the same, and if you are doing this type of research, I would encourage you to think carefully about data quality when selecting a recording system and when developing your electrode application procedures. In particular, I suspect that the use of an electrode gel to make an electrical connection between the electrode and the scalp will lead to more stable impedance levels—and therefore fewer artifactual voltage fluctuations—than will the use of saline to make this connection, especially in subjects who move a lot. Again, I have not formally compared electrode gels and saline, so I cannot make a strong claim here. I would strongly recommend talking to people who use multiple different systems before choosing a particular system (and talk to the people who actually record and analyze the data, not the lab chief who sits in his or her office drinking coffee and writing grants and papers all day long).

If you decide to use a high-impedance system, there are several things you can do to minimize skin potentials and their effects on statistical power. First, many labs ask subjects to wash their hair on the day of the recording, which will decrease the oils on the surface of the skin that contribute to the electrode impedance. Some labs even have subjects comb or brush their hair vigorously immediately prior to electrode application, which may displace some of the dead skin cells on the surface of the epidermis, decreasing the impedance. This seems like a reasonable idea (as long as the comb or brush is disinfected after each use), and my lab has recently started doing it. Second, many high-impedance systems provide a measure (e.g., an impedance value or offset value) for each electrode that you can use to make sure that the impedance is within a reasonable range. Spending some time checking these values and monitoring them over the course of each session will save you a lot of time in the long run. Third, you should make sure that your recording environment is cool, because this will tend to reduce sweating. Chapter 16 provides some specific advice about this. Fourth, if you are running an experiment in which you need extra statistical power, you may want to abrade the skin a little bit to keep the impedances reasonably low (e.g., <20 kΩ). In some cases, you can do this just for the electrodes that will be contributing to the main statistical analyses (along with the ground and reference electrodes). Finally, you should do all the other small and large things that I describe in this book that influence data quality and statistical power. For example, you should track down and eliminate sources of electrical noise and follow the advice that I provide in online chapter 16 about monitoring subjects and keeping them happy.

Amplifying, Filtering, and Digitizing the Signal

Once the EEG has been picked up by the electrodes, it must be amplified and then converted from a continuous, analog voltage into a discrete, digital form that can be stored in a computer. Fortunately, these processes are relatively straightforward, although there are a few important issues that must be considered, such as choosing an amplifier gain and a digitization rate.

Analog-to-Digital Conversion and High-Pass Filters

The EEG is an analog signal that varies continuously over a range of voltages over time, and it must be converted into a set of discrete *samples* to be stored on a computer. This is called *digitizing* the EEG. As illustrated in figure 5.6, the samples are discrete in terms of both voltage and time (i.e., there are a fixed set of possible voltage values at a fixed set of time points). The continuous EEG is converted into these discrete samples by a device called an *analog-to-digital converter* (ADC). In older EEG digitization systems, the ADC had a resolution of 12 bits. A 12-bit ADC can code 2^{12}, or 4096, different voltage values (intermediate values are simply rounded to the nearest whole number). For example, if the ADC has a range of –5 V to +5 V, a voltage of –5 V would be coded as 0, a voltage of +5 V would be coded as 4096,[1] and the intermediate voltages would be coded in discrete steps of 0.00244 V (because the 10-V range of values is divided into 4096 equal steps, and 10 ÷ 4096 = 0.00244).

The EEG is amplified before being digitized, which provides much smaller steps between values. For example, if the EEG is amplified by a factor of 10,000 before being digitized, your ADC will have an effective range of ±500 μV and a step size of 0.244 μV. This amplification factor is called the *gain* of the amplifier. Amplification is necessary to bring the EEG voltage into the appropriate range for the ADC.

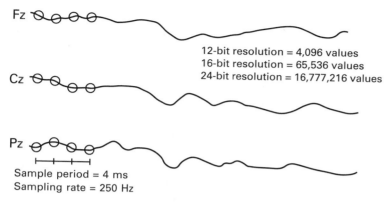

Fz

12-bit resolution = 4,096 values
16-bit resolution = 65,536 values
24-bit resolution = 16,777,216 values

Cz

Pz

Sample period = 4 ms
Sampling rate = 250 Hz

Figure 5.6
Digitization (sampling) of the EEG. A discrete sample is taken from each channel every 4 ms in this example, so the sample period is 4 ms. This is the same as 250 samples per second in each channel, so the sampling rate is 250 Hz.

Most modern EEG recording systems use the gain of the amplifier and the properties of the ADC to recode the discrete ADC values into units of microvolts. In the bad old days, the raw EEG was stored in the original ADC units, and we had to go through a series of extra steps to convert these values into microvolts.

Newer EEG systems typically use ADCs with 16–32 bits. This provides both greater resolution (smaller differences in voltage between each possible ADC value) and a broader range of possible values. The broader range is actually more important than the increased resolution. If your ADC has a resolution (step size) of 0.244 μV, you might think you would be unable to detect an experimental effect of 0.2 μV. But you would be wrong. When you average together many trials to create averaged ERP waveforms, you effectively increase your resolution. For example, even though the set of numbers {1, 2, 2, 2, 1, 1, 1, 1} contains only 1s and 2s, the average of this set is 1.375. Thus, a resolution of a quarter of a microvolt is sufficient for almost any ERP experiment. Most newer EEG recording systems provide a far greater resolution than this. You shouldn't worry about the resolution, and you certainly shouldn't choose one system over another because of differences in voltage resolution.

However, a broad range of values is very useful. If you have a 12-bit ADC with an effective range of ±500 μV, the EEG will often drift past the range of values. This drift can occur because of skin potentials and because of small static charges in the electrodes. If you are using a 12-bit or 16-bit ADC, you will need to filter out low frequencies to make sure that the voltages don't drift outside the ADC range. You will also need to make sure that the gain of your amplifier is low enough that you don't exceed the ADC range very often (and that you reject trials when this happens, as will be discussed in chapter 6). However, you should make sure that the gain of your amplifier is high enough that your voltage resolution is reasonably good (at least a quarter of a microvolt).

Filters are described in more detail in chapter 7 and online chapter 12, but I will say a few words about them in this chapter so that you know how to set the filters in your EEG acquisition system. To avoid drift, you can use a *high-pass* filter (a filter that passes high frequencies and suppresses low frequencies). You will definitely want to use a high-pass filter during data acquisition if you are using a 12- or 16-bit ADC (with a half-amplitude cutoff of between 0.01 and 0.1 Hz; see chapter 7 for details). However, the high-pass filters that are implemented in EEG amplifiers are inferior to the filters that you can apply in software. Consequently, if your ADC has a resolution of at least 24 bits, you can record without a high-pass filter (assuming that your system allows you to turn off the high-pass filter). This is called a *direct coupled* (DC) recording, because the EEG signal is directly coupled to the amplifier rather than being coupled through a capacitor. You can then filter out the low frequencies offline with a superior software filter.

If you are shopping for an EEG recording system, I recommend that you get a system with at least 20 bits of ADC resolution if you can afford it. If you can't afford it, you can live with 12 or 16 bits as long as you filter out the low frequencies prior to digitization. If a system has 12–16 bits of ADC resolution, the gain should be adjustable, with a minimum gain of 1000 or less and a maximum gain of 20,000 or more. In addition, a system like this should allow you to select among several high-pass cutoffs, including 0.01 Hz, 0.1 Hz, and 0.5 Hz (or nearby frequencies). These options will give you the ability to do many different kinds of ERP experiments. Systems with at least 20 bits of ADC resolution do not need to have a user-adjustable amplifier gain because they have enough range and resolution to deal with virtually any imaginable ERP experiment. These systems also do not need to include high-pass filters that operate prior to digitization because the EEG will never exceed the amplifier range (unless an electrode becomes disconnected) and the low frequencies can be filtered offline. I don't think there is any value in going beyond 24 bits of ADC resolution, although I won't be surprised if manufacturers someday start advertising that they are using 64-bit ADCs.

Discrete Time Sampling and Low-Pass Filters

The ADC takes samples of the continuous EEG signal at regularly spaced time points and converts the measured voltage into a number at each time point. As illustrated in figure 5.6, the *sampling period* is the amount of time between consecutive samples (e.g., 4 ms), and the *sampling rate* is the number of samples taken per second (e.g., 250 Hz). When multiple channels are sampled, most systems sample the channels sequentially rather than simultaneously; however, the digitization process is so fast that you can think of the channels as being sampled simultaneously (unless you are examining extremely high frequency components, such as brainstem evoked responses). Some systems contain a separate ADC for each channel, but this seems like an unnecessary expense for the vast majority of experiments.

How do you decide what sampling rate to use? To decide, you need to know about the *Nyquist theorem*, which states that all of the information in an analog signal such as the EEG can be captured digitally as long as the sampling rate is more than twice as great as the highest frequency in the signal. You should avoid sampling at lower rates because you will be losing information.

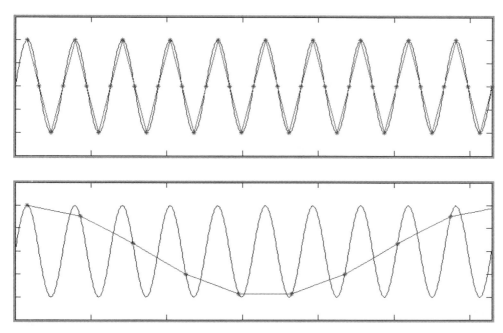

Figure 5.7
Example of aliasing. The black line is the original sine-wave signal, and the gray line simply connects the sampled values together. When the sampling is fast enough (top), the frequency of the samples reflects the frequency of the original signal. When the sampling is too slow (bottom), the sampled data appear to be at a much lower frequency.

In addition, if you sample at lower rates, you will induce artifactual low frequencies in the digitized data (this is called *aliasing*).

Aliasing is illustrated in figure 5.7. The top panel shows a sine wave that is sampled four times for each cycle, and you can see that it captures the frequency of the sine wave quite well. The bottom panel shows the same sine wave, but sampled a little less than once per cycle. When the samples are connected into a waveform, they form a very slow oscillation. Thus, when you sample too slowly, your high frequencies will be transformed into low frequencies.

Figure 5.8 provides a closer look at the idea that sampling allows you to capture all of the information in a continuous signal. When you sample, you are missing little bits of the signal between the samples, so you might think you are losing information. In fact, if you try to reconstruct the signal without any interpolation, you will get a waveform that is quite a bit different from the original signal. However, if you interpolate by simply drawing lines between each sample, you will obtain a fairly close approximation of the original signal. To obtain an exact reconstruction, you would need to use a more sophisticated interpolation method (which is beyond the scope of this book). With the appropriate interpolation method, you can reconstruct the original signal essentially perfectly (within the limits of the physical system). Thus, you will

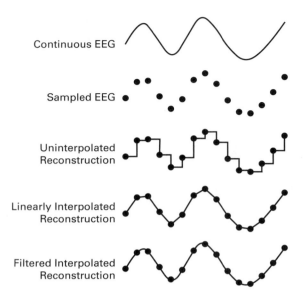

Figure 5.8
Interpolation to reconstruct the continuous EEG from the samples. When appropriate interpolation is used, the reconstruction is essentially perfect. Linear interpolation is imperfect, but good enough for most purposes (assuming that the data were originally digitized at a sufficiently high sampling rate).

not lose any information by sampling (as long as you sample more than twice as fast as the highest frequency in the signal), but you will need to use intelligent interpolation if you ever want to reconstruct the original continuous signal.

To sample at a high enough rate, you need to know the frequency content of the signal that you are recording. If you are interested in ERPs that mainly contain power below 30 Hz, you might think you could just record at any rate higher than 60 Hz. However, the EEG will contain frequencies higher than 30 Hz, even if you're not interested in them, and the Nyquist theorem says that we need to sample more than twice as fast as *any* frequency that is in the signal. You can't really know in advance what frequencies might be present, so the standard approach is to use a hardware filter that eliminates all activity above a certain frequency and then sample more than twice as fast as this cutoff frequency.

Filtering will be discussed in more detail in chapter 7, but I will say a few more words here so that you know how to filter out the high frequencies prior to digitization. A filter that suppresses high frequencies and passes low frequencies is called a *low-pass filter*. Often, the filters that are used prior to digitization are called *anti-aliasing filters*, but they are really just low-pass filters.

In the vast majority of experiments that examine cognitive or affective processes, there is very little information of interest in the EEG above approximately 100 Hz. For such experiments, you can set your low-pass filter cutoff at 100 Hz and sample at 400 Hz or higher. You will want to sample three to five times faster than the cutoff frequency of the filter because some activity above the cutoff frequency will be present in the filter's output (see chapter 7 for details). If you are not interested in high-frequency activity such as gamma oscillations, you could filter at 50 Hz and sample at 200 Hz or higher. If you are interested in very early sensory responses (e.g., <20 ms), you will want to filter and sample at a higher frequency (just look at published papers in your area of interest to see what frequencies other people use). Note that many systems automatically set the low-pass cutoff on the basis of your selected sampling rate (which is a good idea, because it avoids accidental errors).

For almost all cognitive and affective neuroscience experiments, I recommend a sampling rate between 200 and 1200 Hz. If you are near the low end of this range, you will use much less disk space, and all of your data analyses will run faster. Speed can be a significant issue, with some kinds of analyses taking hours or days (especially if you are using a computer with limited RAM). The main advantage of a higher sampling rate (assuming you are looking at typical cognitive or affective ERPs) is that you will have more precision in your latency measurements. But this greater precision may provide no practical advantage. For most analyses, the noise in the data is sufficiently great that a rounding error of ±2 ms (which is the error that you will get with a sampling rate of 250 Hz) is inconsequential. One exception to this arises when you use the jackknife analysis approach (described in chapter 10), which dramatically reduces noise in the measurement process, making small rounding errors meaningful. If you use this approach, you may want a higher sampling rate. However, you don't need a higher sampling rate with this approach if your data analysis system uses intelligent interpolation during latency measurement or if you can resample the data at a higher sampling rate prior to latency measurement (using interpolation during the resampling). I find that a sampling rate of approximately 250 Hz is ideal for most cognitive and affective neuroscience experiments.

Amplifier Gain and Calibration

The signal from each active electrode passes through a separate amplifier before being digitized (so that the range of EEG values is appropriate for the ADC). In the old days, a 16-channel EEG amplifier was a fairly large device, covered with switches for setting the gain and filters for each channel. In most modern systems, the electrodes are attached to a small box that contains both the amplifiers and the ADC, and all the settings are controlled by an attached computer rather than by physical switches. Each channel is typically set to the same gain (amplification factor) and filter settings, although some systems allow different settings for different channels. Some systems do not give you any control over the gain and filter settings; they have a fixed gain and automatically vary the cutoff frequency of the anti-aliasing filter to match whatever sampling rate you specify. Although this gives you less flexibility, it avoids errors. When my lab used a system that allowed custom settings, we would sometimes find that someone had changed the settings without telling anyone, and these new settings were therefore used for a subset of subjects in ongoing experiments. I was happy to give up flexibility to avoid these kinds of mistakes when we switched to our current system, which does not allow the user to directly modify the gain and filter settings.

Whether or not you directly set the gain on your amplifier channels, the actual gains will not be exactly what you specify, and they will not be exactly the same across channels (because analog devices like amplifiers are never perfect). It is therefore important to calibrate your system. The best way to do this is to pass a voltage of a known size through the system and measure the system's output. For example, if you create a series of 10-μV voltage pulses and run them into your recording system, it may tell you that you have a signal of 9.8 μV on one channel and 10.1 μV on another channel. You can then generate a scaling factor for each channel (computed by dividing the actual value by the measured value), and multiply all of your data by this scaling factor. You can apply this multiplication process to the EEG data or to the averaged ERP waveforms; the result will be the same. The gains of the channels of an EEG amplifier may drift over time, so it is a good idea to calibrate the amplifiers for each subject.

Some EEG amplifiers are permanently calibrated, meaning that the manufacturer guarantees that the actual gain of each channel will be within some small range of the specified gain. For example, if you set the gain to be 1000, the manufacturer guarantees that the actual gain will be within X% of this value (e.g., within 5%). In these systems, it's not always obvious to the casual user that the EEG is being amplified before being digitized, because the output of the system is in units of microvolts. However, whatever system you are using, the signal is being amplified in some manner, and imperfect calibration of the amplifiers will have some effect on your data.

The effects of imperfect calibration may be miniscule or enormous, depending on how great the imperfections are and what you are doing with your data. If you are just looking at individual channels, minor imperfections in calibration (e.g., ±5%) are not important because the exact size of an ERP effect does not usually mean very much. For example, it is unlikely to change your conclusions if the average N400 amplitude in a given experiment appeared to be 4.2 μV even though the true value was 4.0 μV. However, if the gain varies from subject to subject, this will add variance to your data and decrease your statistical power. If your system is not permanently calibrated, you should calibrate it for each subject to avoid this kind of variance. You should also calibrate regularly if you are combining data recorded from multiple systems (even multiple instances of the same type of system).

Whether or not your system is permanently calibrated, small differences in gain from channel to channel can distort your scalp distributions. If, for example, some of the channels have an actual gain of 95% of the specified value and others have an actual gain of 105% of the specified value, this could lead to systematic errors in ERP source localization. If you ever want to do any kind of ERP localization, you should definitely calibrate your amplifiers, even if the manufacturer claims that they are permanently calibrated. In my lab, we don't do source localization of ERP data, and we have permanently calibrated amplifiers, so we don't worry much about calibration.

Suggestions for Further Reading

Coles, M. G. H., Gratton, G., Kramer, A. F., & Miller, G. A. (1986). Principles of signal acquisition and analysis. In M. G. H. Coles, E. Donchin, & S. W. Porges (Eds.), *Psychophysiology: Systems, Processes, and Applications* (pp. 183–221). New York: Guilford Press.

Dien, J. (1998). Issues in the application of the average reference: Review, critiques, and recommendations. *Behavior Research Methods, Instruments, & Computers, 30*, 34–43.

Kappenman, E. S., & Luck, S. J. (2010). The effects of electrode impedance on data quality and statistical significance in ERP recordings. *Psychophysiology, 47*, 888–904.

Keil, A., Debener, S., Gratton, G., Junhöfer, M., Kappenman, E. S., Luck, S. J., Luu, P., Miller, G., & Yee, C. M. (in press). Publication guidelines and recommendations for studies using electroencephalography and magnetoencephalography. *Psychophysiology*.

Picton, T. W., Lins, O. G., & Scherg, M. (1995). The recording and analysis of event-related potentials. In F. Boller & J. Grafman (Eds.), *Handbook of Neuropsychology, Vol. 10* (pp. 3–73). New York: Elsevier.

Picton, T. W., Bentin, S., Berg, P., Donchin, E., Hillyard, S. A., Johnson, R., Jr., et al. (2000). Guidelines for using human event-related potentials to study cognition: Recording standards and publication criteria. *Psychophysiology, 37*, 127–152.

Regan, D. (1989). *Human Brain Electrophysiology: Evoked Potentials and Evoked Magnetic Fields in Science and Medicine*. New York: Elsevier.

6 Artifact Rejection and Correction

Overview

The signal that you record in an ERP experiment will consist of the EEG plus a variety of non-neural noise sources. These non-neural sources mainly consist of induced electrical signals from the recording environment (e.g., line noise from lights and computers) and biological signals such as blinks, eye movements, muscle activity, and skin potentials. All of these non-EEG signals are considered artifacts. Some of them are small and constant (e.g., line noise and some types of muscle activity), but others are large and transient (e.g., blinks and eye movements). The large and transient artifacts are often eliminated by discarding the trials on which they occur. The smaller and more constant artifacts cannot be eliminated in this way because they are typically present on every trial. However, they can often be attenuated by data processing procedures (e.g., averaging, filtering).

Artifacts can be problematic in three ways. First, they may decrease the signal-to-noise ratio (SNR) of the averaged ERP waveform, decreasing your ability to find significant differences between groups or conditions. Second, some types of artifacts may be systematic rather than random, occurring in some conditions more than others and being at least loosely time-locked to the stimulus so that they are not eliminated by the averaging process. Such artifacts may lead to erroneous conclusions about the effects of an experimental manipulation. For example, some stimuli may be more likely to elicit blinks than others, which could lead to differences in amplitude in the averaged ERP waveforms. Thus, a difference between conditions might be a result of a difference in a motor behavior (blinking) rather than reflecting a difference in brain activity. Third, the most common artifacts in ERP experiments are ocular artifacts related to blinks and eye movements, both of which change the sensory input. If these artifacts differ across conditions, then the sensory input may also differ across conditions, which can be an important confound. This type of confound is often not obvious to the experimenter, and it's on my list of Top Ten Reasons for Rejecting an ERP Manuscript (see online chapter 15).

There are two main classes of techniques for minimizing the effects of artifacts. First, it is possible to detect large artifacts in the EEG and simply exclude contaminated trials from the averaged ERP waveforms (this is called *artifact rejection*). Alternatively, it is sometimes possible to estimate the influence of the artifacts on the ERPs and use correction procedures to subtract away the estimated contribution of the artifacts (this is called *artifact correction*). In this chapter, I will discuss both approaches, including their advantages and disadvantages. However, I would first like to make a point that should be obvious but is often overlooked. Specifically, it is always better to minimize the occurrence of artifacts rather than to rely heavily on rejection or correction procedures, which always have a cost. This is really just a special case of Hansen's axiom: There is no substitute for good data. In other words, time spent eliminating artifacts at the source will be well rewarded by time saved later. This chapter will therefore also include hints for reducing the occurrence of artifacts.

This chapter provides very detailed and concrete recommendations about artifact rejection. In the first edition of this book, I was frustrated by the fact that most of my recommendations could not be followed in commercial ERP analysis packages, which typically do not provide very sophisticated tools for artifact rejection. This is one of the factors that motivated my lab to create ERPLAB Toolbox, a freely available ERP data analysis package (available at http://erpinfo.org/erplab). All of my recommendations about artifact rejection can easily be accomplished with ERPLAB Toolbox.

The description of artifact correction in this chapter is more theoretical. This is partly because there are many different approaches to artifact correction, and it's not practical to discuss all of them. It is also partly because my preferred approach for artifact correction—which uses independent component analysis (ICA)—is complex and rapidly evolving. However, a detailed description of how ICA-based artifact correction works in general, along with an example of how my lab implements it, can be found in the online supplement to chapter 6.

The General Artifact Rejection Process

Artifact rejection has been used since the earliest ERP studies to minimize the effects of artifacts. It has an obvious drawback: If you throw out some proportion of trials, you end up with fewer trials in your averaged ERPs, which can reduce your statistical power. And if the number of rejected trials differs across conditions or groups, this can make the data noisier in some conditions or groups than in others (see the online supplement to chapter 9 for a discussion of the consequences of differences in the number of trials). In many cases, however, it is a simple procedure that works very well, and the benefits greatly outweigh the drawbacks.

Before I get into the details of how to detect specific types of artifacts, I would like to provide a general framework for conceptualizing the artifact rejection process.[1] Detecting artifacts is, in essence, a *signal detection* problem, in which the artifact is treated as the to-be-detected signal. If you don't know about signal detection theory, I would recommend finding a perception textbook and reading about it. It's a fundamental framework for understanding how the mind works, and it is very useful for understanding artifact rejection.

Here I will provide a simple example of signal detection. Imagine that you have lost a valuable ring on a beach, and you have rented a metal detector to help you find it. The metal detector makes a sound that tells you the extent to which there is evidence of nearby metal, getting louder when there is a lot of mineral content and getting softer when there is not much mineral content. However, this output is quite variable due to random fluctuations in the mineral content of the sand. If you started digging in the sand any time there was a hint of nearby metal, you would make very slow progress because you would start digging every few steps. However, if you only started digging when the metal detector's output was very high, you might miss the ring altogether because it's small and doesn't create a large change in the detector's output. You therefore want to start digging at some intermediate level.

The key aspects of this example are as follows. You are trying to detect something that is either there or not (the ring) based on a noisy, continuously variable signal (the metal detector's output). You select a threshold value, and if the signal exceeds that value, you make a response (digging). In this context, we can define four outcomes for each patch of sand: (1) a *hit* occurs when the sought-after object is present, the signal exceeds the threshold, and you respond (i.e., the ring is under the metal detector, the metal detector's output exceeds a certain value, and you dig); (2) a *miss* occurs when the object is present but the signal fails to exceed the threshold, and you don't respond; (3) a *false alarm* occurs when the object is absent, the signal exceeds the threshold due to random variation, and you respond; (4) a *correct rejection* occurs when the object is absent, the signal doesn't exceed the threshold, and you don't respond. Hits and correct rejections are both correct responses, and misses and false alarms are both errors. Importantly, you can increase the number of hits by choosing a lower threshold (i.e., digging when the metal detector's output is fairly low), but this will also lead to an increase in the number of false alarms. The only way to increase the hit rate without increasing the false alarm rate is to get a better metal detector with an output that better differentiates between the presence or absence of small metal objects and is less influenced by other minerals in the sand.

Now imagine that you are trying to detect blinks in a noisy EEG signal rather than a ring on the beach. When a subject blinks, the movement of the eyelids across the eyeball creates a voltage deflection, and it is possible to assess the presence or absence of a blink by measuring the size of the largest voltage deflection within a given segment of EEG (just like assessing the presence or absence of the ring by examining the output of the metal detector). If the voltage deflection exceeds a certain threshold level, you conclude that the subject blinked and you discard that trial; if the threshold is not exceeded, you conclude that the subject did not blink and you include that trial in the averaged ERP waveform. If you set a low threshold and reject any trials that have even a small voltage deflection, you will eliminate all of the trials with blinks, but you will also discard many blink-free trials, reducing the signal-to-noise ratio of the averaged ERP waveform. If you set a high threshold and reject only trials with very large voltage deflections, you will have more trials in your averages, but some of those trials may contain blinks that failed to exceed your threshold. Thus, simply changing the threshold cannot increase the rejection of true artifacts without also increasing the rejection of artifact-free trials. However, just as you can do a better job of finding a ring in the sand by using a better metal detector, you can do a better job of rejecting artifacts by using a better procedure for measuring artifacts (i.e., a procedure that does a better job of having a high value when the artifact is present and a low value when it is absent).

When you are doing artifact rejection, it is important to be clear about your goal, which is ultimately to learn the true answer to some scientific question. To get the truth from your data, you need to maximize your statistical power (so that you have the best possible chance of getting a significant p value if a real effect is present), and you need to avoid confounds (so that a valid conclusion can be drawn from a significant effect). The goal is not to eliminate all artifacts from your data. You should typically eliminate artifacts only when they reduce your statistical power or create confounds. If you discard every trial that contains a hint of an artifact, you will end up throwing out virtually every trial in the experiment (because there is always at least a hint of muscle activity, induced noise, eye movement, etc.). However, if an artifact differs systematically across conditions or groups, it may create a significant but bogus difference between your conditions or groups. Thus, your goal in artifact detection is to discard trials that contain "problematic" artifacts (i.e., artifacts that reduce your statistical power or create confounds) and retain "good" trials (i.e., trials that improve your statistical power without creating confounds). This is not terribly difficult if you spend some time learning about the artifacts that are present in your data, and you think carefully about how the artifacts might influence your data.

In ERPLAB Toolbox, we make a distinction between artifact *detection* and artifact *rejection*. We use the term *detection* to refer to the process of determining that an artifact is present. That is, when the algorithm determines that an EEG epoch exceeds the threshold for rejection, that epoch is given a special mark. At the time of averaging, the marked epochs are excluded from the averages (unless the user specifies otherwise). The artifact-containing epochs are not permanently discarded, and this makes it easy to change the artifact rejection parameters and try again. However, ERPLAB also includes tools for artifact *rejection*, in which periods of data containing artifacts are deleted from the data file. Most people use the term *rejection* to mean the combination of detecting artifacts and excluding them from the average, and this is how I will use this term. Keep in mind, however, that the artifacts are not really being deleted from the EEG file in most cases; they are just being excluded during the ERP averaging process.

In most systems, all channels are discarded for a given trial if an artifact is detected in any channel. This makes sense because an artifact that is present in multiple channels might be harder to detect in some than in others. For example, if an eyeblink artifact is 100 μV right above the eyes at the Fp1 and Fp2 electrodes, it would be approximately 36 μV at the Fz electrode, approximately 16 μV at the Cz electrode, and approximately 10 μV at the Pz electrode (Lins, Picton, Berg, & Scherg, 1993a). This artifact would not be big enough to be reliably rejected at the Cz and Pz electrode sites, but it would certainly be big enough to impact a 2-μV experimental effect at these sites. Consequently, researchers typically reject all channels for trials with eyeblinks, even though the blinks are detected only at a subset of channels on individual trials. If you excluded trials from the Cz and Pz electrodes only when blinks were detected at those electrodes, you would fail to reject many blinks that are large enough to significantly distort your data. Moreover, if you reject trials only in the electrodes in which the artifact was detected,

Box 6.1
Electrode-Specific Rejection or Interpolation

Some software systems allow you to reject only the channel in which an artifact was detected or to replace the waveform in that channel with interpolated values from the surrounding electrode sites. This approach is particularly popular among people who are recording from very large numbers of electrodes. I have heard people say that they can't throw out all the channels for a given trial when an artifact is detected because they would end up throwing out almost every trial.

In my view, something is deeply wrong with this. If your data quality is so low that at least one channel has an artifact on a high proportion of trials, your data quality is presumably too low to do anything that would require a large number of electrodes (e.g., provide very precise scalp distributions). My recommendation in this situation would be to buy a better EEG recording system or to spend more time making sure that you are getting a good recording from every channel.

If you have one channel that is bad during the entire recording (e.g., because the electrode is broken), it should be okay to use an interpolated value for that channel. However, this should be a rare event: In most recordings, you should be able to make sure that all channels are delivering high-quality data before you start the task.

your averages will be based on different trials at different electrodes, making it difficult to trust the scalp distributions. Moreover, you would end up with different numbers of trials in your averages at different electrode sites, which might make it problematic to compare amplitudes across electrode sites (see the online supplement to chapter 9). See box 6.1 for some further thoughts.

Choosing an Artifact Measure

Many ERP analysis systems provide only a single crude method for rejecting trials with artifacts, in which a trial is rejected if the voltage during the epoch exceeds a user-defined threshold, such as ±75 μV. In essence, this approach uses the absolute value of the maximum voltage in the epoch as the measure of the artifact, rejecting the trial if this voltage exceeds the threshold. I am always shocked that a company will sell an ERP analysis system for thousands of dollars (or thousands of euros, tens of thousands of yuan, hundreds of thousands of yen, etc.) with such primitive artifact detection abilities. This method works okay for blink rejection under some conditions because blinks are very large, but it is totally inadequate for detecting and rejecting more subtle artifacts, such as eye movements. This is one of the reasons that we developed ERPLAB Toolbox, which contains excellent methods for assessing specific types of artifacts. I should note that some commercial packages also have good methods for assessing artifacts, and this should be a factor if you are deciding which package to buy.

If your software forces you to use an absolute voltage rejection approach, it is essential to perform baseline correction prior to artifact rejection. As will be discussed in chapter 8, baseline correction typically involves subtracting the average prestimulus voltage from the entire waveform. Panels A–C of figure 6.1 show why this is necessary if you are rejecting artifacts on the

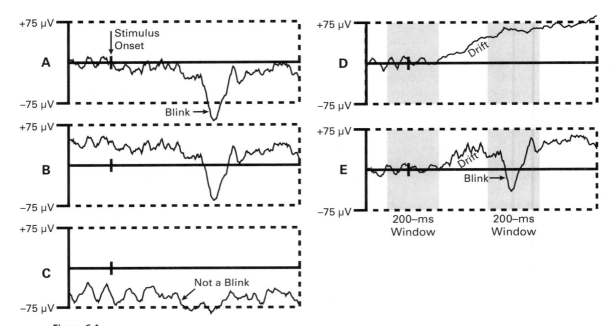

Figure 6.1
(A) Single-trial vertical EOG waveform containing a blink, with baseline correction to eliminate any overall offset in the signal. The dashed lines reflect the voltage threshold for artifact rejection (±75 μV). The blink (the large negative-going voltage deflection) brings the signal beyond the threshold, so the trial would be correctly rejected. (B) Same as (A), but without baseline correction. Now the blink does not exceed the threshold for rejection, and the trial would not be rejected (incorrectly). (C) EOG waveform on a trial without a blink, with no baseline correction. The overall offset of the signal causes the trial to be rejected (incorrectly) even though no blink is present. (D) Baseline-corrected EOG waveform on a trial with some voltage drift, which exceeds the threshold for rejection and is therefore rejected (incorrectly). The gray regions indicate two of the many time windows that would be tested with the moving window peak-to-peak amplitude method. The peak-to-peak amplitude is small within each window, so the trial would not be rejected with this method (correctly). (E) Baseline-corrected EOG waveform on a trial with some voltage drift and a blink. The drift partly counteracts the blink, so the trial would not be rejected if a simple voltage criterion were used (incorrectly). The gray regions indicate two of the many time windows that would be tested with the moving window peak-to-peak amplitude method. The peak-to-peak amplitude is large within the window that contains the blink, so the trial could be rejected with this method (correctly).

basis of the absolute voltage. Panel A shows a typical blink that is large enough to exceed the 75-μV threshold for rejection. Baseline correction was applied to this waveform, so the voltage is centered on 0 μV prior to the onset of the blink. Panel B shows the same waveform without baseline correction. A voltage offset is present over the entire epoch that, by chance, moves the waveform up enough that this blink no longer exceeds the threshold for rejection. Panel C shows a different epoch of data in which baseline correction was not performed. No blink was present, but a voltage offset brought the voltage close enough to the rejection threshold that ordinary voltage fluctuations caused the waveform to exceed the threshold for rejection. Thus, baseline correction is absolutely essential if you will be using this type of rejection algorithm.

Even if you perform baseline correction, drifts in the data may lead to poor detection of eye-blinks. Figure 6.1D shows a baseline-corrected example in which no blink is present but a drift in the data leads the voltage to exceed the threshold for rejection. Figure 6.1E shows a baseline-corrected example in which a blink is present, but an opposite-polarity drift causes it to stay within the voltage threshold and avoid rejection. The former problem could be addressed by choosing a higher rejection threshold, and the latter problem could be addressed by choosing a lower rejection threshold. Although these solutions would work for these specific trials, increasing the threshold would lead to more misses, and decreasing the threshold would lead to more false alarms. A much better solution would be to use a more sensitive blink detection algorithm.

When Javier Lopez-Calderon was developing the artifact detection tools in ERPLAB Toolbox, he came up with a simple but very effective algorithm for more reliably distinguishing between epochs with blinks and epochs without blinks. It's called the *moving window peak-to-peak amplitude* method.[2] With this method, you define a window width, such as 200 ms (see the gray regions in panels D and E of figure 6.1). The algorithm places this window at the beginning of the epoch and finds the peak-to-peak amplitude within this window (i.e., the difference in amplitude between the most positive and negative points in the window). The window is then shifted rightward by a user-defined amount (e.g., 50 ms), and the peak-to-peak amplitude is determined in this new window. This continues until the whole epoch has been tested (or whatever portion of the epoch the user wants to test). The largest of these peak-to-peak amplitudes is then compared with the threshold for rejection.

Because this method uses the difference between the highest and lowest points in each small window, it is completely insensitive to the overall voltage offset of the epoch. It is also relatively insensitive to slow voltage drifts. Consider, for example, the blink-free data in figure 6.1D. Because there are no blinks, there are no sudden changes in voltage, and the peak-to-peak voltage within any given 200-ms window is small. When a blink is present, however, the peak-to-peak voltage within the 200-ms period containing the blink will be large, even if an opposite-polarity drift is present (as in figure 6.1E). I have found this to be a much more sensitive measure of blinks than the absolute voltage method. With an appropriate threshold, it can detect all true blinks while rarely having false alarms. Another good algorithm will be described later in the chapter.

Choosing a Rejection Threshold

Once you have chosen an appropriate measure of an artifact, you must choose a threshold that does a good job of balancing misses and false alarms. One option is to pick a threshold on the basis of experience or previous studies and use this value for all subjects. For example, you may decide that all trials with a peak-to-peak EOG amplitude of 50 µV or higher will be rejected. However, there is often significant variability across subjects in the size and shape of the voltage deflections produced by a given type of artifact and in the characteristics of the EEG in which these voltage deflections are embedded, and a one-size-fits-all approach is therefore not optimal. Instead, it is usually best to tailor the threshold for each individual subject.

Depending on your data analysis system, you may have control over other parameters as well, such as the time window that is scanned for artifacts. Some systems force you to look for artifacts over the entire epoch, but others let you choose a window within the epoch. You will ordinarily want to reject artifacts at any time in the window, but remember that the goal is not to remove all artifacts but instead to maximize statistical power and avoid confounds. In my lab's N2pc experiments, for example, subjects often make small eye movements in the direction of the target, which leads to a lateralized EOG voltage that can distort the N2pc measurements. These eye movements are fairly small compared to the single-trial noise, but they are still large enough to distort the results in the averaged ERP waveforms. In these cases, we are concerned about eye movements that are consistently directed toward the side of the target stimulus and not with random eye movements. The consistent eye movements never occur prior to 150 ms, and they rarely occur after 500 ms. Thus, it's reasonable to reject small eye movements only between 150 and 500 ms. This allows us to use a low rejection threshold without discarding too many trials. In some cases, we still end up either rejecting too many trials or not catching all of the eye movements, but in those cases we take a close look at the individual subjects and tailor the rejection window for each subject. For example, one subject may make consistent eye movements only between 200 and 300 ms poststimulus and another might make them only between 250 and 400 ms, and we would use different rejection windows for these two subjects.

You may be thinking that this kind of experimenter control over the data processing has the potential to bias the results of an experiment. This is not actually a problem if your experiments exclusively involve within-subject manipulations, because the same parameters will be used in every condition for a given subject.[3] If your experiments involve different groups of subjects, however, the artifact rejection parameters should be set by someone who is blind to the group membership of each subject. This is what we do in our schizophrenia research. It can be inconvenient, because it's difficult to look at the results before the experiment is complete if the person doing the analyses is blind to group membership, but I believe it's still the best approach.

If you set the parameters individually for each subject, the best way to do this is usually by means of visual inspection of the raw EEG. You can do this with the following sequence of steps. First, use your prior experience (or published values) to select an initial threshold. You will then have your software package perform a preliminary artifact detection with these parameters. This allows you to determine visually whether trials with problematic artifacts are not being rejected or if trials without problematic artifacts are being rejected. Of course, this requires that you are able to determine the presence or absence of artifacts by visual inspection. In most cases, this is fairly straightforward, and some hints on how to do this are provided in the next section. After you have tested the effects of this initial threshold, you can adjust the threshold (and/or other parameters) and try the artifact rejection again. You can repeat this until you are able to reject all of the trials that clearly have artifacts without rejecting too many artifact-free trials (as assessed visually). Some types of artifacts also leave a distinctive "signature" in the averaged waveforms (as will be discussed below), so it is also possible to evaluate whether the threshold adequately rejected trials with artifacts after the data have been averaged.

It can also be useful to ask the subject to make some blinks and eye movements at the beginning of the session so that you can easily see what that subject's artifacts look like. Keep in mind, however, that voluntarily produced blinks are usually larger than spontaneous blinks.

When you are first learning to do artifact rejection, you may spend an hour with the data from each subject, looking through the entire session and trying lots of different parameters. After you think you have the right parameters, you should ask someone with experience to take a look and make sure you've done it right. This will help you to train your visual system to detect artifacts. Once you have more experience, you will typically spend only 5–10 min per subject, looking at a few parts of the session to make sure that the parameters are appropriate for the whole session. Moreover, you will be able to use the same parameters for most subjects in most experiments.

Visual Inspection

Some investigators visually inspect the EEG on each trial to determine which trials contain artifacts, but this process is conceptually identical to the procedure that I just outlined. The only difference is that the experimenter's visual system is used instead of a computer algorithm to determine the extent to which an artifact appears to be present, and a subjective, informal, unspecified, internal threshold is used to determine which trials to reject. The advantage of this approach is that the human visual system can be trained to do an excellent job of differentiating between real artifacts and normal EEG noise. However, a well-designed computer algorithm may be just as sensitive. In addition, computer algorithms have the advantages of being fast, consistent, and bias-free.

In my experience, the use of manual artifact rejection is often a consequence of commercial data analysis packages that have only primitive automated tools for rejection. If I were using one of those packages, I would also end up rejecting artifacts by visual inspection. In almost all cases, I would recommend using a good automated artifact rejection system rather than spending hours trying to identify artifacts by eye. Your visual system will still be used in setting the rejection parameters, but the computer will do most of the tedious work of checking every single trial for artifacts. If you are dealing with difficult-to-obtain data sets with very small numbers of trials (e.g., data from infants or hard-to-find patients who can only tolerate short recording sessions), you may want to do the automated rejection and then visually verify each trial. If you end up using visual inspection, then it is absolutely crucial that the person doing the rejection is blind to groups and conditions.

Minimizing and Detecting Specific Types of Artifacts

In this section, I will discuss several common types of artifacts and provide suggestions for reducing their occurrence and for detecting them when they do occur.

Blinks

Understanding Blinks Within each eye, there is a large, constant electrical potential between the cornea at the front of the eye and the retina at the back of the eye (the *corneal-retinal potential*), which is like a dipole with positive at the front of the eye and negative at the back of the eye. This potential spreads to the surrounding parts of the head, falling off gradually toward the back of the head. This is one source of the voltage offsets that are seen in EEG and EOG recordings (as in figure 6.1B). Ordinarily, this offset is subtracted by the baseline correction procedure, making it effectively invisible. However, anything that changes this potential rapidly will cause a large voltage deflection across much of the scalp. The voltage recorded on the scalp from this dipole is called the *electrooculogram* (EOG). This should not be confused with the *electroretinogram* (ERG), which is the much smaller neural response of the retina to a visual stimulus.

When the eyes blink, the eyelid moves across the eyes, which acts as a variable resistor that changes the EOG voltage recorded from electrodes near the eyes. Other factors (e.g., ocular rotation) also contribute to this voltage, but the movement of the eyelid is the main contributor (for an excellent review, see Plochl, Ossandon, & Konig, 2012). Figure 6.2 shows the typical waveshape of the eyeblink response at a location below the eyes (labeled VEOG) and at several locations on the scalp (all are referenced to a mastoid electrode). The eyeblink response consists primarily of a monophasic deflection of 50–100 μV with a typical duration of 200–400 ms. Perhaps the most important characteristic of the eyeblink response is that it is opposite in polarity for sites above versus below the eye (compare, for example, the VEOG and Fz recordings in figure 6.2). This makes it possible to distinguish between a blink, which would produce opposite-polarity voltage shifts above versus below the eye, and a true EEG deflection, which would typically produce same-polarity voltage shifts above and below the eye. An example of a true EEG deflection is shown on the right side of figure 6.2, where same-polarity deflections can be seen at the VEOG and Fz sites.

Reducing the Occurrence of Blinks Reducing the occurrence of an artifact is always better than rejecting trials with artifacts, and there are several ways in which the number of blinks can be reduced. The first is to ask subjects who normally wear contact lenses—which cause a great deal of blinking—to wear their glasses instead of their contact lenses. These individuals tend to blink more than average even when wearing their glasses, and it is therefore useful to keep a supply of eye drops handy (although these should be used only by individuals who normally use eye drops, and single-use bottles should be used to avoid infection risks). It is also helpful to use short trial blocks of 1–2 min, thus providing the subjects with frequent rest breaks for blinking (this also helps to keep the subjects more alert and focused on the task).

If you see a lot of blinks (or eye movements), it's important to let the subject know. Don't be shy about telling subjects that they need to do a better job of controlling these artifacts. My students tell me that it took them a long time to become comfortable doing this, but you really need to do it, even if it makes you uncomfortable at first.

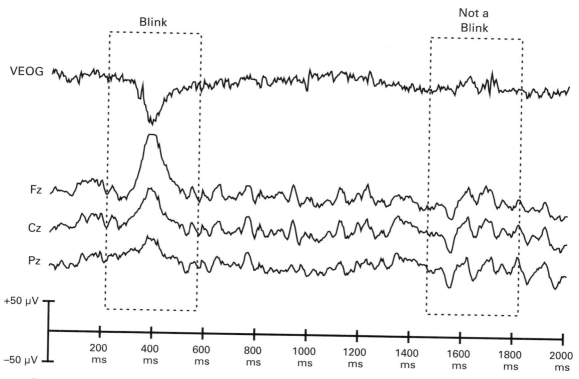

Figure 6.2
Recordings from a vertical EOG (VEOG) electrode located under the left eye and EEG electrodes located at Fz, Cz, and Pz, with a right mastoid reference for all recordings. A blink can be seen at approximately 400 ms, creating a negative deflection at the VEOG electrode and a positive deflection at the scalp electrodes. Note that the deflection is quite large at Fz and then becomes smaller at Cz and even smaller at Pz. The area labeled "Not a Blink" contains moderately large voltage deflections in all of these channels, but these deflections do not reflect a blink because the polarity is not inverted at the under-the-eye VEOG electrode relative to the scalp electrodes.

Some experiments can be designed so that subjects have a well-defined period during the intertrial interval (ITI) in which blinks are allowed. For example, the fixation point might disappear during the ITI, and the subject would be told that blinks are allowed when the fixation point is not present. This can work very well, but you need to be careful about two potential problems. First, if the fixation point appears shortly before the stimulus of interest (e.g., 500 ms before the target), the ERP elicited by the fixation point will overlap the ERP elicited by the stimulus of interest. I have seen many papers in which the prestimulus baseline contained a large and unexpected voltage deflection because of the onset of a fixation point. Second, if a blink occurs immediately before the stimulus on a large proportion of trials, the offset of the blink may be large enough to contaminate the data while being small enough to avoid rejection.

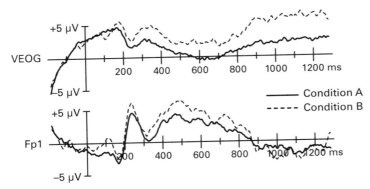

Figure 6.3
Recordings from a vertical EOG (VEOG) electrode under the left eye and the Fp1 electrode, which is over the left eye. The offset of a blink is visible in the prestimulus period, appearing as a negative voltage deflection at the VEOG electrode and a positive voltage deflection at the Fp1 electrode. Two different experimental conditions are shown for a single subject (a patient with a neurological condition).

Figure 6.3 shows an example of this from an actual experiment. The data were recorded by a collaborator, Erik St. Louis, from a patient with a neurological disorder. One day, Erik e-mailed me the averaged ERP waveforms shown in this figure and asked what might be causing the obvious distortion in the baseline. It was clear to me that the subject must have frequently blinked just before the onset of the trial, and the tail end of the blink was present during the prestimulus baseline period. Specifically, the offset of the blink was causing a deflection of approximately 10 μV between −200 and +200 ms (relative to stimulus onset). This was far too small to be detected and rejected on the single trials, but it was large enough and consistent enough to cause significant distortion of the ERP waveforms. I will come back to these waveforms in a little bit and tell you what we did to deal with this artifact (the answer may surprise you).

Detecting Blinks Blinks are relatively easy to detect on single trials. As I mentioned before, a simple voltage threshold works reasonably well if you have clean data, but it's far from perfect even with clean data and is woefully inadequate in some cases. The moving window peak-to-peak amplitude measure works great, and I highly recommend it (if your software implements it). I will describe another measure—the *step function*—in the section on eye movements, and this function also works nearly perfectly for blink detection. I've never done a formal comparison of these two methods, but I suspect the step function method works a little better than the moving window peak-to-peak amplitude method. However, both work a zillion times better than a simple voltage threshold.

Because the blink potential is negative below the eyes and positive above the eyes, it can be isolated from other EEG activity by recording the difference between an electrode below the eyes and an electrode above the eyes. This also increases the size of the blink response (because a negative value minus a positive value is an even bigger negative value). The best way to do this is to start with separate recordings from an electrode below the eye and an electrode above the eye, with both electrodes referenced to a common, distant site (e.g., an EOG electrode located below the left eye and the Fp1 site above the left eye, both referenced to a mastoid electrode). You can then create a new channel offline that is the difference between these values (below-the-eye minus above-the-eye). Any brain activity that is in common to these two channels is removed (e.g., most of the EEG), leaving only the blink response and the little bit of EEG that differs between these nearby sites. The blink detection algorithm is then applied to this new channel.

The polarity inversion also makes it possible to determine whether the artifact rejection worked properly. As described earlier, you can look at the averaged ERP waveform at the electrodes under and over the eye (with their original reference) to see if there are any polarity inversions. You might see something like the waveforms in figure 6.3, in which case you may want to take another look at your artifact rejection parameters. You should be especially concerned if an experimental effect has an opposite polarity under versus over the eyes, because this is likely to mean that more blinks survived artifact rejection in one condition than in the other. It's possible for a real effect to show this kind of polarity reversal if the dipole is located at the frontal pole and oriented vertically. However, you should definitely rule out the possibility that it is a blink artifact before accepting it as a real effect.

When Erik St. Louis sent me the data shown in figure 6.3, there was no realistic way to adjust the rejection parameters to avoid the blink offset that contaminated the data. The subject blinked prior to the stimulus on almost every trial, so a really low threshold for rejection would have caused almost every trial to be rejected. I told Erik not to worry about it because the experimental effect (the difference between the two conditions) had the same polarity above and below the eyes and could not be a blink artifact. If the data had come from a college student in one of my lab's basic science studies, we probably would have replaced the subject. But the data in figure 6.3 came from a hard-to-find neurological patient, and it was not worth throwing out the data given that the artifact did not impact the experimental effect. The contamination from the blink artifact in this case was much like nondifferential overlap, which can usually be ignored (as discussed in chapter 4 and online chapter 11).

Eye Movements

Like blinks, eye movements are a result of the strong dipole inside the eye. When the eyes are stationary, this dipole creates a static voltage gradient across the scalp, which is removed by baseline correction (and by high-pass filters). When the eyes move, the voltage becomes more positive over the side of the head that the eyes now point toward (for an excellent review, see Plöchl et al., 2012). For example, a leftward eye movement causes a positive-going voltage deflection on the left side of the scalp and a negative-going voltage on the right side of the scalp. Horizontal eye movements are most easily observed with bipolar recordings, in which an electrode lateral to one eye is the active site and an electrode lateral to the other eye is the reference site. This is called a *horizontal EOG* (HEOG) recording. Vertical eye movements can be seen in the same bipolar recordings used to isolate blinks from other EEG activity (i.e., the difference between electrodes below and above the eyes). This is called a *vertical EOG* (VEOG) recording. An eye movement that is not perfectly vertical or perfectly horizontal will show a voltage deflection in both the VEOG and HEOG channels, where the amplitude of the deflection in a given channel reflects the degree to which the eyes moved in the corresponding direction.

Unless the subject is viewing moving objects or making gradual head movements, the vast majority of eye movements will be saccades (sudden shifts in eye position). Examples of saccadic eye movements are shown in the bottom three waveforms of figure 6.4. In the absence of noise, a saccade would consist of a sudden step from one voltage level to another voltage level, where it would remain until the eyes moved again (unless the data have been high-pass filtered, in which case the voltage will fall gradually to zero). In most ERP experiments, subjects make a saccade from the fixation point to some other location and then make another saccade to return

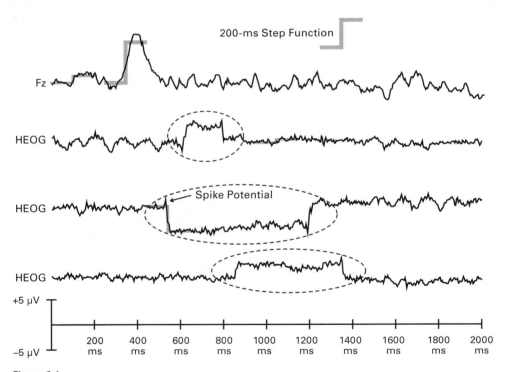

Figure 6.4
The step function and its application to an eyeblink recorded from the Fz electrode and to eye movements recorded in a horizontal EOG (HEOG) channel. The Fz recording was referenced to the right mastoid, and the horizontal EOG was recorded with the active electrode adjacent to the right eye and the reference electrode adjacent to the left eye. The areas indicated by the dashed lines are eye movements that move away from the fixation point and then back again, creating a "boxcar" shape. The step function is overlaid with the EOG waveforms in a few places so that you can easily see how the step function matches or mismatches the EOG. The step function would ordinarily be compared with each point in the EOG waveform, not just the few points shown here.

to the fixation point. This leads to a *boxcar*-shaped voltage deflection in the EOG recording (as in the examples in figure 6.4). The length of the boxcar depends on how long the eyes remain at the peripheral location before moving back to the fixation point. This characteristic shape can help you distinguish between real eye movements and noise in the EOG recording.

The average size of the saccade-related deflection has been systematically measured in bipolar recordings by Hillyard and Galambos (1970) and by Lins et al. (1993a), and these studies yielded the following findings: (a) the voltage deflection at a given electrode site is a linear function of the size of the eye movement, at least over a ±15° range of eye movements; (b) an HEOG recording between electrodes at locations immediately adjacent to the two eyes will yield a deflection of approximately 16 µV for each degree of horizontal eye movement; and (c) the voltage falls off in a predictable manner as the distance between the electrode site and the eyes increases (for a list of the propagation factors for a variety of standard electrode sites, see tables V and VI of Lins et al., 1993a).

It should also be noted that eye movements cause the visual input to slip across the retina, which creates a visual ERP response (saccadic suppression mechanisms make us unaware of this motion). This saccade-induced ERP depends on the nature of the stimuli that are visible when the eyes move, just as the ERP elicited by a moving stimulus varies as a function of the nature of the stimulus. Procedures that attempt to correct for the EOG voltages produced by eye movements—which are discussed at the end of this chapter—cannot correct for these saccade-induced ERP responses.

Because of the approximately linear relationship between the size of an eye movement and the magnitude of the corresponding EOG deflection, large eye movements are relatively easy to detect on single trials, but small eye movements are difficult to detect. If a simple voltage threshold is used to detect and reject eye movement artifacts, and a typical threshold of 100 µV is used, eye movements as large as 10° can escape detection (e.g., if the voltage starts at –80 µV, a 10° eye movement in the appropriate direction will cause a transition to +80 µV, which would be entirely within the allowable window of ±100 µV). Of course, a 10° eye movement greatly changes the position of the stimulus on the retina, which can be an important confound, and the resulting voltage deflection is quite large relative to the size of a typical ERP component, even at scalp sites fairly far from the eyes.

The best method for detecting small eye movements that I have seen uses the *step function* shown in figure 6.4 (see box 6.2 for the story of how this method was originally developed). In general, a step function is a flat period of one voltage level followed immediately by another flat period at a lower or higher voltage level. The step function is like the moving window peak-to-peak amplitude algorithm, insofar as a window (e.g., 200 ms) is slid along the data and something is computed within each window. Instead of finding the peak-to-peak amplitude, the step function finds the difference in mean amplitude between the first half of the window and the second half of the window (e.g., between the first 100 ms and the last 100 ms of a 200-ms window). This is an excellent method for finding artifacts that consist of changes in amplitude between adjacent time periods. As you can see from the middle HEOG waveform in figure 6.4,

Box 6.2
The Step Function

When I was in graduate school, I ran lots of N2pc experiments, in which eye movements can be a major headache, and I needed to develop a better way of detecting trials that contained small horizontal eye movements. As figure 6.4 shows, saccadic eye movements have a distinctive shape in HEOG recordings, with a sudden transition from one relatively flat voltage level to a different relatively flat voltage level. That is, they have a step-like shape. I thought I could detect these step-like transitions by looking at the correlation between the HEOG waveform and an actual step function (a 100-ms series of −1 values followed by a 100-ms series of +1 values, as shown in figure 6.4). More precisely, I computed the *cross-correlation function* between the step function and the HEOG waveform. I started by lining up the step function with the initial 200 ms of the HEOG waveform and calculating the correlation (Pearson's *r*) between the step function and this period of the HEOG waveform. I then temporally shifted the step function by one sample period relative to the HEOG waveform and then calculated the correlation again. I did this over and over again, shifting the step function by one sample point each time. This gave me a set of correlations, one for each 200 ms in the HEOG waveform. I then took the highest correlation value and compared that with a threshold to determine whether the trial should be rejected. I reasoned that a high correlation between the step function and any 200-ms section of the HEOG waveform would mean that an eye movement likely occurred in that portion of the HEOG.

When I finally got the program to work, I tested it and found that it worked fairly well but frequently identified sections of HEOG that were fairly flat. When I took a closer look, I saw that these sections did have a step-like shape, but that it was extremely small. In other words, using the correlation worked well for finding the right shape, but it was completely insensitive to the size of the HEOG deflection. I then remembered an important fact that I had read in one of Manny Donchin's papers years before; namely, that you can use covariance rather than correlation when you care about the magnitude of an effect. I changed my program to calculate the covariance between the step function and the HEOG rather than the correlation, and it worked great!

The only problem was that it was very slow. I spent several hours trying to optimize the program, and then I had a realization: The covariance between a step function and some other function is equal to the difference in mean amplitude between the first and second halves of the period of the step function. That is, the covariance for a 200-ms period is simply the difference between the mean amplitude in the first 100 ms of this period and the mean amplitude in the second half of this period. When I rewrote my program to do it this way, I got exactly the same result in 10% of the time.

Now you know how I spent my time in San Diego while everyone else was out surfing.

a step function matches the shape of a saccade-related HEOG deflection very well. It also works well for detecting blinks. This measure is effective for two reasons. First, averaging together the voltage over each 100-ms half of the window filters out any high-frequency noise. Second, computing the difference between successive 100-ms intervals minimizes the effects of any gradual changes in voltage, which corresponds with the fact that a saccadic eye movement produces a sudden voltage change.

The step function can easily detect eye movements of 2° or larger, and it can work reasonably well to detect eye movements as small as 1° under optimal conditions. If even smaller eye movements are a concern in your research (e.g., if you are measuring the N2pc or CDA components), see box 6.3 for additional advice.

Box 6.3
Detecting Small but Consistent Eye Movements

Eye movements are not a big issue for most researchers. For people who study covert shifts of spatial attention, however, eye movements are a major problem. First, when the eyes rotate laterally, the EOG produces a negative voltage over the contralateral hemisphere that contaminates the N2pc and CDA components. Second, if the eyes move prior to the appearance of a lateralized stimulus, this changes the sensory input produced by the stimulus. Even a small eye movement that is consistently directed to a specific location can create a substantial EOG artifact and a meaningful change in stimulus location.

Using a step function, it is often possible to detect eye movements as small as 1° (a 16-μV step) on individual trials, but the SNR of the EOG signal is not good enough to detect smaller eye movements without an unacceptably large number of false alarms. However, it is sometimes possible to use averaged EOG waveforms to demonstrate that a given set of ERPs is uncontaminated by very small systematic eye movements. Specifically, if different trial types would be expected to elicit eye movements in different directions, virtually unlimited resolution can be obtained by averaging together multiple trials on which the eye movements would be expected to be similar. For example, if an experiment contains some targets in the left visual field (LVF) and other targets in the right visual field (RVF), one can compute separate averaged EOG waveforms for the LVF and RVF targets and compare these waveforms. The same thing can be done in a cuing paradigm by averaging left-cue trials with right-cue trials (time-locked to the cue). Any consistent differential eye movements will lead to differences in the averaged EOG waveforms, and even very small eye movements can be observed (if they are present on a large proportion of trials) due to the improvement in the SNR produced by the averaging process.

This procedure will not allow individual trials to be rejected, nor will it be useful for detecting eye movements that are infrequent or not consistently related to the side that contains the target. However, it can be useful when combined with the rejection of individual trials with large eye movements in a two-tiered procedure. The first tier consists of the rejection of individual trials with large saccades (>1°) by means of the step function. Residual EOG activity can then be examined in the averaged EOG waveforms, and any subjects with differential EOG activity exceeding some criterion (e.g., 1.6 μV, corresponding to 0.1°) can be excluded from the final data set (see, e.g., Woodman & Luck, 2003b).

Eye movements are caused by contraction of the extraocular muscles. This contraction mainly occurs at the onset of the eye movement, and very little muscle activity is needed to maintain the eye position once the saccade has ended. Consequently, a spike of activity from the extraocular muscles (the *spike potential*) is often seen at the onset of each eye movement. You can see this at the beginning and end of each boxcar-shaped deflection in figure 6.4. This artifact is not ordinarily a significant problem, but it can create the appearance of gamma-frequency oscillations in time–frequency analyses (Yuval-Greenberg, Tomer, Keren, Nelken, & Deouell, 2008). This general issue will be discussed more in online chapter 12.

It should be noted that the techniques described above are useful for detecting saccades but are not usually appropriate for detecting slow shifts in eye position or for assessing absolute eye position. To assess these, it is usually necessary to use a video-based eye tracker.

Skin Potentials and Other Slow Voltage Shifts

As described in chapter 5, skin potentials arise when sweat begins to accumulate in sweat glands, changing the impedance of the skin and therefore causing a change in the standing electrical potential of the skin over a period of many seconds. Some examples can be seen in figure 6.5.

Voltage shifts can also be caused by slight changes in electrode position, which are usually the result of movements by the subject. A change in electrode position will often lead to a change in impedance, thus causing a sustained shift in voltage. This type of artifact can be reduced by making sure that the subjects are comfortable and do not move much. If electrodes are placed at occipital sites, the subjects should not lean the back of the head against the back of the chair.

If the voltage shifts are small, slow, and random, they shouldn't distort the averaged ERPs very much, and there is no need to reject the trials. As discussed in chapter 7, these slow potentials can be minimized by high-pass filters. However, a movement in the electrodes will sometimes cause the voltage to change suddenly to a new level, and this can be detected by means of the moving window peak-to-peak amplitude method or a step function method (you'll want to keep the threshold fairly high to avoid rejecting trials with large ERP deflections, like a large P3 wave).

Amplifier and ADC Saturation/Blocking

Slow voltage shifts may sometimes cause the amplifier or ADC to saturate, which causes the EEG to be flat for some period of time (this is also called *blocking*). If this happens frequently, you should use a lower gain on the amplifier. This should happen rarely or never in systems with 24 or more bits of resolution (see chapter 5).

As illustrated in figure 6.5, amplifier blocking is relatively easy to spot visually because the EEG literally becomes a flat line. You could reject trials with amplifier saturation by finding trials in which the voltage exceeds some value that is just below the amplifier's saturation point, but in practice this would be difficult because the saturation point may vary from channel to channel and may even vary over time. Another possibility would be to determine if there are a large number of points with identical voltages within each trial, but this isn't quite optimal

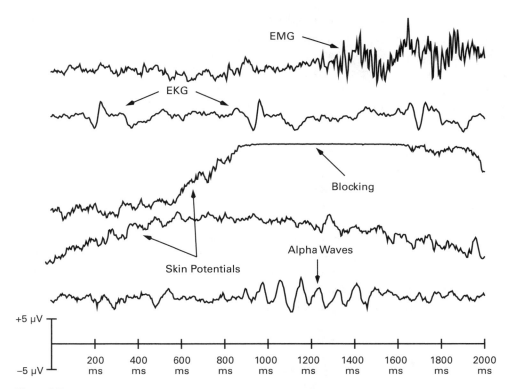

Figure 6.5
Examples of artifacts on different trials in EEG recordings. The EMG, blocking, and skin potential artifacts were recorded at Cz with a right mastoid reference. The EKG artifacts were recorded at the left mastoid with a right mastoid reference. The alpha waves were recorded at O2 with a right mastoid reference.

because the voltages might not be exactly the same from moment to moment. A better procedure is to use an algorithm that I call the *X-within-Y-of-peak* method, which was developed by Jon Hansen at UCSD. This method first finds the maximum EEG value within a trial (the peak) and then counts the number of points that are at or near that maximum. X is the number of points, and Y defines how close a value must be to the peak to be counted. For example, you might want to reject any trial in which 30 or more points are within 0.1 μV of the peak (i.e., X = 30 and Y = 0.1 μV). Of course, the same function must be applied for both the positive peak voltage and the negative peak voltage, and it should be applied to every channel if your system is prone to blocking. The same method can be used to detect other causes of flat-lined data, such as when an electrode becomes intermittently disconnected.

Alpha Waves   Alpha waves are EEG oscillations at approximately 10 Hz that are typically largest at posterior electrode sites and occur most frequently when subjects are tired or have their eyes closed (see the bottom waveform in figure 6.5). The best way to reduce alpha waves is to use well-rested subjects and give them interesting tasks to perform, but some individuals have substantial alpha waves even when they are fully alert. Alpha waves can be particularly problematic when a constant stimulus rate is used, because the alpha rhythm can become entrained to the stimulation rate such that the alpha waves are not reduced by the averaging process. Thus, it is useful to include a jitter of at least ±50 ms in the intertrial interval (as discussed in chapter 4).

It is not usually worthwhile to reject trials with alpha waves because you will typically end up rejecting almost all trials in some subjects and few or no trials in other subjects. Moreover, alpha oscillations are not always noise, and they may contribute to ERP effects in an important way (see Mazaheri & Jensen, 2008; Bastiaansen, Mazaheri, & Jensen, 2012; van Dijk, van der Werf, Mazaheri, Medendorp, & Jensen, 2010).

Figure 6.6 shows what alpha looked like after averaging across many trials in a subject who had a very large alpha artifact (subject 1 in the figure). As is common, the alpha in this subject was large during the intertrial interval (and therefore during the prestimulus baseline) and became temporarily suppressed within 200 ms of stimulus onset. In addition, the alpha was largest at the back of the head and was quite small at the Fp1 and Fp2 electrode sites. This is a fairly extreme case (the subject with the largest alpha from an experiment with 16 subjects).

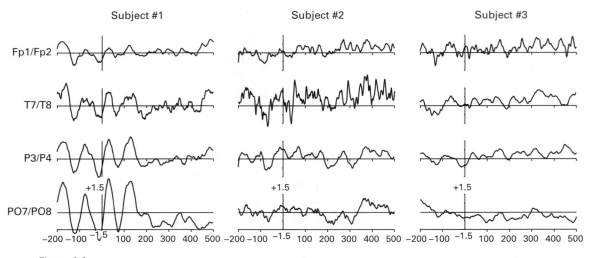

Figure 6.6
Examples of artifacts that are still visible after averaging in three electrode sites for three different subjects. These are contralateral-minus-ipsilateral difference waves in an N2pc experiment (data courtesy of Emily Kappenman). You can see alpha activity in the posterior electrodes in subject 1, EMG activity at the T7/T8 electrodes in subject 2, and EMG activity at the Fp1/Fp2 electrodes in subject 3.

Muscle and Heart Activity

The electrical potential created during the contraction of a muscle is called the electromyogram, or EMG (see top waveform in figure 6.5). The EMG typically consists of rapid voltage fluctuations, and much of the EMG can be eliminated by using a low-pass filter with a half-amplitude cutoff somewhere between 30 and 100 Hz. The best way to reduce EMG is to have subjects relax the relevant muscles. You therefore need to know which muscles cause EMG artifacts at which electrode sites.

The temporalis muscles are powerful muscles that we use to contract our jaws, and they are located right under the T7 and T8 electrodes (see figure 6.7). If you see a lot of EMG in this region of the head, you can ask the subject to relax his or her jaw and avoid teeth-clenching. The temporalis muscles are so large, however, that you will see small but consistent high-frequency EMG artifact at T7 and T8 throughout the session in some subjects, even when they try to relax. Subject 2 in figure 6.6 shows this pattern, with high-frequency noise localized to the T7 and T8 electrodes. And if a subject is chewing on something, you will see enormous bursts of EMG activity at T7, T8, and surrounding electrodes.

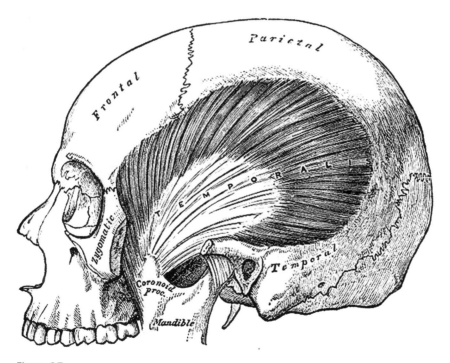

Figure 6.7
Temporalis muscles. From Gray, H. (1918). *Anatomy of the Human Body*, 20th ed. Philadelphia: Lea & Febiger (now in the public domain).

The muscles of the forehead are also a common source of EMG noise, largest over anterior electrode sites. These muscles can be activated when the electrode cap pulls at the forehead or if the subject is furrowing his or her brow in concentration. Subject 3 in figure 6.6 shows this pattern, with high-frequency noise localized to the Fp1 and Fp2 electrodes. EMG from the forehead muscles can usually be minimized by asking the subject to relax these muscles.

The muscles of the neck are the remaining common source of EMG noise. If a mastoid reference is used, this activity may be picked up by the reference electrode and therefore appears in all channels that use this reference. If a different reference site is used, EMG noise arising from the neck appears at the most inferior occipital and temporal electrode sites. It can usually be minimized by asking the subject to sit straight upright rather than leaning the head forward. Neck EMG can also be minimized by having the subject sit back in a recliner with the head leaning against the recliner, but this can cause artifacts in the occipital electrodes.

It is not usually necessary to reject trials with EMG, assuming that appropriate precautions have been taken to minimize the EMG. However, if it is necessary to reject trials with EMG activity, EMG can be detected by calculating the amount of high-frequency power in the signal (e.g., power above 100 Hz).

It should also be noted that some stimuli will elicit reflexive muscle twitches, and these are particularly problematic because they are time-locked to the stimulus and are therefore not attenuated by the averaging process. These also tend to be sudden, high-frequency voltage changes, but they are usually limited to a very short time period and are therefore difficult to detect by examining the high-frequency power across the entire trial. To reject these artifacts, it is best to look for sudden shifts in voltage during the brief time period during which they are likely to occur (usually within 100 ms of stimulus onset).

The beating of the heart (the electrocardiogram, or EKG) can also be observed in EEG recordings in some subjects, and its distinctive shape is shown in figure 6.5. The EKG propagates to the head through the carotid arteries and is usually picked up by mastoid electrodes. If a mastoid is used as a reference, the EKG is seen in inverted form in all of the electrode sites. The EKG can sometimes be reduced by slightly shifting the position of the mastoid electrode, but usually there is nothing that can be done about it. In addition, this artifact usually occurs approximately once per second during the entire recording session, so rejecting trials with EKG deflections will usually lead to the rejection of an unacceptably large proportion of trials. Fortunately, this artifact is almost never systematic, and it will simply decrease the overall SNR. Because you can't realistically reject trials with EKG artifacts, artifact correction is usually the best approach if the EKG is causing serious problems with your data.

The previous paragraph raises an important point: If you see an artifact or some type of noise equally in all of your EEG channels, it is probably being picked up by the reference electrode. Most artifacts and noise sources will be more prominent at some electrodes than at others, but any signals picked up by the reference electrode will appear in inverted form in all electrodes that use that reference. However, if you are using bipolar recordings for some of your channels (e.g., for EOG recordings), these recordings will not have artifacts or noise arising from the main reference electrode. This can help you identify and eliminate the sources of noise and artifacts.

Speech-Related Artifacts

I had always heard that you cannot record clean EEG data while the subject is speaking, but I didn't know why. I thought it must be a result of the muscle activity caused by moving the mouth. Several years ago, Emily Kappenman and I designed an experiment in which we hoped to use a verbal response. To minimize mouth movement, we used /d/ and /t/ as the verbal responses. The EEG was hopelessly contaminated with weird-looking artifacts. When Emily presented the data in a psychology department brown-bag seminar, one of the language-oriented graduate students explained why our data were so noisy. It turns out that there is a strong electrical gradient between the base of the tongue and the tip of the tongue. Consequently, when the tongue moves up and down in the mouth, it creates large voltages that propagate to the surface of the head. These voltages are called *glossokinetic artifacts*. The /d/ and /t/ sounds involve substantial tongue movements, and this is why we were seeing large artifacts.

This places rather unfortunate limitations on language production studies. It is possible to use verbal word-speaking paradigms if the analyses focus on the waveform prior to the spoken output, but it is not usually practical to look at the ERPs after speech onset. It seems plausible that artifact correction could remove this artifact, but I haven't yet seen anyone do it.

Sporadic Artifacts of Unknown Origin

From time to time, you will see periods of "crazy" voltage fluctuations. This often happens at the end of a break when you tell the subject that you are about to start a trial block. You then see all kinds of large voltage deflections in many channels for a few seconds, and then all the channels return to normal. This also happens occasionally in the middle of a block, usually for just a few seconds. In most cases, you will have no idea why the artifact occurred, and it may be from a combination of things (e.g., a yawn and a stretch followed by a clearing of the throat).

In my lab, we call these *commonly recorded artifactual potentials* (C.R.A.P.; see box 6.4). They are easy to reject with almost any algorithm.

Some Practical Advice about Artifact Rejection

Before transitioning to artifact correction, I'd like to provide some very concrete advice about performing artifact rejection.

First, as described earlier in the chapter, I recommend setting the artifact rejection parameters individually for each subject on the basis of visual inspection of the EEG and adjusting the parameters as necessary to optimize rejection. Once you have become practiced at doing this, it will not take much time, and it will be very worthwhile. If your experiments involve between-group comparisons, however, you need to make sure that the parameters are set by someone who is blind to group membership.

Box 6.4
Commonly Recorded Artifactual Potentials

When describing miscellaneous artifacts to students many years ago, I would refer to them as "crap" in the data. One day, I was giving a guest lecture in a grad course on ERPs, and I said something like "*crap* is a technical term" (just to be funny). I then suggested that someone needed to figure out how C.R.A.P. could be an acronym for some reasonable phrase. Tamara Swaab, who was teaching the class, came up with "commonly recorded artifactual potentials." I thought this was great, and now I regularly use this acronym. My lab integrated this into the design for the T-shirt we created for the very first UC-Davis ERP Boot Camp.

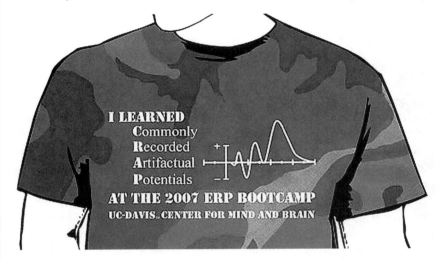

Second, look at the averaged ERPs to determine whether the rejection is working (especially with blinks and eye movements, as described earlier). It is also useful to create averages that include only the trials with artifacts. This will allow you to see how the artifacts differ across conditions, which may give you hints about how artifacts that escaped detection may be affecting your data.

Third, don't worry that you are reducing your SNR by rejecting trials with artifacts unless you are throwing out more than about 20% of your trials. As was discussed in chapter 5, the square root law means that a 20% reduction in the number of trials leads to only an 11% reduction in the SNR.

Fourth, test for specific artifacts in the channels in which the artifacts are most easily observed (e.g., test for blinks in the VEOG channel and test for eye movements in the HEOG channel). In addition, test for C.R.A.P. in all channels by setting some reasonably high rejection criterion (e.g., a 200-μV threshold for the moving window peak-to-peak test).

Fifth, keep track of the percentage of trials that was rejected for each subject, and report this in your journal articles. I like to give the average percentage across subjects and the range (separately for each group, when appropriate). In most cases, you can collapse across conditions when giving these percentages. In experiments where the number of remaining trials may be very small (e.g., studies of infant cognition), you may want to report this information separately for each condition.

Finally, you will probably want to completely exclude subjects for whom a large percentage of trials has been rejected, but you should do this in a way that avoids biasing your results. For example, you will occasionally have a subject who blinks on almost every trial. When this happens in my lab, I tell my students to send the subject home before the experiment is complete, and then the subject is automatically excluded from the final data set. However, if the blinks are happening during the intertrial interval but end up contaminating the baseline, you might still end up rejecting 90% of trials. Obviously you don't want to include a subject like this in your final data analyses. But what if you reject 80% of trials? Or 60%? Or 30%? Or 5%?

Imagine the following scenario. You spend 18 months collecting data for a very important experiment. You do all of the data processing and finally run your statistical analysis. Your conclusions (and your future in academia) hinge on getting a significant three-way interaction. You nervously scan through the output of your statistics program, and your heart sinks when you see that the p value for this three-way interaction is the most hated number in all of science, namely 0.06. What do you do? You might take a look at your single-subject ERPs to see if something went wrong with one of the subjects. Imagine that you notice that one subject has noisy and strange-looking waveforms. You then realize that this subject had a very large number of blinks and that 48% of trials were rejected. Obviously, this subject should be excluded from the statistical analyses. You then repeat your statistical analysis, and now the p value for your three-way interaction is 0.04. A big smile crosses your face. You write a paper describing your results, noting that you excluded one subject who had artifacts on 48% of trials. Your paper gets published, because reviewers agree that a subject with artifacts on 48% of trials should be excluded, and you get a promotion and a big new grant.

This is a realistic example of how some subjects end up being excluded. It falls within the norms of the field, and it is not a case of ethical misconduct. In some sense, it is obvious that you should exclude a subject with artifacts on 48% of trials. The problem is that the exclusion of the subject was a result of looking at the overall results of the experiment (i.e., it was post hoc). If you exclude subjects only when you find that your effects are not significant, you will produce a systematic bias in your results and end up with more than a 5% false positive rate (i.e., you will reject the null hypothesis when it is true more than 5% of the time). This is obviously a problem, and it does not serve the goal of finding the truth. A p value should reflect the probability of a Type I error (a false positive), and anything that makes the p values better (smaller) than the actual probability of a Type I error is a misuse of statistics (see box 6.5 for a brief rant about the exclusion of "outlier" subjects).

Box 6.5
Outliers

Many years ago, I reviewed a journal submission in which the authors were looking at the correlation between two variables. I have no memory of what the variables were, so let's just call them X and Y. The correlation between these variables was very important for the conclusions of the paper, but it was not significant unless one "outlier" subject was removed. This outlier subject had X and Y values that were within the typical X range and the typical Y range, but the X,Y combination was quite different from the rest of the subjects. In other words, the subject was an "outlier in multidimensional space." The authors computed some statistic indicating that this X,Y combination was truly beyond the range one would typically expect (e.g., the analog of something like 2 standard deviations away from the mean). Of course, in a truly random sample from an infinite distribution, there is always the chance of obtaining an outlier like this.

The problem with removing an outlier like this is that, in practice, it biases the results and increases the probability of a Type I error (i.e., rejecting the null hypothesis when the null hypothesis is true). I say this with considerable certainty because I have been reading journal articles for about 30 years, and I have never EVER seen a study in which the authors looked at the data and then decided to remove an outlier that would have helped their data if they had kept it. That is, I have never seen a case where the effect was significant (in the predicted direction) when all the data were included, and yet the authors decided to remove an outlier even though the effect was no longer significant with the outlier excluded.

This can and does happen when a criterion for excluding subjects is established prior to the experiment and carried out automatically. However, I have never seen someone do this post hoc. In contrast, I have seen many cases of people excluding outliers that hurt their effects post hoc.

There is growing awareness that practices such as post hoc outlier removal can dramatically increase the rate of false positives in psychology, neuroscience, and related areas (see, e.g., Simmons, Nelson, & Simonsohn, 2011). You can help solve this problem by not letting people get away with this kind of post hoc outlier removal (and, of course, I'm sure you will never do it yourself).

To avoid this problem, I strongly recommend that you adopt a criterion for excluding subjects that you apply to all of your experiments in an *a priori* manner. In my lab's basic science experiments with college students, for example, we automatically exclude any subject for whom more than 25% of trials were rejected, and we automatically include any subject for whom 25% or less of trials were rejected (collapsed across all trial types). We do this irrespective of the way the waveforms look and irrespective of the outcome of the statistical analyses. In our experiments on schizophrenia, we see a lot more artifacts (in both the patients and the control subjects), so we exclude subjects for whom more than 50% of trials were rejected. By adopting a preset criterion, we are just as likely to exclude subjects who "help" or "hurt" our effects, so excluding subjects does not bias us toward obtaining significant results. I encourage you to adopt a standard criterion of this nature and to report it in the methods sections of your papers (e.g., "We always exclude subjects for whom more than 25% of trials were rejected because of artifacts; three subjects were excluded for this reason in the present study").

Basics of Artifact Correction

In this section, I will describe the reasons why it is sometimes better to subtract away the voltages produced by artifacts (artifact correction) rather than to reject trials with artifacts. I will then describe the rationale behind three major classes of artifact correction approaches. Finally, I will discuss some of the potential pitfalls with artifact correction. An online supplement to this chapter provides a detailed description of one specific method, based on independent component analysis.

Why Artifact Correction May Be Useful

Artifact rejection is a relatively crude process because it completely eliminates a subset of trials from the ERP averages. As discussed by Gratton, Coles, and Donchin (1983), there are three potential problems associated with rejecting trials with ocular artifacts. First, in some cases, discarding trials with eye blinks and eye movements might lead to an unrepresentative sample of trials (e.g., you might end up including only trials on which subjects are in a particularly alert state). Second, there are some groups of subjects (e.g., children and psychiatric patients) who cannot easily control their blinks and eye movements, making it difficult to obtain a sufficient number of artifact-free trials. Third, there are some experimental paradigms in which blinks and eye movements are integral to the task, and rejecting trials with these artifacts would be counterproductive. Under these conditions, it would be useful to be able to subtract away the voltages due to eye blinks and eye movements rather than to reject trials with these artifacts.

Ochoa and Polich (2000) identified a related problem: If you instruct subjects not to blink while they are performing a task, you are essentially giving them two tasks to perform at the same time (the task you are explicitly studying and the task of suppressing blinks). If you have ever been a subject in an ERP task in which you were asked to minimize blinking, you know that this takes considerable mental effort. My own experience is that trying not to blink makes me focus attention on the sensations arising from my eyes, which makes me want to blink more rather than less. Ochoa and Polich directly tested the possibility of dual-task interference by giving subjects an oddball task and either telling them to avoid blinking or saying nothing at all about blinking. The P3 had a smaller amplitude and a longer latency when subjects were asked to avoid blinking. The subjects in this experiment were college students, and it seems likely that an even larger effect would be observed in subjects who have lower levels of overall cognitive ability. For example, any differences in P3 between a patient and a control group could reflect the interactive effect of the disorder and the instruction to minimize blinking.

It can be even more difficult for subjects to control their eye movements. Subjects don't have very direct information about where their eyes are pointing and when their eyes move, and this makes it difficult for them to avoid making eye movements. In some cases, it is useful to train subjects to maintain fixation, which can be done with a very simple task that does not involve an eye tracker (Guzman-Martinez, Leung, Franconeri, Grabowecky, & Suzuki, 2009).

General Approaches to Artifact Correction

To deal with these issues, various procedures have been developed to estimate the artifactual potentials generated by blinks and eye movements and to subtract them from the EEG. Some of these procedures work for other types of artifacts as well. Artifact correction procedures fall into three categories: (1) regression-based procedures, (2) dipole localization procedures, (3) statistical component isolation procedures. Before you read about these procedures, you might want to take a look at box 6.6, which provides some general reasons to be cautious about artifact correction.

Gratton et al. (1983) developed a regression-based procedure to estimate the artifactual potentials generated by blinks and eye movements and to subtract them from the EEG. The basic idea is that an artifact generated in the eyes propagates to each scalp site in a very predictable and

Box 6.6
Why Artifact Correction Is Scary

I find artifact correction to be a bit scary. Any procedure that subtracts an estimated value from my data makes me nervous. For example, baseline correction and re-referencing are extremely simple procedures that merely subtract a constant value from the entire waveform, and this book contains many, many pages describing the problems that can result from these subtractions (see chapters 5 and 8). Artifact correction procedures are much more complicated than baseline correction and re-referencing procedures, making it more difficult to anticipate exactly how they might distort the ERP waveforms. I am therefore concerned that the cure might be worse than the disease.

Moreover, I have been unimpressed by the evidence that has been provided about the accuracy of artifact correction procedures. Mathematical justifications are unsatisfying because it is hard to know how well the assumptions of the procedure are met by real data. Many attempts at validation simply apply the procedure to existing data and show that the results look sensible. However, these studies do not have a *ground truth* that can be used to assess the accuracy of the correction. In other words, there is no way to know what the true, artifact-free waveforms should look like in these studies, and consequently there is no way to tell how much distortion is produced by the correction. The best studies use real EEG data but add experimenter-created artifact waveforms to these data. This makes it possible to see how well the known artifact is removed (for a really great example of this approach in a different domain, see Kiesel, Miller, Jolicoeur, & Brisson, 2008).

I haven't been fully convinced by any of the validation studies of artifact correction that I've read (although it's possible that I missed some). One problem is generality: The fact that a procedure works with one set of subjects, conditions, and artifacts doesn't mean that it will work with other subjects, conditions, and artifacts. In particular, there are some situations that might be particularly difficult to correct, and most studies don't try to "break" the procedure by testing these challenging situations. In many cases, the studies were conducted by the people who developed the procedure, and so it is natural that they did not try to "break" the procedure. However, I am naturally skeptical about such studies, and I much prefer studies that were conducted by independent third parties. The bottom line is this: Take a skeptical perspective when reading studies claiming to provide evidence of the validity of an artifact correction procedure, and don't assume that their results will generalize to your studies unless their subjects, conditions, and artifacts are similar to yours.

quantifiable manner. Consequently, the artifact-related voltage recorded at a given electrode site will be equal to the size of the artifact recorded at the eyes multiplied by a propagation factor. To correct for eye artifacts, therefore, you can simply estimate the propagation factor between the eyes and each of the scalp electrodes and subtract a corresponding proportion of the recorded EOG activity at the eyes from the ERP waveform at each scalp site. For example, Lins et al. (1993a) found that 47% of the voltage present in a bipolar EOG recording propagated to the Fpz electrode, 18% to the Fz electrode, and 8% to the Cz electrode. To subtract away the EOG contribution to the averaged ERP waveforms at these electrode sites, it would be possible to subtract 47% of the EOG waveform from the Fpz electrode, 18% from the Fz electrode, and 8% from the Cz electrode.

Although the development of this procedure was an important step forward, it has a significant shortcoming. Specifically, the EOG recording contains brain activity in addition to true ocular activity, and, as a result, the subtraction procedure ends up subtracting away part of the brain's response as well as the ocular artifacts (see, e.g., Lins, Picton, Berg, & Scherg, 1993b; Plochl et al., 2012). Consequently, I recommend against using the Gratton et al. (1983) method or similar regression-based methods. Thirty years have passed since the introduction of this method, and better techniques are now widely available.

A second approach uses dipole modeling to create a more detailed biophysical model of the artifact and its propagation through the head (Berg & Scherg, 1991a, 1991b). Although I am not usually a big fan of dipole modeling (see online chapter 14), the locations of the ocular dipoles are already known, eliminating much of the uncertainty that is ordinarily involved in dipole modeling. However, the accurate use of this approach may require considerable effort. For example, Lins et al. (1993b) recommended that recordings should be obtained from at least seven electrodes near the eyes, and a set of calibration runs must be conducted for each subject. In addition, this approach typically assumes that a vertical eye movement and the effect of the eyelid passing over the eye have the same scalp distribution, which is not true (see, e.g., Plochl et al., 2012).

A third approach uses the statistical properties of the data to identify a set of components, each of which is characterized by a scalp distribution, and then uses these components to isolate and then subtract the artifact-related voltages. Different statistical methods can be used estimate these components, including principal component analysis (PCA), independent component analysis (ICA), and second-order blind inference (SOBI).

These methods assume that each artifact has a fixed scalp distribution in a given subject. For example, eyelid closure would have one scalp distribution, an eye movement would have another scalp distribution, the EKG would have yet another scalp distribution, and so forth. For many artifacts, this is a valid assumption. The scalp distribution of the EEG at any moment is assumed to consist of the weighted sum of the scalp distribution of the artifact plus the scalp distributions of all the other brain signals. By examining the relationships of the voltages at each electrode site at each time point, these techniques are able to find a small number of scalp distributions that, when summed together, can account for the EEG at each time point. ICA has become the most commonly used of these statistical methods, largely because it is widely available and relatively easy to use in the free EEGLAB Toolbox package (Delorme & Makeig, 2004). You can find a more detailed description of how ICA-based artifact rejection works, along with an example from a real experiment, in the online supplement to chapter 6.

These methods are not limited to ocular artifacts. They can work with any artifact that has a consistent scalp distribution, including EKG artifacts and induced line-frequency noise from nearby electrical devices. However, they will not work with artifacts that have a variable scalp distribution (as may be the case with skin potentials and movement artifacts).

Potential Pitfalls and General Advice

No matter what method you use, you should be aware of two potential pitfalls with artifact correction.

First, no artifact correction technique has been demonstrated to work perfectly in all situations. I read several reviews and validation studies to prepare for writing this section, and I found evidence for and against virtually every major artifact correction approach. Different methods appear to work best in different situations. If you decide to use artifact correction in your own research, you should maintain an appropriate level of skepticism about whatever method you use, and you should take steps to assess the validity of the results.

One way to do this is to compare artifact correction with artifact rejection. If your data look similar with both approaches—although noisier with rejection owing to the smaller number of trials—then the correction procedure probably didn't distort your data very much. Carly Leonard, a postdoc in my lab, took this approach the first time we tried using ICA to correct blink and eye movement artifacts. She found comparable results with correction and rejection, although the data were noisier with rejection, and she said so in the methods section of the paper (Leonard et al., 2012). If your results look different between correction and rejection, this may mean that the correction is distorting the data. Alternatively, it could mean that the brain was working somewhat differently on trials with versus without artifacts, and removing trials with artifacts therefore led to an incomplete view of what the brain actually does. If you find yourself in this situation, you will need to think carefully about how to proceed. There is a nice saying in Spanish for difficult situations like this, in which you might need some divine intervention: *Vaya con Dios*.

A second potential pitfall is that artifact correction, even if it works perfectly, cannot account for the changes in sensory input caused by blinks and eye movements. For example, if a subject blinks or makes an eye movement at the time of a visual stimulus, then this stimulus may not be seen properly, and this cannot be accounted for by artifact correction techniques. It should be obvious that closing or moving the eyes will change the processing of a stimulus. An experiment has actually quantified this, showing that reaction times were increased by approximately 200 ms when a blink occurred at the time of stimulus onset (Johns, Crowley, Chapman, Tucker, & Hocking, 2009). Stimulus duration was 400 ms, so the subjects were able to detect the stimuli even when they blinked. If the stimulus duration had been shorter, however, subjects presumably would have completely missed the stimuli.

The obvious solution to this problem is to reject trials with blinks or eye movements that occur within a few hundred milliseconds of stimulus onset prior to performing artifact correction. An alternative solution would be to demonstrate that very few blinks or eye movements actually occurred during this time period. Although these solutions seem obvious, I have seen very few published studies that addressed the problem of blinks changing the sensory input. This may partly reflect the fact that it is not always easy to do both correction and rejection in commercial ERP analysis software. When we were designing ERPLAB Toolbox (which relies on EEGLAB for ICA-based artifact correction), one of our goals was to make this possible.

Should you use artifact correction, or should you just stick with the old-fashioned rejection approach? In the first edition of this book, I said the following: "I would recommend against using artifact correction procedures unless the nature of the experiment or subjects makes artifact rejection impossible." Since that time, my lab has used ICA-based correction in a number of experiments, and we have looked very closely at the results for evidence of problems. As a result, my advice has now changed. I now think it's reasonable to use one of the newer correction techniques to correct for blinks in almost any experiment. Blinks are very large, which tends to make these methods work well, and I have seen no evidence of substantial distortions of the data with ICA-based blink correction.

There are three situations in which I would recommend caution with blink correction. The first is when your experimental effect consists of a subtle effect with a blink-like scalp distribution and a relatively long duration (>=200 ms). A subtle distortion from the blink correction algorithm could be problematic and difficult to detect in this situation. The second is when the blinks are highly consistent in their timing (e.g., when the subject blinks at the time of the response or in reaction to an intense stimulus); the statistical methods for finding blink-related activity may lump brain activity with the blinks in this situation. If you are in either of these situations, you might try using correction but testing it very carefully for signs that it is distorting your data (e.g., by comparing rejection and correction).

The third and more common situation in which caution is advised for blink correction is when the blinks differ significantly across groups or conditions, especially during time periods when significant ERP effects are found. How will you know if you are in this situation? One way to determine this is to do the opposite of artifact correction: Instead of removing the component(s) associated with the blink, you can remove all of the other components. If you then average the data, the resulting waveform will show you the artifact-related activity in each group and condition. This seems like something that should be done for every experiment that uses artifact correction. If you find that the artifacts do differ across groups or conditions, you can look at the data in greater detail and determine whether the major effects in your experiments could be explained by inaccuracy in the artifact correction (*Vaya con Dios, amigo!*).

I am less enthusiastic about using correction for other types of artifacts. For example, we have found that ICA works imperfectly for removing eye movement artifacts, as described in the online supplement to this chapter. In Carly Leonard's study, which I mentioned earlier (Leonard et al., 2012), we ended up using ICA to remove eye movement artifacts despite the fact that it worked imperfectly. It was "good enough," meaning that several aspects of the results convinced us that the imperfections did not impact our conclusions (e.g., the time course of the eye movements was different from the time course of the ERP effects). But it would not have been good enough if we had been looking for more subtle effects or if we needed precise scalp distribution information. If you use correction for eye movements, make sure that you have a way to determine whether it might be leading you to incorrect conclusions. Plochl et al. (2012) found that ICA worked well for correcting eye movements, but I suspect this reflects the particular nature of their experiments, in which the task involved making large eye movements to a small number of locations. In addition, the nature of the study (which involved real eye movements rather than simulated data) made it difficult to quantify how well the correction worked.

There is controversy about whether these approaches are adequate for removing EMG artifacts (see, e.g., McMenamin, Shackman, Maxwell, Greischar, & Davidson, 2009; McMenamin et al., 2010; Olbrich, Jodicke, Sander, Himmerich, & Hegerl, 2011). The problem may be that the scalp distribution is not sufficiently constant, especially if multiple muscle groups are involved. Similarly, skin potentials may not have a consistent scalp distribution, making them difficult to correct with these methods. I would expect that line-frequency noise and EKG artifacts would be relatively easy to correct, and previous studies have provided some evidence (Jung et al., 2000; Ille, Berg, & Scherg, 2002).

I have sometimes seen people remove dozens of different artifact-related components from the EEG, without a clear understanding of what these components were. "They just looked like artifacts," the experimenter told me. This seems like a bad idea. As noted in box 6.6, it is a dangerous thing to subtract complex information from your data, and you should do it only when you understand exactly what you're removing.

In general, I find that remarkably little attention is paid to the details of artifact correction in journal articles. Authors tend to say very little about how they did the correction, and reviewers let them get away with this. Readers of journal articles seem to just accept the assertion that the artifact correction worked. I think we all need to think a little more critically when we read articles that use artifact correction, and editors and reviewers should force authors to give the details of what they did and, in many cases, provide evidence that it did not distort their results.

Suggestions for Further Reading

Berg, P., & Scherg, M. (1994). A multiple source approach to the correction of eye artifacts. *Electroencephalography & Clinical Neurophysiology, 90*(3), 229–241.

Frank, R. M., & Frishkoff, G. A. (2007). Automated protocol for evaluation of electromagnetic component separation (APECS): Application of a framework for evaluating statistical methods of blink extraction from multichannel EEG. *Clinical Neurophysiology, 118*, 80–97.

Gratton, G., Coles, M. G. H., & Donchin, E. (1983). A new method for off-line removal of ocular artifact. *Electroencephalography and Clinical Neurophysiology, 55*, 468–484.

Groppe, D. M., Makeig, S., & Kutas, M. (2008). Independent component analysis of event-related potentials. *Cognitive Science Online, 6.1*, 1–44.

Hillyard, S. A., & Galambos, R. (1970). Eye movement artifact in the CNV. *Electroencephalography and Clinical Neurophysiology, 28*, 173–182.

Lins, O. G., Picton, T. W., Berg, P., & Scherg, M. (1993). Ocular artifacts in EEG and event-related potentials I: Scalp topography. *Brain Topography, 6*, 51–63.

Lins, O. G., Picton, T. W., Berg, P., & Scherg, M. (1993). Ocular artifacts in recording EEGs and event-related potentials. II: Source dipoles and source components. *Brain Topography, 6*, 65–78.

Jung, T. P., Makeig, S., Humphries, C., Lee, T. W., McKeown, M. J., Iragui, V., et al. (2000). Removing electroencephalographic artifacts by blind source separation. *Psychophysiology, 37*, 163–178.

Jung, T. P., Makeig, S., Westerfield, M., Townsend, J., Courchesne, E., & Sejnowski, T. J. (2000). Removal of eye activity artifacts from visual event-related potentials in normal and clinical subjects. *Clinical Neurophysiology, 111*, 1745–1758.

Plochl, M., Ossandon, J. P., & Konig, P. (2012). Combining EEG and eye tracking: identification, characterization, and correction of eye movement artifacts in electroencephalographic data. *Frontiers in Human Neuroscience, 6*, 278.

Talsma, D. (2008). Auto-adaptive averaging: Detecting artifacts in event-related potential data using a fully automated procedure. *Psychophysiology, 45*, 216–228.

Verleger, R., Gasser, T., & Moecks, J. (1982). Correction of EOG artifacts in event-related potentials of the EEG: Aspects of reliability and validity. *Psychophysiology, 19*(4), 472–480.

7 Basics of Fourier Analysis and Filtering

Overview

When you record the EEG and construct averaged ERP waveforms, you represent the data as a voltage value at each time point. However, it is also possible to use *Fourier analysis* to represent EEG and ERPs in terms of the sum of a set of oscillating voltages at various frequencies. This frequency-based representation is essential for understanding how filters work, and it is also useful for understanding time–frequency analyses. Unfortunately, many people misunderstand what frequency-based representations really tell us, and this also leads to the misuse of filters and time–frequency analyses. The goal of this chapter is to provide a basic overview of Fourier analysis and filtering. This overview will be sufficient for most conventional ERP studies. If you are interested in doing more sophisticated frequency-based analyses, you can find more detailed information in online chapters 11 and 12.

I begin this chapter with a simple overview of Fourier analysis. I then describe the conventional approach to filtering EEG and ERP data, which is based on Fourier analysis. This is followed by a mathematically equivalent approach to filtering that is based entirely on time, which is often more appropriate for EEG and ERP data. I then describe how filters can distort your data, and then I finish with some concrete recommendations for how and when you should filter your data. I use very little math in presenting this material (nothing beyond arithmetic and averaging). My goal is to give you the conceptual background to understand frequency-based representations and filtering, along with some practical advice about how you should filter your data.

Filtering is essential, but it can lead to severe distortions if applied incorrectly. I will show examples of these distortions near the end of the chapter, and then I'll provide you with some simple recommendations that will help you avoid them. You may want to deviate from these recommendations, but I would strongly recommend following them until you have read and fully understand online chapters 11 and 12 (which provide the math).

You may already know a bit about Fourier analysis and filtering, so you might be tempted to skip part or all of this chapter. However, even though this chapter focuses on the basics, you should at least skim it, because you will probably discover some important principles. For example, if you think that the presence of power at 18 Hz in a Fourier analysis means that a brain oscillation was present at 18 Hz, you should definitely read this chapter. Similarly, you should read this chapter if you don't know that a high-pass filter with a half-amplitude cutoff of 0.5 Hz can produce artifactual peaks in your data or if you don't know why high-pass filters should ordinarily be applied to the continuous EEG rather than to averaged ERPs.

Basics of Fourier Analysis

Transforming ERPs into Frequency-Domain Representations

Figure 7.1A shows an ERP waveform that is contaminated by 60-Hz line noise. You can tell that it's 60 Hz because there are six peaks in this noise every 100 ms (and therefore 60 peaks every 1000 ms). This waveform is equivalent to the sum of the "clean" ERP waveform and a 60-Hz sinusoidal oscillation (see the glossary if you need a reminder about sine waves). Fourier analysis, named after the French mathematician Joseph Fourier, provides a way of determining the amplitude and phase of the 60-Hz oscillation in this complex waveform. This then allows us to design a filter that can remove the 60-Hz oscillation, leaving us with the clean ERP waveform.

Although it's fairly obvious that a 60-Hz oscillation is present in the waveform shown in figure 7.1A, we can actually decompose the entire waveform into sine waves. Joseph Fourier proved that any waveform—no matter how complex—can be re-created by summing together a set of sine waves of various frequencies, amplitudes, and phases. This is truly a stunning and counterintuitive idea: The waveform shown in figure 7.1A certainly doesn't look like it could be re-created by summing together a bunch of sine waves. Well, it can, and this forms the basis of many widely used techniques in mathematics, engineering, and science.

A mathematical procedure called the *Fourier transform* is used to determine the frequencies, amplitudes, and phases of the sine waves that need to be added together to re-create a given waveform.[1] For example, figure 7.1B shows the result of applying the Fourier transform to the ERP waveform in figure 7.1A. This is called transforming the data from the *time domain* (which just means that the *X* axis is time) into the *frequency domain* (which just means that the *X* axis is frequency). The frequency-domain representation shown in figure 7.1B shows the amplitude needed for each frequency of sine wave in order to reconstruct the ERP waveform shown in figure 7.1A. There is also a phase value for each frequency (not shown here). If you took a set of sine waves of the frequencies and amplitudes shown in figure 7.1B (with the appropriate phases), and you added them together, you would get exactly the same waveform shown in figure 7.1A. If you did this, you would be performing the *inverse Fourier transform*, which is the mathematical procedure that transforms a frequency-domain representation back into a time-domain representation.

Note that the amplitude in figure 7.1B is higher at 60 Hz than at the surrounding frequencies. This tells us that we would need to include a fairly large sine wave at 60 Hz to reconstruct the ERP waveform, which is consistent with the fact that the waveform obviously contains a 60-Hz

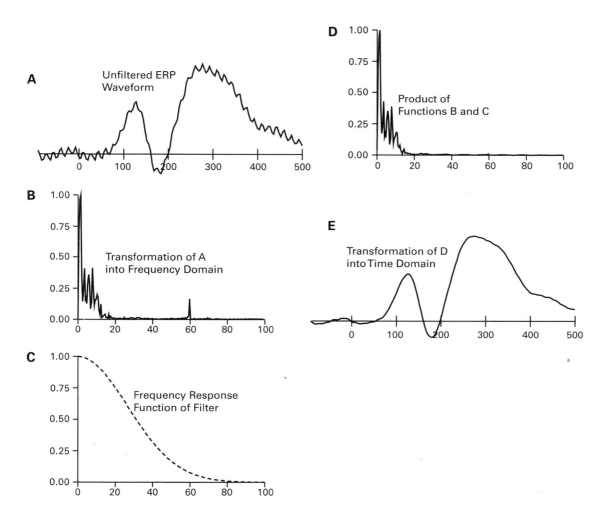

Figure 7.1
Frequency-based conceptualization of filtering. (A) Unfiltered ERP waveform, contaminated by substantial noise at 60 Hz. (B) Transformation of the waveform from panel A into the frequency domain by means of the Fourier transform. Note the clear peak at 60 Hz. (C) Frequency response function of a low-pass filter that can be used to attenuate high frequencies, including the 60-Hz noise. (D) Product of the waveforms in panels B and C; for each frequency point, the magnitude in panel B is multiplied by the magnitude in panel C. Note that panel D is nearly identical to panel B in the low-frequency range, but it falls to zero at high frequencies. (E) Transformation of the waveform from panel D back into the time domain by means of the inverse Fourier transform. The resulting waveform closely resembles the original ERP waveform in panel A, except for the absence of the 60-Hz noise. Note that the phase portion of the frequency-domain plots has been omitted here for the sake of simplicity. Also, Y-axis units have been omitted to avoid the mathematical details, except for panel C, in which the Y axis represents the gain of the filter (the scaling factor for each frequency).

oscillation. There are also many frequencies under 15 Hz that have a relatively high amplitude in figure 7.1B, but it is less obvious that we need these frequencies. This is the "magic" of Fourier analysis: Even though it isn't visually apparent that we need these frequencies, these are in fact the frequencies we would need to reconstruct the ERP waveform from sine waves.

People often use power rather than amplitude to represent the strength of each sine wave. Power is simply amplitude squared, and amplitude is simply the square root of power. Using power instead of amplitude has some advantages, but it's really the same information expressed on a different scale.

You need to be aware of one small caveat about applying the Fourier analysis to ERP waveforms. Specifically, sine waves have an infinite duration, whereas ERP waveforms do not. Consequently, if you reconstructed the original waveform by summing together a bunch of infinite-duration sine waves, the ERP waveform would repeat over and over again, both backward and forward in time. That is, the 600-ms waveform that we see from −100 to +500 ms would repeat again from 500 to 1100 ms, from 1100 to 1700 ms, and so forth. It would also repeat from −700 to −100 ms, from −1300 ms to −700 ms, and so forth. This isn't really a problem in practice, and it doesn't mean that something is wrong with Fourier analysis. After all, we've given the Fourier transform no information about the voltages that were present outside this narrow time window, so we can't expect it to come up with anything sensible beyond this window. If we gave it an infinite-duration waveform as the input (e.g., by making the prestimulus interval extend to negative infinity and the end of the ERP extend to positive infinity), the reconstruction of this waveform from sine waves would not show this repetition. Nonetheless, this is a warning sign that we need to be careful about how we interpret the results of a Fourier analysis. We'll return to this issue soon.

You may have heard people talk about the "sine and cosine components" of a Fourier transform. These are sometimes called the "real and imaginary components." These are just alternative ways of talking about amplitude and phase, and there is nothing "imaginary" about them. If you are interested in understanding these terms, see the online supplement to chapter 7.

The Fourier Transform of a Simple Square Wave

Figure 7.2 shows the application of the Fourier transform to a simpler time-domain waveform, a repeating square wave. You might think that it would be impossible to re-create a square wave—which consists of straight lines and sharp corners—from a set of smoothly varying sine waves. However, this example shows that you can, in fact, reconstruct the square wave from sine waves, as long as you're willing to add together enough sine waves. Unlike the Fourier transform of the ERP waveform in figure 7.1, which requires sine waves of many consecutive frequencies, it turns out that the Fourier transform of a square wave requires a distinct subset of possible frequencies. Specifically, if the square wave repeats at a frequency of F times per second, the sine waves needed to reconstruct the square wave are at frequencies of F, 3F, 5F, 7F, and so forth, without any of the in-between frequencies. These frequencies are called the *fundamental frequency* (F) and the *odd harmonics* (3F, 5F, 7F, etc.).

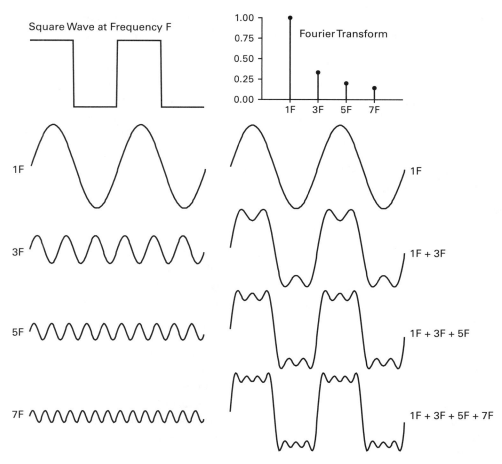

Figure 7.2
Example of the application of Fourier analysis to a repeating square wave.

If you add the 1F and 3F sine waves (see 1F + 3F in figure 7.2), the result looks a little more like a square wave, but it still lacks straight lines and sharp corners. Once you add 1F, 3F, 5F, and 7F, however, the result starts to approximate the square wave. I hope you can imagine that we would get closer and closer to a perfect square wave by adding more and more of the odd harmonics together. It takes an infinite number of these odd harmonics to perfectly reconstruct a square wave, but you can get extremely close with a relatively modest number.

What a Fourier Transform Really Means
People often make a fundamental error in interpreting Fourier transforms (and related processes, such as time–frequency analyses). The problem is that they mistake *mathematical* equivalence

for *physiological* equivalence. The essence of Fourier analysis is that any waveform can be reconstructed by adding together a set of sine waves. This does not mean, however, that the waveform actually *consists* of a set of sine waves. For example, the Fourier transform shown in figure 7.1B has some power at 18 Hz, and this means that we would need to include an 18-Hz sine wave if we wanted to add together a set of sine waves to re-create the ERP waveform shown in figure 7.1A. It does not mean that the brain was generating an 18-Hz sine wave when it produced this ERP waveform. This is just like saying that a $1 bill is equal in value to 100 pennies. You can exchange a $1 bill for 100 pennies, but the bill is not physically made of 100 pennies. You can also exchange a $1 bill for nine dimes and two nickels. These are all mathematically equivalent, but they are not physically equivalent. This brings us to a key concept that you should commit to memory:

Fundamental principle of frequency-based analyses Power at a given frequency does not mean that the brain was oscillating at that frequency.

When I run ERP boot camps, I always ask the participants whether they find this fundamental principle surprising. At least half of them say yes. And this is consistent with my experience reading journal articles that discuss frequency-based analyses. That is, at least half of the papers I read implicitly assume that power at a given frequency means that the brain was actually oscillating at this frequency. For example, the paper might say something like "We observed a neural oscillation between 25 and 60 Hz that differed across conditions." This conclusion usually goes well beyond the actual data, because they did not actually *observe* an oscillation; they merely saw power in a particular frequency range in a Fourier-based analysis. In essence, these papers are *assuming* that the data consist of the sum of a set of sine waves when they apply the frequency-based analysis technique, and the conclusion is simply a restatement of this assumption. No matter whether the data contain sine wave oscillations or not, there will always be power at some frequencies. Power at a frequency in a Fourier-based (or wavelet-based) analysis is therefore a mathematical inevitability, not evidence for physiological oscillations.

After I describe this fundamental principle at an ERP Boot Camp, the first question I am always asked is this: "If power at a given frequency doesn't mean that the brain was oscillating at that frequency, then what does it mean?" It just means that we would need some power at that frequency if we were trying to reconstruct the observed data by summing together a set of sine waves. The summing of sine waves is something the *experimenter* is doing to reconstruct the waveforms, not necessarily something the *brain* is doing when it is generating the waveforms. The brain may or may not be oscillating at this frequency, and Fourier analysis cannot tell us whether the brain is actually oscillating.

Of course, a true oscillation at a given frequency will lead to power in that frequency in a Fourier transform. For example, the ERP waveform in figure 7.1A contains a 60-Hz oscillation that was induced by oscillating electrical activity in the environment. Consequently, power at 60 Hz in the Fourier transform shown in figure 7.1B really was a result of a 60-Hz oscillation in the data. Similarly, you can often see alpha band oscillations (approximately 10 Hz) in the raw EEG. The alpha oscillations appear to reflect, at least in part, recurrent connections between the cortex and the thalamus that produce a loop, and brain activity oscillates at 10 cycles per second as signals travel around this loop (Schreckenberger et al., 2004; Klimesch, Sauseng, & Hanslmayr, 2007). The alpha activity typically leads to high power in a narrow band of frequencies around 10 Hz in a Fourier transform of the EEG. This is another case in which a true oscillation can be measured by examining the power at a specific frequency in a Fourier transform. But the power at 18 Hz in the Fourier transform of the ERP waveform shown in figure 7.1 probably does not reflect a neural oscillation at 18 cycles per second. The 18-Hz power reflects a mathematical equivalence, not a physiological equivalence.

There is an asymmetry here: A neural oscillation at a given frequency will virtually always produce power at that frequency in a Fourier analysis, but power at a given frequency in a Fourier analysis does not entail that the brain was oscillating at that frequency.

The next question I'm asked at the ERP Boot Camp is this: "How can you tell when power at a given frequency reflects a true oscillation at that frequency?" In most cases, strong power in a narrow range of frequencies in a Fourier analysis reflects a true oscillation in that range of frequencies. For example, there is strong power from 58 to 60 Hz in figure 7.1B, with very low power from 40 to 58 Hz and from 62 to 100 Hz, and this reflects the 60-Hz oscillation that you can see in the ERP waveform in figure 7.1A. In contrast, the power at 18 Hz is not much different from the power at 16 Hz or 20 Hz, and this activity probably doesn't reflect a true oscillation. Thus, if you see a narrow band of high power, it is probably an oscillation, but if it is a broad range of frequencies, it is probably not an oscillation. In particular, a transient brain response (i.e., a short-duration, non-oscillating voltage deflection) usually contains power all the way down to 0 Hz. There are exceptions to this, but it's a good heuristic in most cases. The reasoning behind this heuristic is provided in online chapter 12.

If you're reading a paper that shows some kind of frequency-based analysis, and it doesn't show you what happens at the lower frequencies (e.g., <10 Hz), you should be skeptical of any conclusions the paper draws about oscillations. True oscillations may be present, but it's hard to be certain without seeing the low frequencies. For example, figure 7.3 shows a simulated time–frequency analysis. The X axis is time relative to stimulus onset, the Y axis is frequency, and the darkness of the shading represents change in power relative to baseline. Periods of gamma-band power (30–40 Hz) are present from approximately 100 to 300 ms and from approximately 400 to 800 ms. The gamma-band power from 400 to 800 ms is likely to reflect true neural oscillations, because it is isolated to a fairly narrow band of frequencies (30–40 Hz), with no power at lower frequencies. In contrast, the power from 100 to 300 ms extends down to the lowest frequency shown (20 Hz), making it impossible to know whether this is a real oscillation or whether it is instead some kind of transient, non-oscillating response. I won't name names, but I have seen many papers in which activity like that shown from 100 to 300 ms was interpreted as an oscillation, even though it is quite likely that it reflected a transient, non-oscillating brain response.

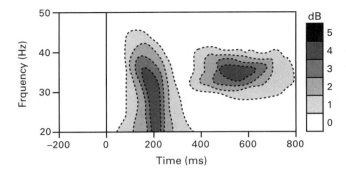

Figure 7.3
Simulated time–frequency data. The *X* axis shows time, the *Y* axis shows frequency, and the darkness indicates the change in power relative to the prestimulus baseline (in units of decibels, dB).

The final question that I get from ERP Boot Camp participants is usually this: "Why does it matter whether something is truly an oscillation?" The answer is that the presence of an oscillation in a specific frequency band has implications for the kind of neural circuitry that is producing the brain activity. For example, gamma-band oscillations are often assumed to involve short-range recurrent connections between fast-spiking, parvalbumin-expressing interneurons and pyramidal cells (see, e.g., Knoblich, Siegle, Pritchett, & Moore, 2010; Volman, Behrens, & Sejnowski, 2011). Moreover, oscillations are often a sign of neural synchrony, which may play an important role in representing and processing information in neural circuits (see, e.g., Singer, 1999). Thus, it can be useful to know if the brain is oscillating at a given frequency. Unfortunately, researchers often assume that they have detected an oscillation without very good evidence that the brain is indeed oscillating.

Basics of Filtering in the Frequency Domain

Now that I've explained the basics of Fourier analysis, it's time for you to learn the basics of filtering. Here I will explain the most common approach to filtering, which focuses on suppressing specific frequency bands. This approach treats ERPs as if they actually consist of the sum of a set of sine waves, which sometimes leads to interpretive errors. Nonetheless, the most common uses of filters involve reducing sources of noise that can be well approximated by sine waves, so there is considerable value to this approach. And once you understand this approach, you will be ready to understand the time-domain approach to filtering.

Why Are Filters Necessary?
A key message of this chapter is that filters can substantially distort your data. It is therefore useful to ask why it's worthwhile to use filters. There are two answers to this question, the first of which is related to the *Nyquist theorem* (see chapter 5). This theorem states that it is possible

to convert a continuous analog signal (like the EEG) into a set of discrete samples without losing any information, as long as the rate of digitization is more than twice as high as the highest frequency in the signal being digitized. This means that we can legitimately store the EEG signal as a set of discrete samples on a computer. However, this theorem also states that if the original signal contains frequencies that are more than twice as high as the digitization rate, the high frequencies will appear as artifactual low frequencies in the digitized data (this is called *aliasing*). Consequently, EEG recording systems use hardware-based *anti-aliasing* filters to suppress high frequencies, and these filters are generally set to eliminate frequencies at or above the *Nyquist frequency* (one-half of the sampling rate). For example, a typical cognitive ERP experiment might use a digitization rate of 250 Hz, and it would therefore be necessary to make sure that everything greater than or equal to 125 Hz is eliminated.

The second main goal of filtering is noise reduction, and this is considerably more complicated. The basic idea is that the EEG consists of a signal plus some noise, and some of the noise can be suppressed simply by attenuating certain frequencies. For example, most of the relevant portion of the ERP waveform in a typical cognitive neuroscience experiment consists of frequencies between approximately 0.1 Hz and 30 Hz, whereas EMG activity primarily consists of frequencies above 100 Hz; consequently, suppressing frequencies above 100 Hz will greatly reduce the EMG activity while producing very little change to the EEG signal. However, if the frequency content of the signal and the noise are similar, it is difficult to suppress the noise without significantly distorting the signal. For example, alpha waves can provide a significant source of noise, but because they are within the range of frequencies of the ERP, it is difficult to filter them without significantly distorting the ERP waveform.

In addition to suppressing high frequencies, filters are also used in most experiments to attenuate very low frequencies, which typically arise from the electrodes and the skin rather than from the brain (as was discussed in chapter 5). It is usually a good idea to remove these slow voltage shifts by filtering frequencies lower than approximately 0.1 Hz. This is especially important when recordings are obtained from patients or from children, because head and body movements are one common cause of these sustained shifts in voltage. You can reduce these slow voltage shifts even more by filtering with a higher cutoff, such as 0.5 or 1.0 Hz, but these frequencies also contribute in an important way to the ERP waveform. Filtering with such a high cutoff may cause substantial distortion of the ERP waveforms and may decrease the statistical power (as described later in this chapter).

I should also note that filters are often useful for helping you visually inspect your average ERP waveforms. If these waveforms contain a lot of high-frequency activity, this will make it difficult for you (and the people reviewing your journal submissions) to see the differences between conditions or groups. For this reason, it can be useful to filter out the high-frequency noise before plotting the data.

Filters are usually described in terms of their ability to suppress or pass various different frequencies. The most common classes of filters are (1) *low-pass filters*, which attenuate high frequencies and pass low frequencies; (2) *high-pass filters*, which attenuate low frequencies and pass high frequencies; (3) *band-pass filters*, which attenuate both high and low frequencies, passing only an intermediate range of frequencies; and (4) *notch filters*, which attenuate some narrow band of frequencies (e.g., 60 Hz) and pass everything else. Personally, I find it very confusing to describe a filter in terms of the frequencies that it passes, and I would find it much more natural to refer to a *high-block* filter than a *low-pass* filter. But this is the convention, and we need to live with it. Note that a band-pass filter is equivalent to filtering once with a low-pass filter and then filtering a second time with a high-pass filter (or vice versa).

Filtering as Multiplication in the Frequency Domain

The effects of a filter are usually expressed by the filter's *frequency response function,* which describes how each frequency is influenced by the filter. Usually, we focus on how the filter changes the amplitude at each frequency, which is determined by the *gain* at each frequency. The gain is a multiplication factor; each frequency in the input of the filter is multiplied by the gain for that frequency. A gain of 1.0 means that the amplitude of that frequency is unchanged by the filter. A gain of 0.25 means that the amplitude is multiplied by 0.25 and is therefore reduced by 75%. A gain of 1.5 means that the amplitude is multiplied by 1.5 and is therefore increased by 50%. In the typical filters used with EEG/ERP data, the gain is between 0 and 1 for each frequency.

The frequency response function of a typical low-pass filter is shown in figure 7.1C. It has a gain near 1 for frequencies below 10 Hz, so these frequencies are not influenced much by the filter. It has a gain near zero for frequencies above 80 Hz, so these frequencies are almost completely eliminated by the filter. There is a gradual decline in gain between 5 and 80 Hz, and these frequencies are partly suppressed and partly passed by the filter. The gain is approximately 0.10 at 60 Hz, so 90% of the 60-Hz noise will be suppressed by the filter.

The actual filtering process is quite simple when conceptualized in this manner. The original ERP waveform (panel A of figure 7.1) is transformed into the frequency domain with the Fourier transform (panel B). The amplitude of each frequency in the frequency-domain representation of the data is then multiplied by the corresponding gain value in the frequency response function (panel C), leading to a filtered version of the frequency-domain data (panel D). The filtered frequency-domain data are then transformed back into the time domain with the inverse Fourier transform, giving us a filtered version of the time-domain ERP waveform (panel E).

If you compare the original and filtered versions of the frequency-domain data (panel B versus panel D), you can see that the pattern of amplitudes in the low frequencies (under 10 Hz) is very similar for the filtered and unfiltered data. This is because the gain values are near 1.0 for these frequencies. However, the spike at 60 Hz in the unfiltered data (panel B) is nearly eliminated in the filtered data (panel D), because the gain at 60 Hz is approximately 0.10. Consequently, the rapid up-and-down noise oscillations that you can see in the unfiltered ERP waveform (panel A) are largely eliminated in the filtered ERP waveform (panel E).

You can use any function you want for the frequency response function. For example, you could use a function that has a gain of 1 for every frequency between 0 and 50 Hz and a gain of 0 for every frequency above 50 Hz. You might think this would be a much better frequency response function, because you wouldn't have the partial attenuation of intermediate frequencies that is produced by the filter in figure 7.1. However, as I will describe in more detail later, this sort of very sudden transition in the frequency response function can have negative side effects.

Filtering Twice

What happens if you filter the same data twice? For example, you might use a hardware low-pass filter during data acquisition to avoid aliasing, and then you might want to apply a low-pass filter offline in software to further reduce high-frequency noise in the data. You can understand how this would work in terms of the sequence of operations shown in figure 7.1. Filtering twice just repeats this sequence of operations twice.

However, there is a much more convenient way to think about filtering twice. Specifically, you can just multiply the frequency response functions of the two filters together, and this will give you the frequency response function that results from applying both filters. This is illustrated in figure 7.4, which shows the frequency response functions of two filters, labeled "filter A" and "filter B." Filtering the data twice, once with filter A and then once with filter B, is equivalent to filtering the data once with a frequency response function that is created by simply multiplying the frequency response functions for the two individual filters. This is true even if one of the filters is implemented in hardware and the other is implemented in software.

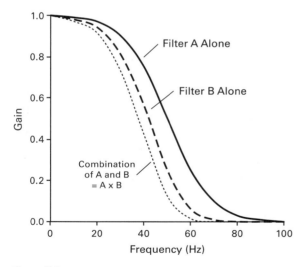

Figure 7.4
Frequency response functions of two filters, A and B, along with the effective frequency response function that would be obtained by filtering the data with both filters in succession. Filtering the data with filter A and then filtering with filter B is equivalent to filtering once with a frequency response function that is the product of the frequency response functions of filter A and filter B. That is, the frequency response function of the combination of the two filters is computed by multiplying the gains of the two individual functions at each frequency.

You might think that filtering a second time with exactly the same filter wouldn't change your data. After all, if you've already eliminated a frequency, how could filtering a second time change that frequency? But filtering a second time will change your data. Imagine, for example, that we applied filter A from figure 7.4 twice. This filter has a gain of approximately 0.80 at 38 Hz. The first time we filter the data, we would remove 20% of the signal at 38 Hz. The second time, we would remove 20% of the remaining signal at 38 Hz. Similarly, filter A has a gain of approximately 0.71 at 41 Hz. If we filter twice with this filter, the gain of the double filtering will be 0.5 at 41 Hz (because $0.71 \times 0.71 = 0.5$). Thus, whereas filter A has a half-amplitude cutoff at approximately 50 Hz, filtering twice with this filter gives us a half-amplitude cutoff at approximately 41 Hz.

Phase Shifts

When we use the Fourier transform to create a frequency-domain representation, as in figure 7.1B, there is a phase as well as an amplitude for each frequency (although this is frequently omitted from figures, including the figures in this book). Similarly, a frequency response function specifies a phase shift as well as a gain for each frequency. In most offline (software) filters, the phase shift is zero for every frequency, so you don't need to worry about phase with these filters. Hardware filters, however, always have a phase shift, which moves the ERP waveform rightward in a complicated manner. The amount of shift is negligible for most antialiasing filters, which have a relatively high cutoff frequency, so you don't usually need to worry about the phase shift. However, if you use hardware filters with a relatively low cutoff frequency (e.g., under 50 Hz), you might produce a meaningful latency shift. This is one reason why it is best to do most of the filtering offline, using filters that do not shift the phase or latency.

As I will explain in more detail later in the chapter, filters create a filtered value at a given time point by means of a weighted combination of the surrounding time points. Most offline filters do this in a symmetrical manner, giving equal weight to the previous and subsequent time points. As a result, they do not produce a phase shift. In contrast, hardware filters only "know" about the past time points, because the future time points haven't yet happened, so only previous voltages are used to create the filtered value at a given time point. The voltage at a given time point influences the value at subsequent time points, and this effectively "pushes" this value later in time, causing a phase shift. Such filters are called *causal* filters, because they follow the basic principle of causation that the past can influence the future but the future cannot influence the past.

Offline filters use both previous and subsequent voltages to compute the filtered value at a given time point. Thus, they violate the unidirectionality of time and are therefore called *noncausal* filters. Although it might sound like a bad idea to use something that does not obey a basic principle of causation, this is not usually a problem in practice. In almost all cases, you will want to use noncausal filters for your offline filtering to avoid latency shifts.

A

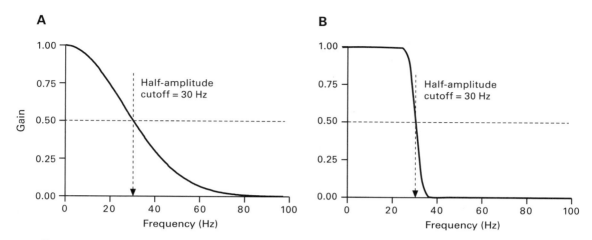

Figure 7.5
Examples of the frequency response functions of two filters that have the same half-amplitude cutoff (30 Hz) but with a gentle roll-off (A) or a steep roll-off (B).

Cutoff Frequency, Roll-off, and Slope

In ERP research, a filter is often described solely in terms of its *half-amplitude cutoff*, which is the frequency at which the amplitude is reduced by 50%. However, this does a poor job of describing the entire frequency response function. For example, figure 7.5 shows two frequency response functions that both have a half-amplitude cutoff at 30 Hz. The function in panel A falls off very gently between 10 and 80 Hz, whereas the function in panel B is very flat from 0 to 25 Hz, falls rapidly between 25 and 35 Hz, and is then very flat again.

The steepness of the decline in gain is called the *roll-off* of the filter, and it is typically quantified by the *slope* of the frequency response function at the cutoff frequency. The frequency response function in figure 7.5A has a shallow roll-off, whereas the frequency response function in figure 7.5B has a steep roll-off. The slope is usually specified in dB/octave, which is a logarithmic scale. A decrease of 3 dB is a 50% drop in power, and a change of 6 dB is a 50% drop in amplitude. An octave is a doubling of frequency. Thus, a filter with a half-amplitude cutoff at 30 Hz and a slope of 6 dB/octave has a 50% drop in amplitude (6 dB) between 20 and 40 Hz (one octave).[2]

Note that the cutoff frequency is sometimes specified in terms of power rather than amplitude. Whereas the half-amplitude cutoff is the point at which the gain is reduced by 6 dB, the half-power cutoff is the point at which the gain is reduced by only 3 dB. A half-power cutoff might be described like this: "The data were low-pass filtered (−3 dB at 30 Hz)." The amplitude is reduced by only 29% at the point where the power is reduced by 50%. If the slope of the filter is fairly gentle, the half-power cutoff might be quite different from the half-amplitude cutoff. For example, a filter with a slope of 12 dB/octave that had a half-power cutoff of 30 Hz would

Box 7.1
What Kind of Filter Was That?

I was an associate editor at *Cognitive, Affective, and Behavioral Neuroscience* for several years, and in that role I handled almost all of the ERP manuscripts that were submitted to the journal. In addition to making big-picture decisions about whether each submission was publishable, I made sure that every ERP paper in the journal met a set of minimal standards for describing the methods and results. Much to my surprise, I found that the vast majority of submitted manuscripts failed to provide a remotely adequate description of the filters that were applied. For example, there was often no mention of the hardware filter that was used to eliminate aliasing during data acquisition. In addition, most papers that mentioned filter cutoffs failed to specify whether the specified cutoffs were half-amplitude or half-power values. And most papers neglected to indicate the slope of the filter. In fact, one paper involved comparing experiments conducted in different laboratories, and it wasn't clear that the filters used for the two data sets were equivalent. The graduate student who wrote the paper had to learn a lot about filtering to satisfactorily revise the paper (which was eventually accepted).

I also heard a story about two labs that tried to replicate each other but kept failing, even though they thought they were processing their data in exactly the same way. They eventually discovered that the discrepancy occurred because one lab was specifying the cutoff frequency in terms of the half-amplitude point and the other was specifying the cutoff frequency in terms of the half-power point. Once they used equivalent filters, the discrepancy was eliminated.

The bottom line is that you need to describe your filters in enough detail so that the reader knows what you did to your data and can replicate your data processing procedures. This should be an obvious point, but most ERP researchers don't seem to realize how much filtering can influence the results. Don't make this mistake in your own papers. At a minimum, you should indicate the cutoff frequency, whether this was the half-amplitude or half-power point, and the slope of the roll-off. It also helps to describe the general class of filter (e.g., Gaussian, Bessel, Butterworth), but that is a more advanced topic that will be described in online chapter 12.

have a half-amplitude cutoff at approximately 46 Hz. That's a 50% higher cutoff frequency when expressed in terms of amplitude instead of power! See box 7.1 for a brief rant about the importance of fully specifying your filters when you write a paper for publication.

High-Pass Filters and the Time Constant

Up to this point, we have mainly considered low-pass filters. In most cognitive experiments, low-pass filters help to reduce induced electrical noise and EMG noise. High-pass filters are used to reduce slow changes in voltage caused by skin potentials and other gradual changes in the voltage offset, which can improve statistical power. An example of the application of a high-pass filter to a 30-s period of continuous EEG is provided in figure 7.6. The unfiltered data exhibit a clear upward drift over this time period, and this drift is eliminated by the filter, which had a half-amplitude cutoff of 0.1 Hz and a slope of 24 dB/octave. You may not even notice the drift if you only look at short EEG epochs (e.g., 5 s or less), and it's a good idea to look at longer time periods (e.g., 60 s or more) so that you can see the slow drifts in your data. As

Unfiltered

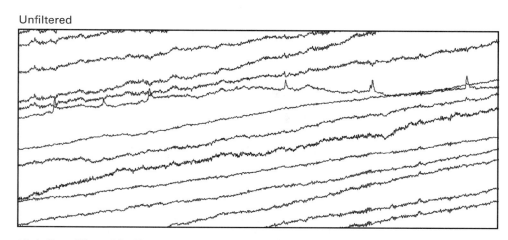

High-Pass Filtered (half-amplitude cutoff = 0.1 Hz, slope = 24 dB/octave)

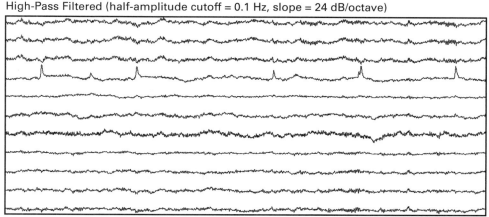

Figure 7.6
Example of a 30-s period of EEG without any filtering (top) and after the application of a high-pass filter with a half-amplitude cutoff of 0.1 Hz and a slope of 24 dB/octave (bottom). The filter eliminates the gradual upward drift in the data.

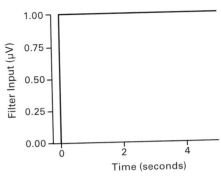
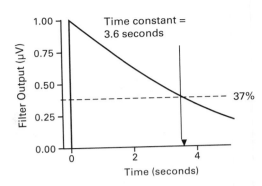

Figure 7.7
Example of the time constant of a high-pass filter. If the input to the filter is a constant voltage (left), the output of the filter will fall gradually toward zero according to an exponential function. The time constant of this particular filter is 3.6 s, which means that the output of the filter at a given time point will be $1/e$ (37%) of the value 3.6 s previously. As long as the input to the filter is a constant voltage, this percentage-based drop will be true over any 3.6-s period.

discussed in the section "When Should You Filter?" near the end of the chapter, high-pass filters are usually applied to the continuous EEG rather than to the epoched EEG or to averaged ERPs.

High-pass filters are sometimes described in terms of the *time constant* rather than the half-amplitude cutoff. As shown in figure 7.7, if the input to a high-pass filter is a constant voltage, the output of the filter will start at this voltage and then gradually fall toward zero. The filter's time constant is a measure of the rate at which it causes the voltage to fall. The decline in output voltage over time is exponential, which means that the voltage drops by a particular percentage between the beginning and end of a time period of a given length (e.g., it might drop by 50% between the beginning and end of an 8-s period, and then by 50% of the remaining voltage by the end of the next 8-s period). Because the drop in voltage over a period of time is always a percentage of the voltage at the beginning of that period, the voltage never quite reaches zero. Consequently, the time constant is expressed as the time required for the filter's output to reach a particular proportion ($1/e$, or 37%) of the starting value. It's called a "constant" because, for any starting point, the voltage will drop to $1/e$ of the starting value in that amount of time.

As the half-amplitude cutoff becomes higher, the time constant becomes shorter. If you know the half-power cutoff frequency of a high-pass filter (f_c, the frequency at which the filter's output is reduced by 3 dB), the time constant can be computed as $1/(2\pi f_c)$.

Basics of Filtering in the Time Domain

In conventional ERP analyses, frequency per se is not an important variable for analysis. Instead, ERP studies usually focus on the time course of the brain activity. However, ERP researchers almost always talk about the frequency-domain properties of filters rather than their time-domain properties. I find this a little odd. In this section, I will therefore explain how filters operate as a time-based process, without any frequency-domain representations. I will focus on a particularly simple type of time-domain filter called a *running average* filter, which won't require any math beyond simple averaging. Online chapter 12 provides a more generalized description of how filters operate in the time domain, which I encourage you to read if you want to go beyond the filtering recommendations provided at the end of this chapter (or if you someday hope to become an ERP guru).

Low-Pass Filtering with a Running Average Filter

Let's start by looking at the ERP waveform in figure 7.8A. It has a lot of high-frequency noise (small rapid changes in voltage that make the waveform look a bit fuzzy). A simple way to

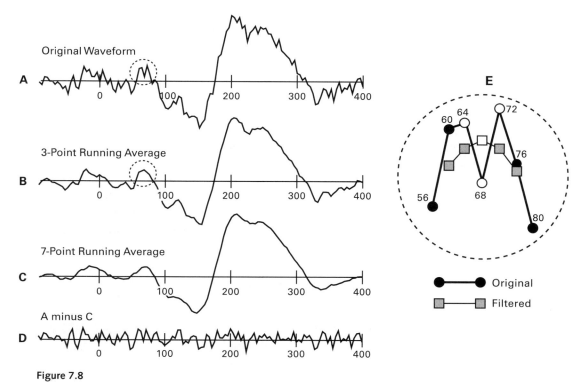

Figure 7.8
Example of filtering an ERP waveform with a running average filter, which works by averaging together the voltages surrounding each time point. (A) Unfiltered ERP waveform, contaminated by substantial high-frequency noise. (B) Result of filtering the waveform in panel A by averaging the voltage at each time point with the voltages at the immediately adjacent time points (a three-point running average filter). (C) Result of filtering the waveform in panel A by averaging the voltage at each time point with the voltages at the three time points on either side (a seven-point running average filter). (D) High-pass–filtered waveform, constructed by subtracting the filtered waveform in panel C from the unfiltered waveform in panel A. (E) Close-up of the original and filtered data between 56 and 80 ms. The original values are indicated by circles, and the filtered values are indicated by squares. Each number represents the time point of the value. The three white-filled circles were averaged together to create the white-filled square.

reduce these rapid changes would be to replace each point in the waveform with the average of the voltage at that point and the voltages at the time points immediately before and immediately after. For example, if we have one sample every 4 ms, the filtered value at 68 ms would be equal to the average of the unfiltered voltages at 64, 68, and 72 ms. Similarly, the filtered value at 72 ms would be equal to the average of the unfiltered voltages at 68, 72, and 76 ms. This would be called a *three-point running average filter* because the filtered waveform is computed by taking a running average of every three points in the original waveform. The result of applying this simple procedure is shown in figure 7.8B. The filtered waveform looks smoother than the original waveform in figure 7.8A, but it is otherwise the same basic waveform.

Just to make sure this is clear, figure 7.8E shows a blown-up view of seven samples from the original waveform (the circled part of the waveform in figure 7.8A, from 56 to 80 ms) and five of the corresponding samples from the filtered waveform (the circled part of the waveform in figure 7.8B, from 60 to 76 ms). The filtered value at 68 ms (the white square) is simply the average of the points at 64, 68, and 72 ms in the original waveform (the three white circles). Similarly, the filtered value at 64 ms is the average of the original values at 60, 64, and 68 ms.

I am showing you seven values from the original waveform, but I have shown you only five points of the filtered values, leaving out the first and last points in this time period. This is because filters run into a problem at the beginning and end of the waveform. To compute the filtered value at 56 ms, you would need to know the original value at 52, 56, and 60 ms. Similarly, to compute the filtered value at 80 ms, you would need to know the original value at 84 ms (and average this with the original values at 76 and 80 ms). Because I haven't shown you the original values at 52 and 84 ms, it's hard to show you the filtered values at 56 and 80 ms. Of course, if we look at the entire original waveform, we can find the values at 52 and 84 ms and compute the filtered values at 56 and 80 ms. However, this problem will arise at the very beginning and very end of the overall waveform (at −100 and +400 ms), and the filtered waveform becomes undefined at these time points. I will return to this later, because the same underlying problem arises with frequency-domain filters as well.

Figure 7.8C shows the same approach to filtering, but using a seven-point running average rather than a three-point running average. For example, the filtered value at 68 ms was computed as the average of the original values at 56, 60, 64, 68, 72, 76, and 80 ms. This waveform is even smoother than the waveform created with the three-point running average. The longer the running average, the more heavily you are filtering the data. As the next section describes in more detail, increasing the number of points in a running average is mathematically equivalent to decreasing the half-amplitude cutoff of a frequency-domain filter.

Relationship to Frequency-Domain Filtering

You are probably wondering how this type of filtering is related to frequency-domain filtering. As will be described in detail in online chapter 12, any time-domain filter has an equivalent

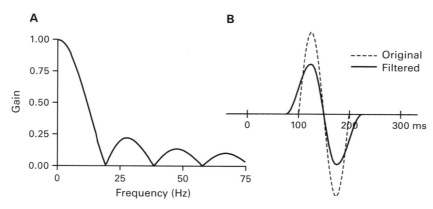

Figure 7.9
(A) Frequency response function of a 13-point running average filter (assuming a sampling rate of 250 Hz). (B) Artificial waveform consisting of one cycle of a 10-Hz sine wave beginning at 100 ms (dashed line), and the result of filtering this waveform with a 13-point running average filter (solid line).

frequency-domain filter, and vice versa. Figure 7.9A shows the frequency response function of a 13-point running average filter (assuming a sampling rate of 250 Hz). It passes low frequencies perfectly, and then the gain drops off as the frequency increases, falling to zero at 20.83 Hz. The gain then rises up a little and falls back to zero again at 41.67 Hz. This up-and-down pattern continues infinitely, getting smaller and smaller as the frequency increases. If you filtered an ERP waveform using the frequency-domain approach to filtering (e.g., converting the ERP waveform to the frequency domain, multiplying it by this frequency response function, and then converting it back into the time domain), the result would be exactly the same as applying a 13-point running average filter.

The frequency response function of a running average filter is a little strange because it goes up and down multiple times. However, it is possible to make a slight change to the running average filter and get a much nicer, monotonically decreasing frequency response function. This change also addresses a time-domain oddity of the running average filter. Specifically, when we compute the filtered value at a given time point with a 13-point running average filter, we give equal weight to all six time points on either side of the current time point. It might make more sense to give a higher weight to the time points that are closer to the current time point. With a three-point running average filter, for example, we could give the current point a weight of 0.5 and each of the surrounding points a weight of 0.25. The filtered value at 68 ms, for example, would be 0.25 times the voltage at 64 ms plus 0.5 times the voltage at 68 ms plus 0.25 times the voltage at 72 ms. By choosing an appropriate set of weights, you can create a "weighted" running average filter that has a monotonically decreasing frequency response function. In fact, by choosing appropriate weights, you can create any frequency response function you desire. The details of this are described in online chapter 12.

Filtering and Temporal Smearing

When we filter with a running average filter, it's clear that we are losing some temporal resolution. That is, because the filtered voltage at a given time point is computed by averaging the surrounding time points, the filtered value represents the average activity over a range of time points rather than representing activity at a single instant. This is illustrated in figure 7.9B, which shows what happens when we apply a 13-point running average filter to an artificial waveform (a single cycle of a 10-Hz sine wave that begins at 100 ms). The filtered waveform has a lower overall amplitude than the original waveform, which makes sense given that the gain at 10 Hz is well below 1.0 in the frequency response function shown in figure 7.9A. However, the most notable thing about the filtered waveform is that it has an earlier onset time than the original waveform (and a later offset time as well). This makes sense given how a running average is computed. For example, the filtered value at 92 ms is the average of the voltages between 68 and 116 ms (this is 13 points given that we have one point every 4 ms). Because the original waveform starts rising above zero in this time range, the average of these voltages is above zero. This causes the filtered value to be above zero at 92 ms, even though the original waveform does not deviate from zero until after 100 ms.

You might think that this would be a reason not to filter with a running average filter and instead use a frequency-domain filter. However, time-domain and frequency-domain filters are mathematically equivalent, and this same kind of "smearing" of temporal information is produced when ERPs are filtered in the frequency domain. I will return to this later in the chapter.

High-Pass Filtering with a Running Average Filter

You now know how to perform a simple low-pass filtering operation. You can perform high-pass filtering in the same way by adding a simple trick. This trick takes advantage of the fact that the overall ERP waveform is equivalent to the sum of the low frequencies plus the high frequencies, whereas the low-pass–filtered waveform contains just the low frequencies. If we take the original waveform (the high frequencies plus low frequencies) and subtract the low-pass–filtered waveform (just the low frequencies), the result is a waveform that contains only the high frequencies. In other words, we can create a high-pass–filtered waveform by subtracting a low-pass–filtered waveform from the original data. This is shown in figure 7.8D, which is the original waveform minus the waveform that was filtered with the seven-point running average filter. If you look closely, each little upward or downward blip in the high-pass–filtered waveform corresponds to a little upward or downward blip in the original waveform. This particular high-pass filter, which used a seven-point running average, has such a high half-amplitude cutoff that you would never want to use it. However, you could achieve a more useful half-amplitude cutoff by subtracting a waveform that was filtered with, for example, a 101-point running average.

The unfiltered waveform shown in figure 7.8A contains only 125 points (500 ms at a sampling rate of 250 Hz). If you applied a 101-point running average filter, you would be unable to compute the filtered value for the first and last 50 points in the waveform (because you need 50 points on either side of a given point to compute the average of 101 points). That would leave us with only the middle 25 points in the filtered waveform, which is obviously a problem. Consequently, a filter with a lot of points should ordinarily be applied to the continuous EEG (i.e., prior to epoching). Losing the first 50 and last 50 points of a 5-min period of continuous EEG is not much of a problem. In fact, you could just start the recording a little bit before the stimuli start and end it a little after the stimuli end, and then you wouldn't lose any important data by filtering.

Distortions Produced by Filtering

Although filters are extremely useful, they are really a form of systematic distortion. The more heavily you filter the data, the more you are distorting your data. The distortions caused by filtering can be summarized by a key principle that you should commit to memory and recall every time the topic of filtering comes up.

Fundamental principle of filtering Precision in the time domain is inversely related to precision in the frequency domain.

In other words, the more tightly you constrain the frequencies in an ERP waveform (i.e., by filtering out everything but a narrow band of frequencies or by using steep roll-offs), the more the ERP waveform will become spread out in time. Conversely, the more temporal precision you have, the broader the range of frequencies will be in your data. Given that temporal resolution is one of the main virtues of the ERP technique, this principle makes it clear that we give up a lot if we try to be highly precise about the frequency information.

To make this principle more concrete, figure 7.10 shows three types of temporal spreading that can be produced by filtering a realistic ERP waveform. These are all fairly extreme examples,

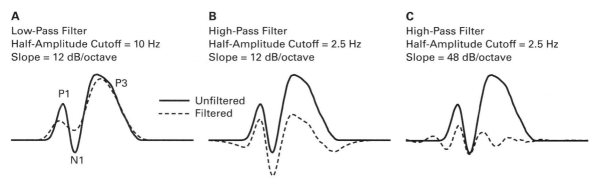

A
Low-Pass Filter
Half-Amplitude Cutoff = 10 Hz
Slope = 12 dB/octave

B
High-Pass Filter
Half-Amplitude Cutoff = 2.5 Hz
Slope = 12 dB/octave

C
High-Pass Filter
Half-Amplitude Cutoff = 2.5 Hz
Slope = 48 dB/octave

Figure 7.10
Examples of distortions caused by filtering. (A) Effects of low-pass filtering on the onset and offset times of an ERP waveform. (B) Effects of a high-pass filter with a relatively gentle roll-off. Note the artificial negative peaks at the beginning and end of the filtered waveform. (C) Effects of a high-pass filter with a relatively steep roll-off. Note the artificial oscillations in the filtered waveform.

and you should not see this kind of distortion in your own data if you follow the recommendations at the end of the chapter. Figure 7.10A shows how low-pass filtering an ERP waveform (half-amplitude cutoff = 10 Hz) causes the filtered waveform to start earlier and end later than the unfiltered waveform. This is similar to the smearing that we saw with a 13-point running average filter in figure 7.9B. Whether implemented in the time domain or in the frequency domain, low-pass filters always have this smearing effect, which may affect the onset and offset times of the ERP components and experimental effects. For example, if you see an experimental effect beginning at 120 ms in a heavily low-pass–filtered waveform, the effect may have actually started quite a bit later (e.g., 150 ms).

Figure 7.10B shows the effect of a high-pass filter with a half-amplitude cutoff at 2.5 Hz and a relatively gradual slope of 12 dB/octave. This panel shows that the spreading of voltage produced by high-pass filtering is inverted in polarity. The spread is inverted because a high-pass filter is equivalent to subtracting a low-pass–filtered waveform from the original data (as described earlier in the chapter). The smearing produced by the low-pass filter is inverted by the subtraction. In the example shown here, the positive P1 and P3 peaks at the beginning and end of the unfiltered waveform lead to artifactual negative peaks at the beginning and end of the filtered waveform. The N1 peak in the middle of the waveform also induces artificial positive activity at surrounding time points, but this just blends into the P1 and P3 peaks and is therefore difficult to see in this particular example. However, if two conditions differed only in N1 amplitude, this type of filtering could lead to artificial positive effects during the P1 and P3 peaks.

Figure 7.10C shows the distortion caused by a high-pass filter with the same half-amplitude cutoff (2.5 Hz) but a sharper slope (48 dB/octave). Instead of individual artifactual peaks at the beginning and end of the filtered waveform, we now have artifactual oscillations that extend even further in time. This sort of artifactual oscillation could also cause an experimental effect to appear to oscillate. As you can imagine, the use of this sort of filter might cause someone to completely misinterpret the results of an ERP experiment. Box 7.2 describes an influential study that reported oscillations that may have been a result of this kind of filter artifact.

Figure 7.11A shows how increasing the high-pass cutoff can attenuate the P3 wave and create an artificial peak at the beginning of the waveform (from the study of Kappenman & Luck, 2010). We found that a high-pass filter with a 0.1-Hz cutoff attenuated the P3 wave only slightly, and it did not produce any discernible artifacts. A filter with a 0.5-Hz cutoff clearly attenuated the amplitude of the P3 wave, and it also produced an artifactual negative deflection from approximately –50 to +100 ms. A filter with a 1.0-Hz cutoff reduced the amplitude of the P3 wave even more, and it also increased the amplitude of the artifactual negative deflection at the beginning of the waveform. I hope this makes it clear that high-pass filters can be dangerous when the half-amplitude cutoff is above approximately 0.1 Hz.

If you have already filtered some data, you may be wondering if your filters produced significant artifacts in your waveforms. Similarly, you may be wondering how you can tell if a given filter produces artificial peaks or oscillations. The best way to determine how a filter distorts the data is to apply the filter to a known, artificial signal. Practical directions for accomplishing this

Box 7.2
An Example from the Literature

Luu and Tucker (2001) published an influential study of the ERN in which they used relatively severe filtering in an attempt to demonstrate that the ERN is a modulation of an ongoing oscillation rather than being a discrete, transient neural response. The original waveforms for correct and error trials (time-locked to the response) are shown on the left side of the figure in this box, and the filtered data (band-pass of 4–12 Hz) are shown on the right side. When I first saw these waveforms, I thought to myself, "the filtered data look a lot like the examples of artificial oscillations that I show in my book" (see, e.g., figure 7.10C). In other words, the oscillations in the filtered data may have been an artifact of the filters rather than a true feature of the underlying brain activity.

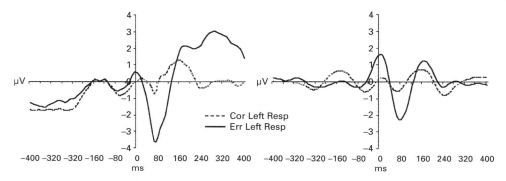

Figure 1 from Luu and Tucker (2001). Reprinted by permission. Copyright 2001 by Elsevier Science Ireland Ltd.

I planned to run some simulations to demonstrate that the pattern of results observed by Luu and Tucker (2001) could reflect a filter artifact, but then I discovered that Nick Yeung and his colleagues had already published a paper making this point (Yeung, Bogacz, Holroyd, Nieuwenhuis, & Cohen, 2007). That is, they conducted simulations demonstrating that the apparent oscillations in the filtered data of Luu and Tucker (2001) could be explained either by a true oscillation or by a filter artifact. This doesn't mean that the conclusions of Luu and Tucker (2001) were incorrect; it just means that their filtering procedure didn't provide any real evidence about whether an oscillation was present. If you filter EEG or ERP data with a narrow band-pass, you will almost always see oscillations, whether or not the underlying brain signal is oscillating.

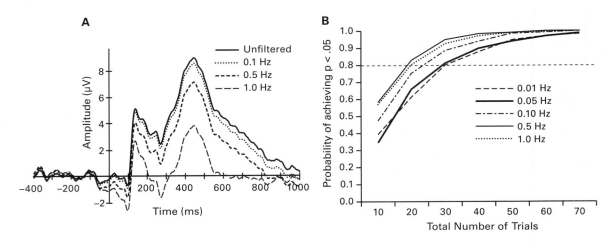

Figure 7.11
(A) Effects of different high-pass filters on the ERP waveform elicited by a visual oddball stimulus. The cutoffs indicate the half-amplitude value of a noncausal Butterworth filter with a slope of 24 dB/octave. (B) Effect of number of trials and high-pass cutoff frequency on the probability of obtaining a significant difference in amplitude between the rare targets and the frequent standards (from high-impedance recordings in a cool and dry recording environment). Adapted from the study of Kappenman and Luck (2010).

are provided in box 7.3. I strongly recommend that you give this a try. You will either learn that everything is okay (which will make you feel better) or that you need to reanalyze your data with different filters (which will allow you to avoid incorrect conclusions).

Do You Really Want to Use a High-Pass Filter?

Under most realistic conditions, high-pass filters are more likely than low-pass filters to produce incorrect conclusions. This is because the spreading of the onset and offset produced by low-pass filters is relatively modest (as long as the cutoff isn't too low) and typically influences all groups and conditions equally. In other words, the absolute onset time of an effect may appear to be slightly early, but this shift should be similar for all the waveforms in a given study. High-pass filters, in contrast, are more likely to be used with cutoffs that produce noticeable distortion, and they can create artificial peaks in the waveforms.

You may therefore wonder if you should avoid using high-pass filters. If you use a high-pass filter with a half-amplitude cutoff of 0.1 Hz or less, it's unlikely that the filter will cause any significant distortion, but it may substantially increase your statistical power, especially if you are looking at relatively late components such as P3 and N400. This is illustrated in figure 7.11B, which shows the results from a set of Monte Carlo simulations that Emily Kappenman ran with the P3 data shown in figure 7.10A (Kappenman & Luck, 2010). Statistical power (the probability of obtaining statistical significance when there truly is an effect) increased as the number of trials per subject increased, which was no surprise. The key finding was that statistical power

Box 7.3
Are Your Filters Distorting Your Data?

If you've already filtered the data for previous experiments, you may be wondering whether you significantly distorted your data. Similarly, if your advisor insists that you filter your data in a way that is inconsistent with my recommendations, you might be worried that you will distort your data. Fortunately, there is an easy way to determine how much your filters are distorting your data (assuming that the filtering in question is being conducted offline by a reasonably flexible software package). Specifically, you can create artificial waveforms that represent what you think your data might look like without any noise, and you can see how your filters distort these artificial waveforms.

To do this, you can create the artificial waveforms in a spreadsheet program (e.g., Microsoft Excel), save them as text files, import them into your analysis system, and apply your filters. To make this a little easier for you to do, I've provided an Excel spreadsheet, the corresponding text file, and an ERPLAB data file (available online at http://mitpress.mit.edu/luck2e). If you are using ERPLAB, you can just load the ERPLAB file and start filtering. If you are using a different analysis system, you can probably import the text file directly into your system. It will probably take you less than an hour of work to see how your filters distort the data. I strongly recommend that you take the time to do this. At a minimum, you may be relieved to see that your filters are producing negligible distortion. But you might find that your filters are producing severe distortion, and this exercise may prevent you from publishing an incorrect conclusion or might help you convince your advisor to follow my filtering recommendations.

for a given number of trials was substantially improved by applying a high-pass filter with a half-amplitude cutoff of 0.1 Hz compared to a lower cutoff frequency (0.01 Hz) or to unfiltered data (not shown here). Power was even better for higher cutoff frequencies (0.5 or 1.0 Hz). However, these higher cutoffs led to substantial distortion of the waveforms (as shown in figure 7.10A). Thus, 0.1 Hz appears to provide the best balance between statistical power and waveform distortion for the P3 wave.

Two additional findings from this study should be noted. First, the effects shown in figure 7.11B were obtained with data from high-impedance recordings. As discussed in chapter 5, low-frequency noise is typically much more of a problem when the electrodes impedances are high than when they are low. Emily found that filtering had much less of an effect on power in low-impedance recordings, presumably because there was less low-frequency noise to be filtered. Second, high-pass filtering had very little effect on statistical power when N1 amplitude was analyzed instead of P3 amplitude. As will be discussed in chapter 8, baseline correction serves as a form of high-pass filtering, but it becomes progressively less effective at filtering low frequencies later in the epoch (see especially figure 8.2D). The N1 wave occurs very soon after the baseline period, so additional filtering of low frequencies does not make much difference. The P3 is farther away from the baseline period, and it is therefore more affected by slow voltage drifts. Thus, high-pass filters are especially useful when you are measuring later components such as P3, N400, and LPP.

Recommendations for Filtering

Now that I've discussed why filters are used, how they work, and how they can distort your data, I will provide some concrete recommendations about filtering. These recommendations may not be appropriate for every experiment, but they will work well for the vast majority of ERP experiments in cognitive and affective neuroscience. If you want to filter your data more heavily than I recommend here, you need to understand the details of how filters work (e.g., by reading online chapter 12 and making sure that you understand all the math), and you should try filtering artificial waveforms to see the distortions produced by your filters (see box 7.3).

Before you start filtering your data, you should remind yourself of Hansen's axiom: *There is no substitute for clean data* (see chapter 5). Some minor filtering is necessary when the data are first being collected, and a modest amount of additional offline filtering is usually a good idea. However, filters cannot help very much if your data are noisy because of variability across subjects, variability across trials, a small number of trials in your averages, and so forth. Filters may make the data *look* better under these conditions, but this may be an illusion that could lead you to draw incorrect conclusions.

Online Filter Recommendations

In general, you should do only minimal filtering online (i.e., hardware filtering during data acquisition). It's always possible to filter the data more offline, but you can't "unfilter" data that have already been filtered. Moreover, the filters that can be applied in software offline are superior to online filters (e.g., online filters will produce a latency shift, but most offline filters will not).

No matter what kind of EEG recording system you are using, you will definitely need to use a low-pass filter to prevent aliasing during data acquisition. The half-amplitude cutoff should be between one-third and one-fifth of the sampling rate so that the gain of the frequency response function is near zero at the Nyquist frequency (one-half the sampling rate). Many systems will automatically select an appropriate anti-aliasing filter when you select the sampling rate.

In many situations, an anti-aliasing filter is the only online filter I would recommend using. Some systems allow you to apply additional filters to the EEG that you are viewing during data acquisition, without actually applying these filters to the data that are being saved to disk. This is a really great option because it's easier for you to monitor the data for artifacts and bad connections if you are viewing filtered data, but it's always better to do the final filtering in software offline. If your system has this ability, I would recommend viewing the EEG with a band-pass of 0.1–30 Hz (i.e., a high-pass filter with a half-amplitude cutoff of 0.1 Hz and a low-pass filter with a half-amplitude cutoff of 30 Hz). You should also look at the unfiltered data occasionally to see if you have large skin potentials or high levels of line noise or EMG activity.

If your data acquisition system has fewer than 20 bits of resolution, you should apply a hardware high-pass filter prior to digitization (see chapter 5 for the rationale). Otherwise, large voltage offsets (due to skin potentials, movement artifacts, etc.) may cause your system to saturate. If you are in this situation, you should use a half-amplitude cutoff somewhere between 0.01 and 0.1 Hz. If your subjects are prone to skin potentials and movement artifacts (e.g., children, neurological patients), I would recommend 0.1 Hz to avoid excessive data loss. With highly cooperative subjects (e.g., healthy young adults), I would recommend 0.01 Hz (with additional high-pass filtering offline).

If your data acquisition system does not allow you to view filtered data while saving the unfiltered data, you may need to apply a notch filter at the line frequency (50 or 60 Hz, depending on where you live). Notch filters are not ideal, but if you have high levels of line noise, you may not be able to monitor the EEG adequately without one. Of course, it's much better to minimize sources of electrical noise in the environment so that the EEG is not contaminated by line noise (as described in online chapter 16). However, this is not always possible (e.g., if you are recording from the bedside in a hospital).

Offline Filter Recommendations

For typical experiments on cognitive and affective processes (or perceptual processes that begin after approximately 50 ms poststimulus), I recommend adding offline filters, if necessary, to achieve a final band-pass of approximately 0.1–30 Hz. That is, you should end up with a half-amplitude high-pass cutoff of 0.1 Hz to attenuate skin potentials and other slow voltage changes and a half-amplitude low-pass cutoff of 30 Hz to attenuate line noise and EMG noise. You don't need to use these exact values; anything between 0.05 and 0.2 Hz for the low end and between 20 and 50 Hz for the high end will be fine. And I recommend a slope of between 12 and 24 dB per octave (a steeper slope is usually acceptable for the low-pass filter but not for the high-pass filter). As mentioned earlier, you should use noncausal rather than causal filters for your offline filtering, which will avoid latency shifts.

I would again like to stress that you should not use a narrower bandwidth than this unless you really know what you're doing. Otherwise you may end up drawing unjustified conclusions (as in the study described in box 7.2).

If you filter your data during data acquisition with a band-pass of approximately 0.1–30 Hz, there is no need to filter again offline. However, if you use a broader band-pass during data acquisition (e.g., 0.01–100 Hz), you can filter again offline to achieve a final band-pass of 0.1–30 Hz.

There are always exceptions to any rule, and here are some common exceptions to my basic filter recommendations:

• If you are looking at very slow or late components, like the CDA or LPP, the conventional advice would be to set your high-pass filter at a lower frequency (e.g., 0.01 Hz) to avoid attenuating the amplitude of the component you are trying to measure. However, the benefits of filtering out low-frequency noise will also be greatest for these late components, and a modest attenuation of the signal might be more than offset by a large attenuation of noise. Thus, you might want to try a lower cutoff, but I suspect you will ultimately find that 0.1 Hz (or perhaps 0.05 Hz) is the best compromise between attenuation of the signal and noise reduction.

• If you are quantifying component amplitudes by measuring the mean amplitude over a time window of at least 50 ms (e.g., quantifying P3 amplitude as the mean voltage from 300 to 500 ms), there is no need to apply a low-pass filter prior to the measurement. The use of a wide window will already attenuate the high-frequency noise (see chapter 9 for details).

• If you are measuring the onset latency of an ERP component (or some other highly noise-sensitive feature of the waveform), you may need an even lower cutoff for your low-pass filter (e.g., 10 Hz). However, you should not do this unless you first gain a fuller understanding of filtering (e.g., by reading online chapter 12 and by filtering some artificial waveforms).

• If you are interested in very fast sensory responses that occur within 50 ms of stimulus onset (e.g., the auditory brainstem responses), you will want to use higher cutoff frequencies for both the low-pass and high-pass filters. To choose the precise frequencies, you should read papers in your area to see what other people use.

When Should You Filter?

During ERP Boot Camps, I am almost always asked whether filters should be applied before or after epoching, before or after artifact rejection, before or after averaging, and so forth. The answer usually depends on the particulars of an experiment, but you can make the right choice if you understand a fundamental principle of data processing: *the order of operations does not matter for linear operations*.

This principle is fully explained in the appendix of this book (including the meaning of the term *linear operations*). The basic idea is that some sorts of mathematical procedures give you the same result no matter the order in which they are applied. For example, (A + B) + C is equal to A + (B + C). With more complicated operations, such as filtering, the order of operations does not matter if the operations are linear. For example, if X and Y are linear operations, you can apply operation X to an ERP waveform and then use the resulting waveform as the input to operation Y, and this will yield the same end result as applying operation Y to the ERP waveform and then using the resulting waveform as the input to operation X.

Many of the filters that are used with EEG and ERP data are *finite impulse response* filters, and these are linear. Others are *infinite impulse response* filters (e.g., Butterworth filters), and these are nonlinear. However, infinite impulse response filters are approximately linear when the band-pass is fairly broad and the slope is fairly gentle. Thus, if you follow my recommendations for filtering, you can apply filtering and other linear operations in any order and get exactly (or almost exactly) the same result. However, you will need to keep in mind one caveat, described in the next section.

A very simple example of this principle arises when you filter the data twice. In the example shown in figure 7.5, it doesn't matter if you apply filter A and then filter B or if you apply filter B and then filter A. Either way, the result is equivalent to a single filter with a frequency response function that is the product of the frequency response functions of A and B. Similarly, it doesn't matter if you first apply a low-pass filter and then a high-pass filter or if you first apply a high-pass filter and then a low-pass filter. And applying a low-pass filter and a high-pass filter sequentially is equivalent to applying a band-pass filter in a single step.

Averaging is also a linear operation. Consequently, it doesn't matter if you filter the EEG immediately before averaging or if you filter the ERP immediately after averaging. Re-referencing is also a linear operation (except in unusual cases), so you can filter either before or after you re-reference the data.

In contrast, artifact rejection is a nonlinear operation, and filtering the EEG and then performing artifact rejection will lead to different results than performing artifact rejection and then filtering. The simple rule is that you should filter prior to artifact rejection if the filtering makes it easier for you to detect artifacts, and you should filter after artifact rejection if the filtering makes it more difficult to detect artifacts. For example, if you have a large amount of 60-Hz noise in the data, this may make it difficult to detect blinks by means of an absolute voltage threshold or by means of the moving window peak-to-peak amplitude approach (see chapter 6). In this case, you might want to apply a low-pass filter with a half-amplitude cutoff of 30 Hz prior to artifact rejection. If your data are reasonably clean, you will get the same results (or nearly the same results) if you filter before or after artifact rejection.

Edge Artifacts

There is one important caveat about the order of operations, which is that filters may work improperly when applied to short segments of data (e.g., epoched EEG segments or averaged ERP waveforms). In particular, filters may produce *edge artifacts* at the very beginning and very end of the waveform. These artifacts occur because the filtered value at a given time point is computed by using the surrounding time points; some of these time points may not exist near the beginning and end of the waveform, which may cause the filter to work incorrectly. We already discussed this issue in the context of running average filters, but the same principle applies to any filter (whether applied in the time domain or in the frequency domain).

This problem is especially severe when the cutoff frequency is low (for both low- and high-pass filters) and when the roll-off is steep, because these factors require the use of more time points to compute the filtered data. A filter with a cutoff frequency of 30 Hz and a slope of 12 dB/octave does not require many time points surrounding the current time point, whereas a filter with a cutoff frequency of 0.1 Hz and a slope of 48 dB/octave requires many time points. In addition, it is difficult to accurately estimate low frequencies from short time periods. For example, you get only half a cycle of an 0.5-Hz sine wave in a 1000-ms averaged ERP waveform, making it difficult to accurately estimate and remove this frequency from the data. Thus, you should avoid filtering short epochs of data with low cutoff frequencies. It is usually fine to apply low-pass filters to epoched EEG or averaged ERPs, but high-pass filters should usually be applied to the continuous EEG. In general, I recommend applying a 0.1-Hz high-pass filter to the continuous EEG, and then applying a 30-Hz low-pass filter after averaging.

Low cutoff frequencies work reasonably well when applied to long periods of continuous EEG because the beginning and end of the EEG waveform represent a small proportion of the overall signal and because there will be many cycles of the very low frequencies. For example, if you filter a 5-min continuous segment of EEG (i.e., the EEG recorded during a single block of trials), any edge artifacts will be a very small part of the overall EEG. To be extra careful, I recommend starting the EEG recording about 20 s prior to the beginning of the trial block and ending it about 20 s after the end of the trial block (making sure that you don't have a lot of movement or other artifacts during these periods. That way, any edge artifacts will occur during a period that does not have any event codes and will not contribute to your ERPs.

Infinite impulse response filters (e.g., Butterworth filters) do not need to use as many time points as finite impulse response filters. Consequently, it can be advantageous to use an infinite impulse response filter when you are filtering epoched EEG data or averaged ERP data. For this reason, ERPLAB Toolbox mainly implements Butterworth filters.

8 Baseline Correction, Averaging, and Time–Frequency Analysis

Overview

This chapter mainly focuses on baseline correction and averaging, which are very simple processes. Baseline correction is achieved by subtracting the average prestimulus voltage from the waveform, and averaging simply consists of summing together a set of EEG epochs and then dividing by the number of epochs. Simple, right? Well, these processes are simple, but their effects can be quite complex, and they can lead to important misinterpretations of the results. I will therefore spend quite a bit of time explaining how they really work and how they can lead you to an incorrect conclusion if you're not careful.

I will begin the chapter by discussing the epoch-extraction and baseline-correction processes that typically precede averaging. I will then describe how averaging decreases the noise in your data and how the number of trials will influence the *p* value you get at the end of your experiment. In this section, I will provide some general advice about how many trials you will need to average together. I will also demonstrate how different subjects can have very different-looking average ERP waveforms and discuss how these differences arise.

I will then discuss trial-to-trial variations in the amplitude and latency of an ERP component, focusing on how latency variability can lead to incorrect conclusions and providing some solutions to this problem. This issue will be described in greater detail in online chapter 11, which also introduces a useful mathematical concept called *convolution*.

The chapter will end with an introduction to time–frequency analysis, in which the time course of specific frequency bands is extracted from the data. Time–frequency analysis will be explored in more detail in online chapter 12, but it's really just averaging with an additional preprocessing step.

Extracting Epochs of EEG Data

As discussed in chapter 5, the EEG is usually recorded as a continuous signal for an entire trial block, and event codes are present to indicate the occurrence of stimuli and responses. Prior to averaging, it is necessary to extract fixed-length *epochs* or *segments* of data from the continuous

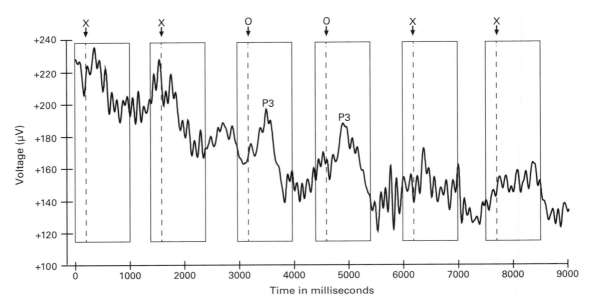

Figure 8.1
Epoch extraction. This example shows 9000 ms of EEG from the Pz electrode in an oddball experiment, with frequent X standards and rare O targets. By chance, there were two consecutive targets in this particular EEG segment. A large P3 component is visible after each target. Prior to averaging, 1000-ms segments of the EEG are extracted. Each segment starts 200 ms prior to an event code and continues until 800 ms after the event code. Note that the EEG has a substantial voltage offset (which is common when the data have not been high-pass filtered), and it also drifts downward due to a changing skin potential. Baseline correction is used to remove this drift, usually at the time of epoch extraction.

EEG, time-locked to the event codes of interest. Each epoch includes a baseline period prior to the event code (typically 100–200 ms) as well as a period after the event code (typically 500–1500 ms, depending on which components will be examined). This is illustrated in figure 8.1, which shows a 9000-ms period of continuous EEG from an oddball experiment in which the letter X was frequent (90%) and the letter O was rare (10%). This particular period of EEG happened to contain four Xs and two Os, which may not seem to fit with the 90%/10% probabilities, but that's the sort of thing that happens with a random sequence. You can see that each of the two rare O stimuli elicited a large P3 wave that is visible in the raw EEG.

Prior to averaging across trials, epochs of data were extracted from the continuous EEG, beginning 200 ms prior to each stimulus and ending 800 ms after each stimulus. Most current EEG recording systems will ordinarily save the entire continuous EEG onto the hard drive, but some systems give you the option of saving only discrete epochs around the event codes. Saving the epochs rather than the continuous EEG usually saves disk space, but I generally recommend against it during recording because this limits your options during analysis. For example, a reviewer may ask for a longer baseline period or for response-locked averages, and you will be out of luck if you did not save the continuous EEG. Also, as described in chapter 7, it is usually better to apply high-pass filters to long periods of continuous EEG rather than to EEG epochs. Note that when the epochs are long relative to the rate of stimulation, one epoch may begin before the previous one has ended (which is not a problem).

Baseline Correction

After you extract the epochs from the continuous EEG, you will typically perform a baseline correction procedure. This is necessary because factors such as skin hydration and static charges in the electrodes may cause an overall vertical offset in the EEG (as was discussed in detail in chapter 5). Figure 8.1 illustrates this offset, showing that the EEG voltage is well over 100 μV throughout the entire 9000-ms interval shown in the figure. If we did not do some kind of correction procedure, the voltage offset would have a huge impact on our ERP amplitude measurements. For example, if we averaged the EEG epochs for the rare stimuli and then measured the P3 amplitude without performing some kind of baseline correction, the measured value would be approximately 180 μV for this subject. For another subject or another period of time, the measured value might be something like –270 μV (if there was a negative voltage offset for that subject or that time period). These differences in voltage offset across time periods and subjects are completely random and are unrelated to brain activity. If we did not remove these offsets, they would add tremendous uncontrolled variance to the data, making it nearly impossible to see significant differences between conditions or groups.

Factors like skin hydration, skin potentials, and static electrical charges often vary slowly over time, causing the offset voltage to drift gradually upward and downward. In the EEG shown in figure 8.1, for example, the voltage gradually drifts downward across the 9000-ms interval. We need to correct for this drift or else it will add substantial error variance to our amplitude measurements, reducing statistical power.

Baseline correction is a simple procedure that can minimize these offsets and drifts. In most cases, the time-locking event for the epoch is a stimulus, and we can assume that the voltage during the prestimulus period can provide a good estimate of the voltage offset for that trial (because it contains the offset but does not contain any stimulus-elicited ERP activity). If we simply subtract this estimate of the offset from the entire epoch, this will eliminate the offset. This is shown in figure 8.2A. In this figure, my data-analysis program computed the average voltage during the prestimulus period to estimate the voltage offset, and then it subtracted this value from every point in the epoch. This just shifts the waveform upward or downward (depending on whether the voltage offset is positive or negative), centering the prestimulus period around the 0-μV line (hint: if this is working correctly, the area of the waveform above the zero line should be the same as the area below the line during the prestimulus period).

In many cases, this baseline correction works beautifully to minimize voltage offsets and gradual drifts. However, it assumes that the prestimulus period contains only the voltage offset, and this assumption is not always correct. For example, the prestimulus period may contain ERP activity from the end of the preceding trial or it may contain preparatory activity that occurs

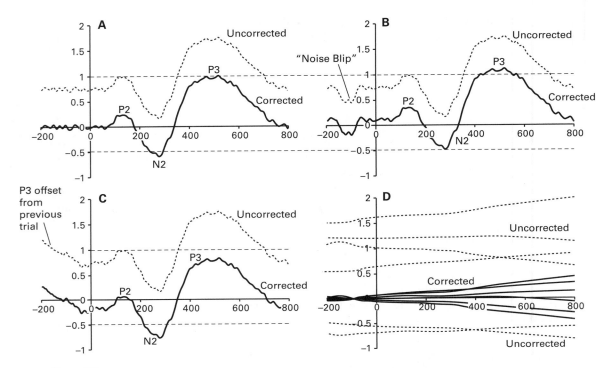

Figure 8.2

Baseline correction. (A) Without baseline correction, the averaged ERP waveform is shifted vertically (reflecting the overall voltage offset of the EEG). The vertical shift is corrected by computing the average voltage during the prestimulus interval and subtracting this voltage from each point in the waveform. This is ordinarily done on the single-trial EEG epochs, but it can also be done on the averaged ERP waveforms. (B) In this example, a negative-going voltage deflection (noise blip) is present during the prestimulus period, and the baseline correction therefore fails to "push" the waveform down as far as it should go. This affects the apparent amplitude of the P2, N2, and P3 components. Horizontal dashed lines at –0.5 and +1.0 µV are provided to facilitate comparison of the amplitudes in panels A and B. (C) In this example, overlapping P3 activity from the previous trial is present during the prestimulus interval, causing the ERP waveform to be pushed too far downward by the baseline correction procedure. (D) Illustration of how trial-to-trial variance tends to increase as time passes from the center of the baseline period. Each dashed line represents slow drift in a single trial without any baseline correction (and without any stimulus-elicited ERP activity, just to make the drift clearer). The solid lines represent the baseline-corrected single trials. The baseline correction shifts each waveform vertically so that they are all together during the baseline period. As time moves away from the baseline period, the signal tends to drift farther and farther away from the baseline voltage.

when the subject anticipates the onset of a stimulus. As we shall see, baseline correction is necessary, but it may lead to unanticipated consequences, so you should think carefully about what it is doing to your data.

Baseline correction is usually performed on the EEG immediately after the data are epoched. This is necessary to avoid problems with some types of artifact rejection (see chapter 6), and it can make the epoched data easier to view. Thus, most software systems perform baseline correction at this point. Some systems allow you to perform baseline correction again after averaging so that you can change the baseline period (assuming that the new baseline period falls within your epochs). You will typically get the same result by performing baseline correction before or after averaging, except when baseline correction affects artifact rejection (see the appendix of this book for a discussion of the factors that influence the order in which you perform operations such as baseline correction, artifact rejection, averaging, and filtering).

High-pass filters will eliminate much of the EEG offset, and you might think that baseline correction would not be needed if you've already applied a high-pass filter. However, filtering is a fairly blunt instrument for removing the offset. In contrast, baseline subtraction is based on the very precise assumption that the voltage during the prestimulus period should contain nothing except the offset and noise. When this assumption is met, baseline correction is the best way to remove the offset. When it is not met, it is still better than filtering alone, because filtering can create systematic differences across conditions if used as the only method of baseline correction. An explicit baseline correction procedure is therefore necessary in almost all conventional ERP studies.

Effects of Baseline Correction on Amplitude Measurements

Although baseline correction is usually performed at the time of epoching, it has a large (and often unappreciated) effect on the amplitude measurements that you make on your averaged ERP waveforms at the end of your data processing pipeline. Specifically, baseline correction involves subtracting the mean baseline voltage from the entire waveform, which therefore affects the amplitude at each point in the waveform. Once you've subtracted the baseline, the voltage at each time point in the waveform represents the difference between that point and the average baseline voltage, and anything that influences the baseline therefore influences your poststimulus amplitude measurements. To illustrate this, figure 8.2B shows how a "noise blip" (a small, random voltage transient) during the prestimulus baseline period influences the amplitude of every component in the ERP waveform. These types of voltage blips are very common in ERP data (and usually reflect EEG noise that remains after averaging a finite number of trials). The negative-going voltage blip in figure 8.2B causes the mean of the prestimulus voltage to be an underestimate of the actual voltage offset, and the voltage subtracted from the ERP waveform is therefore too small. As a result, the waveform isn't "pushed" far enough downward by the baseline correction procedure. This causes the measured amplitude of the N2 to be smaller (less negative) and the measured amplitude of the P3 to be larger (more positive) than they should be. Horizontal lines are drawn at −0.5 µV and +1.0 µV in the figure so that you can see how the peaks are shifted upward in the presence of the noise blip.

Thus, when you measure the amplitude of some poststimulus time period, you should always realize that you are actually measuring the difference between this period and the baseline voltage. Consequently, any noise in the baseline will create noise in your measurements, thus decreasing your statistical power.

Figure 8.2C shows how overlap from the previous trial can influence the amplitude measurements of the ERP components on the present trial. This example simulates the effect of a very short interval between stimuli (~600 ms), which causes the final portion of the ERP from one trial (in this case, the P3 wave) to overlap with the prestimulus period of the next trial. This causes the prestimulus voltage to be an overestimate of the actual offset voltage (greater positivity), and this causes the waveform to be shifted too far downward when the mean prestimulus voltage is subtracted from the waveform. This in turn causes the measured N2 to be artificially large and the measured P3 value to be artificially small. Unlike random noise, this overlap would be consistent across subjects, biasing the voltage measurements toward more negative values for everyone. The same is true of preparatory activity and the offset of blinks during the prestimulus interval (see figure 6.3 in chapter 6). Overlap confounds are on my list of *Top Ten Reasons to Reject an ERP Paper* (see online chapter 15), and overlap is discussed in more detail in online chapter 11. I also encourage you to read a paper on overlap by Marty Woldorff (1993), which is on my list of *Papers Every New ERP Researcher Should Read* (see the end of chapter 1).

As discussed in detail by Urbach and Kutas (2006), baseline correction can have large effects on your ERP waveforms, and you should not assume that it is a benign process that simply centers your waveform on some kind of true zero voltage. All of your poststimulus measures are essentially difference scores between the baseline period and your poststimulus measurement period. This is analogous to the no-Switzerland principle that was described in the context of the reference electrode in chapter 5. That is, just as the voltage at a given electrode site is really the difference between that site and the reference, the voltage at a given time is really the difference between that time and the baseline period.

Figure 8.2D shows an important principle about baseline correction; namely, that slow drifts in voltage tend to cause more and more deviation away from zero the farther you get away from the baseline period. In the first 100 ms after the baseline period, for example, there has not been much time for the voltage to drift away from the mean voltage during the baseline period. By 600 ms after the baseline period, however, there has been more time for the voltage to drift away from the baseline. The drift will sometimes be positive and sometimes be negative, leading to trial-to-trial variation in the amplitude that increases as you get farther and farther away from the baseline period. This increase in variance will make your amplitude measurements less reliable later in the waveform than they are earlier in the waveform (all else being equal). This will tend to reduce statistical power for measurements made very late in the waveform. For this reason, ERPs tend to be especially good for assessing brain activity in the first second or so after an event, but they provide a relatively poor measure of activity many seconds or minutes after an event (assuming you are using the pre-event period as the baseline). If you are interested,

Box 8.1
Accumulation of Variance over Time

The effects of drifts after the baseline period can be understood in terms of a fundamental principle of statistics. Specifically, if you add two uncorrelated random variables to create a new random variable, the variance of the new random variable is equal to the sum of the variances of the two original variables (this is called the Bienaymé formula). For example, if you measure the weight and the annual salary of each individual in a population, and you calculate the weight plus the annual salary of each individual, the variance of this sum will be equal to the variance of the weight alone plus the variance of the salary alone. To apply this principle to EEG drift, imagine that the voltage drifts up and down randomly over time, creating trial-to-trial variations in the mean amplitude measured from 0 to 100 ms after a 100-ms baseline period. If the voltage continues to drift up and down randomly from 100 to 200 ms after the baseline period, this will add even more variation from trial to trial. If we assume that the drift is truly random, then the variance that occurs from 100 to 200 ms will simply add to the variance that occurred from 0 to 100 ms. Thus, if the trial-to-trial variance is X μV in each 100 ms period, then the variance after 200 ms will be $2X$ μV, the variance after 300 ms will be $3X$ μV, and so forth. Note that this is only an approximation because the Bienaymé formula is only true for uncorrelated variables, and the EEG in one time period will be correlated with the EEG in the next time period. Even in this case, however, the variance grows as more and more time passes.

box 8.1 provides a more precise way of describing how the variance across trials increases as time passes from the baseline period.

The fact that the measurements become less reliable as you move farther away from the baseline period has important implications for how you should choose the baseline period. In many language experiments, for example, subjects see or hear sentences containing many words, and the ERP elicited by the last word of each sentence is of interest. You might think it would be best to use the voltage prior to the onset of the sentence as the prestimulus baseline, because it is not contaminated by overlapping activity from prior words (see, e.g., figure 3.13 in chapter 3). However, researchers often use the interval just prior to the final word as the baseline—even though it is contaminated by overlap from the preceding word—because this reduces the amount of slow drift between the baseline period and the ERP elicited by the final word, thus increasing the reliability of the measurements. Although this baseline is contaminated by overlap from the previous word, this overlap is not usually a problem if it is identical across conditions (see online chapter 11 for a detailed discussion). Consequently, it may be worth using a baseline that is contaminated by overlap if it allows the baseline period to be closer in time to the measurement period.

The Importance of Looking at the Baseline

Here's an important bit of practical advice: Whenever you look at a set of ERP waveforms—whether your own or someone else's—you should look at the prestimulus baseline activity before

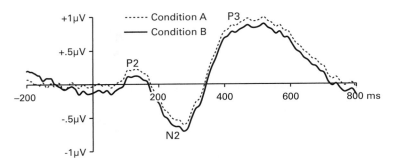

Figure 8.3
Example of how differences between conditions in prestimulus activity can lead to the artifactual appearance of differences in poststimulus activity. The only true differences between these waveforms are in the prestimulus baseline period.

looking at anything else. The baseline interval will tell you several important things. First, it will tell you how much noise remained in the data after averaging. Specifically, if the voltage fluctuations in the prestimulus baseline are as big as the experimental effects in the poststimulus period, you should be very skeptical about whether the effects are real (even if they are statistically significant, which will happen for bogus effects one out of every 20 times).

Second, the baseline will tell you if the waveforms are contaminated by overlapping activity from the previous trial or preparatory activity that occurred before the onset of the stimulus in the present trial. Such effects are not always problematic, but you should definitely think about them.

Third, if the differences between conditions or groups begin in the prestimulus interval or shortly thereafter, they are likely some kind of artifact. This is illustrated in figure 8.3, which shows data from a simulated oddball experiment in which the voltage in the prestimulus baseline period differs between the two experimental conditions. Because the differences in the prestimulus period are subtracted from the entire waveform during the baseline correction process, the prestimulus difference leads to an artifactual difference between conditions during the poststimulus period. This makes the P3 in condition B appear to be smaller than the P3 in condition A, even though the difference is actually in the prestimulus activity. If you did not look carefully at the baseline, you might reach the incorrect conclusion that the P3 differed across conditions. However, if you look at the baseline, you can see that the "experimental effect" begins at time zero, which is of course impossible. Thus, it is important to look at the baseline and be very suspicious about any effects (differences between groups or conditions) that begin near time zero. You should also be a little suspicious if you're reading a paper and it doesn't show a prestimulus baseline in the ERP waveform plots (although I have to admit that I am a co-author on a paper in which this happened by accident—see Hopf, Vogel, Woodman, Heinze, & Luck, 2002).

In addition, you should be highly suspicious about any effects that begin within 100 ms of stimulus onset unless they reflect differences in the stimuli (e.g., larger sensory responses for brighter stimuli). It is very rare that a purely cognitive manipulation influences activity this early. Attention can influence ERP responses within the first 100 ms when attention was focused prior to the onset of the stimulus (reviewed by Luck & Kappenman, 2012a), but most early effects turn out to be a result of noise, artifacts, or confounds.

From the examples shown in figures 6.2 and 6.3 of chapter 6, you might think that baseline correction is a bad idea. After all, baseline correction causes noise and overlap in the prestimulus period to distort the poststimulus voltages. However, if you tried to look at your ERPs without baseline correction, you would be faced with enormous variability from one subject to the next owing to random differences in EEG offset (as illustrated in figure 8.2A). You could try to filter out the EEG offset with a high-pass filter, but this ends up having all the same problems as baseline correction and also creates additional distortions as described in chapter 7 and online chapter 12. Consequently, baseline correction using the mean of the prestimulus voltage is the best option in 99.9% of experiments, and the problems of noise blips and overlap are best solved by recording clean data and by designing experiments in which the overlap is identical across conditions (or using one of the strategies described in online chapter 11 to minimize overlap).

Specific Recommendations for Baseline Correction

The optimal length of a baseline period reflects a balance between the benefits and costs of using a long baseline period. Under ideal conditions, using a longer baseline period will give you a more accurate estimate of the true voltage offset (because little noise blips will tend to cancel out if you're averaging over more points in your baseline). This in turn gives you a more precise measure of the poststimulus amplitudes.

However, if your baseline period is too long, this can have three negative consequences. First, a longer baseline means that you will be performing artifact rejection over a longer time period, which may substantially increase the number of trials that you reject (especially if subjects blink during the intertrial interval). Second, it will move much of the baseline interval farther away from the time period of the components you will be measuring, therefore increasing the gradual drift that occurs as time passes from the baseline interval (see figure 8.2D). Third, the longer your prestimulus baseline interval, the more likely it will be that ERP activity from the previous trial will contaminate your baseline.

In most cases, I recommend a baseline period that is at least 20% of the overall epoch duration (e.g., you could use a 100-ms prestimulus baseline with a 500-ms epoch or a 400-ms prestimulus baseline with a 2000-ms epoch). I typically use a 200-ms prestimulus baseline with an 800-ms poststimulus period when I want to see late components such as P3, and I use a 100- or 200-ms prestimulus baseline with a poststimulus period of 300–500 ms if I'm focusing only on earlier components such as P1 and N2pc.

In most cases, it's a good idea for the baseline period to be a multiple of 100 ms, because this will tend to cancel out alpha-frequency (10 Hz) EEG oscillations. That is, an equal number of the negative and positive portions of an alpha cycle will be present within a given 100-ms period,

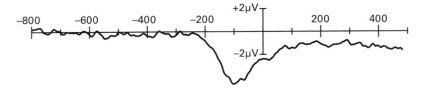

Figure 8.4
Response-locked lateralized readiness potential difference wave, formed by a contralateral-minus-ipsilateral subtraction, collapsed across the C3 and C4 electrode sites. Prior to averaging, the individual epochs were baselined from −800 to −600 ms with an epoch that extended from −800 to +500 ms (relative to the time of the behavioral response).

and these portions will therefore cancel each other out. A 200-ms baseline period includes two full cycles of the alpha oscillation, and I find this does a good job of minimizing the effects of alpha activity.

Up to this point, I have been limiting the discussion to stimulus-locked averages because baseline correction is a little more complicated for response-locked averages. Figure 8.4 shows an example of a response-locked LRP difference wave, formed by a contralateral-minus-ipsilateral subtraction (see the section on "Response-Related ERP Components" in chapter 3 for details). Time zero is the onset of the response. You can see from the waveform shown in figure 8.4 that substantial ERP activity was present in the 200-ms period immediately prior to time zero, so this period does not provide a good estimate of the EEG offset.

To determine an appropriate baseline period, the first step is usually to create an average with a long pre-response interval and then use this to determine when the waveform deviates from a flat line. For example, the data shown in figure 8.4 were initially baseline corrected from −800 to −600 ms, with an overall epoch that extended from −800 to +500 ms. It was clear that the waveform was essentially flat from −800 ms until approximately −250 ms and that the LRP was mostly complete by +200 ms. It would therefore be reasonable to re-epoch the EEG from −500 to +200 ms (relative to the response) and use a new baseline period of −500 to −300 ms.

An alternative solution would be to compute the average voltage over the 200 ms prior to each stimulus on each trial and subtract this prestimulus-defined voltage from each response-locked epoch. However, the onset time of the stimulus relative to the response varies quite a bit from trial to trial (because the reaction time varies from trial to trial), so it is not easy in practice to use the prestimulus interval as the baseline in response-locked data.

Averaging

Basics of Signal Averaging

After you have epoched and baselined the EEG, the next step is usually artifact rejection (as described in chapter 6), and then you are ready to average the data. The actual averaging process is quite simple. The EEG epochs are aligned with respect to the time-locking event, and the voltages from all the EEG epochs at a given time point are averaged together. This type of averaging is sometimes called *signal averaging*. Note that this is equivalent to summing the single-trial EEG waveforms and then dividing by the number of trials.

Signal averaging is used to attenuate noise so that the event-related brain activity can be seen more easily. The EEG data collected on a single trial consists of event-related brain activity (i.e., the ERP) plus other activity that is unrelated to the event (e.g., other EEG activity, skin potentials, muscle artifacts, ocular artifacts, induced environmental noise, etc.). The ERP is assumed to be largely identical on each trial, whereas the other activity is assumed to be completely random with respect to the time-locking event. That is, the noise at a given time point is positive-going on some trials and negative-going on other trials, and the positives and negatives will cancel each other out when the values are averaged together across trials, leaving just the consistent ERP activity.

When I was in graduate school, Steve Hillyard taught me some "advanced" terminology for describing positive and negative voltage deflections, which I would like to share with you. Upward-going deflections are called *uppies*, and downward-going deflections are called *downies*.[1] Because of noise, some trials will have an uppie at a given time point and others will have a downie, and these uppies and downies cancel out when many trials are averaged together. Given a finite number of trials, the uppies and downies will not be perfectly equal and will not cancel out perfectly, so some noise will remain in the averaged ERP waveform. However, the uppies and downies that remain in the averaged waveform tend to become smaller and smaller as more and more trials are averaged together.

Figure 8.5 illustrates how increasing the number of trials leads to reduced noise in the averaged ERP waveforms for the targets in the oddball experiment shown in figure 8.1. The left column in figure 8.5 shows the EEG epochs for eight different target trials. The P3 wave for this subject was quite large, and it can be seen in every trial as a broad positivity between 300 and 700 ms. However, there is also quite a bit of variability from trial to trial in the exact shape of the P3 wave, and this is at least partly due to random EEG fluctuations (although the P3 itself may also vary from trial to trial). The other components are too small to be clearly visible on the single trials.

The right column in figure 8.5 shows how the effects of the random EEG fluctuations are minimized as more and more trials are averaged together. Each row shows the average of the single trials up to that row (e.g., row 3 shows the average of the first three targets and row 6 shows the average of the first six targets). If you look at the prestimulus baseline period, you can see that the uppies and downies in the averaged waveforms get progressively smaller as more trials are averaged together. The poststimulus waveshape also becomes progressively more stable as more trials are averaged together. Note that the P3 is quite nice looking in figure 8.5 when only eight trials were averaged together. This is because the single-trial EEG noise was fairly small and the P3 wave was very large. You will typically need far more trials than this in your averages. I will provide some specific advice about this later in the chapter.

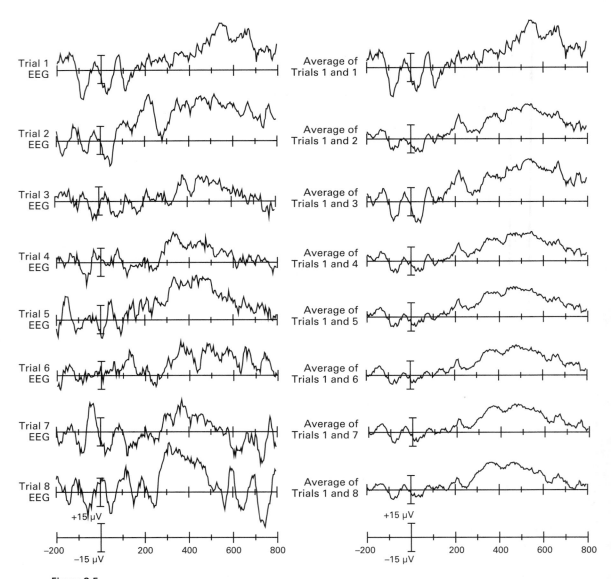

Figure 8.5
Example of signal averaging. The left column shows segments of EEG for each of several target trials in the oddball paradigm shown in figure 8.1, time-locked to stimulus onset. The right column shows the effects of averaging 1, 2, 3, 4, 5, 6, 7, or 8 of these EEG segments.

Signal-to-Noise Ratio

Now that I've provided an informal description of how averaging reduces noise, it's time to get more precise. The absolute amount of noise is not as important as the size of the noise relative to the size of the signal. Thus, researchers typically focus on the *signal-to-noise ratio* (SNR), which is simply the size of the signal divided by the size of the noise. For example, if the signal of interest is a 20-μV P3 wave, and the peak-to-peak EEG noise on a typical trial is 50 μV, we would say that the SNR is 20:50, or 0.4. If we assume that the signal is the same on every trial, then averaging will keep the size of the signal the same but will reduce the noise, thereby increasing the SNR.[2]

The noise in an average decreases progressively as the number of trials increases, and this leads to an increase in the SNR. However, this increase is not linear. For example, doubling the number of trials does not double the SNR. Instead, the SNR increases in proportion to the square root of the number of trials, so doubling the number of trials increases the SNR by the square root of 2. Let's make this more precise by using S to denote the size of the signal, N to denote the size of the noise on a typical single trial, and T to denote the number of trials. The SNR on a single trial is simply S/N (the signal divided by the noise), and the SNR in an average of T trials is equal to $(S / N)\sqrt{T}$ (the single-trial SNR multiplied by the square root of the number of trials). This is a very important fact that you should commit to memory, so I will repeat it: *The SNR increases in proportion to the square root of the number of trials.*

To make this more concrete, consider an experiment in which we are measuring a P3 wave that has an amplitude of 20 μV. If the noise in the EEG is typically 50 μV on a single trial, then the SNR on a single trial will be 20:50, or 0.4. If you average two trials together, then the SNR of the average will be the single-trial SNR multiplied by $\sqrt{2}$ (which is approximately 1.41), leading to an SNR of 0.566. To double the SNR from 0.4 to 0.8, it is necessary to average together four trials (because $\sqrt{4} = 2$). To quadruple the SNR from 0.04 to 1.6, it is necessary to average together 16 trials (because $\sqrt{16} = 4$). Thus, doubling the SNR requires quadrupling the number of trials, and quadrupling the SNR requires increasing the number of trials by a factor of 16. This relationship between the number of trials and the SNR is rather sobering because it means that achieving a substantial increase in the SNR requires a very large increase in the number of trials. You can only quadruple the number of trials so many times before the length of the recording session becomes impractical. This leads to a very important principle: It is often easier to improve the SNR of your averaged ERP waveforms by decreasing sources of noise than by increasing the number of trials.

Although the square root principle makes it difficult to achieve a good SNR, this principle has a positive side as well. Specifically, if you need to decrease the number of trials for some reason, the SNR does not decrease linearly with the decrease in the number of trials. For example, if you reject 20% of trials due to artifacts, this does not reduce your SNR by 20%. Instead, with 80% of the original number of trials, your new SNR will be $\sqrt{0.8}$ times your original SNR (or approximately 89% of your original SNR).

You might be wondering what an acceptable SNR would be. This is not an easy question to answer. In fact, the concept of SNR is not quite as simple as I've led you to believe because the signal of interest and the noise of interest will depend on exactly how you are quantifying your components (e.g., with a mean amplitude measure, an onset latency measure, etc.). Moreover, your SNR doesn't need to be as high if you have a lot of subjects or are looking for a large effect. In general, you shouldn't worry about the specific value of the SNR. Instead, you should just do everything you can (within reason) to make your SNR as good as it can be (by increasing the signal and decreasing the noise).

How Many Trials Do You Need?

One of the most common questions I get from new ERP researchers is, "How many trials do I need to include in an average?" This is related to the question of what an acceptable SNR would be. These are simple questions, but they do not have a simple answer. A detailed discussion of the underlying theoretical issues is provided in the online supplement to this chapter. Here I will provide concrete advice based on my own experience. This reflects the particular kinds of experiments that my lab conducts as well as the details of how we record and analyze our data. Your results should be similar to my lab's results if you conduct similar experiments and follow the advice about recording and analysis provided in this book. If your experiments are quite different, you will need to look at the literature and see what other people in your area of research do. But when you read papers to get this information, take a look at the waveforms and see how noisy they are. This may make you want to increase the number of trials in your experiments. You should also keep in mind that you will want even more trials if you are attempting to look at individual differences.

In my lab's basic science experiments, the experimental manipulations are almost always within subjects rather than between groups. This allows the statistical analyses to factor out the effects of overall differences between subjects (which can be large, as discussed in the next section), and it generally increases the statistical power. You will likely need more trials and/or more subjects if your research involves between-group comparisons or explicit analyses of individual differences.

Our basic science experiments usually have an N of 12–16 subjects per experiment (this is the total after a few subjects have been automatically excluded because of excessive artifacts, as discussed in chapter 6). The subjects in these experiments are college students and are therefore more homogeneous in terms of cognitive abilities than the broader population. They are also generally cooperative, able to understand the instructions, and stay focused on the task. These factors minimize variance and therefore increase our statistical power. You will need more trials and/or more subjects if you test more heterogeneous or less cooperative subjects.

The number of trials we include in these experiments depends on the expected size of the effect, which is related to the size of the ERP component we are measuring. When we are looking at small components (e.g., the visual P1 wave), we typically test 300–500 trials per condition in each subject. When we are looking at somewhat larger components (e.g., N2pc), we typically test 150–200 trials per condition. When we are looking at very large components (e.g., P3 or N400), we typically test 30–40 trials per condition. This is the minimum number of trials per subject in the conditions with the fewest trials (e.g., we might test 200 trials in an oddball experiment to get 40 targets and 160 standards for each subject). In addition, this is the number of trials in the averaged waveforms after combining across theoretically unimportant factors that were manipulated for the purpose of counterbalancing (e.g., we might combine 20 trials in which X was the target with 20 trials in which O was the target to achieve our 40 target trials).

In our experiments comparing schizophrenia patients with control subjects, we are typically trying to find a group × condition interaction (e.g., a smaller difference between two conditions in patients than in control subjects). Statistical power for these interactions is usually substantially lower than the power for detecting a main effect in a within-subjects design. In addition, we typically observe greater variance across individuals than in our basic science studies due to greater true score variance (because the patients and control subjects are more heterogeneous than the college students in our basic studies) and due to greater measurement error (because the EEG is noisier and more trials are rejected because of artifacts). We therefore test more subjects than in our basic science studies, with 20–30 subjects per group. When possible, we also double the number of trials per subject compared to our basic science studies. This means that we usually test fewer conditions per experiment in our schizophrenia studies than in our basic science studies.

All of these numbers reflect the number of trials that we initially include in the experiment, assuming that artifacts will lead to the rejection of 10% to 25% of these trials in our basic science studies and 20% to 50% of these trials in our schizophrenia studies. You can adjust upward or downward if you anticipate a different rate of artifacts or if you will be using artifact correction instead of artifact rejection.

Individual Differences in Averaged ERP Waveforms

Most ERP waveforms shown in journal articles are *grand averages*, which is the term used by ERP researchers to refer to waveforms created by averaging together the averaged waveforms of the individual subjects (see, e.g., figures 1.2 and 1.5 in chapter 1). Single-subject waveforms are shown only rarely in published papers (although they were more common in the early days of ERP research because primitive computer systems made it difficult to create and plot grand averages). The use of grand averages masks the variability across subjects, which can be both a good thing (because variability across subjects makes it difficult to see differences across groups or conditions) and a bad thing (because the grand average may not accurately reflect the pattern of individual results). When you look at your own single-subject data, you may be surprised at how different the subjects look from each other and from the grand averages you have seen in published papers.

Figure 8.6 shows the single-subject waveforms from an N2pc experiment (similar to the experiment shown in figure 3.8 of chapter 3). The left column shows waveforms from a lateral occipital electrode site in five individual subjects (from a total set of nine subjects). As you can see, there is tremendous variability in these waveforms. Everyone has a P1 peak and an N1 peak,

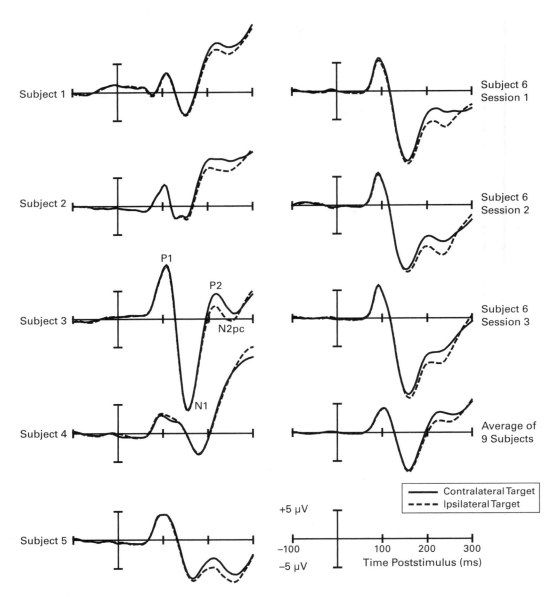

Figure 8.6
Example of individual differences in averaged ERP waveforms. Data from an N2pc experiment are shown for six individual subjects (selected at random from a total set of nine subjects). Subjects 1–5 participated in a single session, and subject 6 participated in three sessions. The grand average of all nine subjects from the experiment is shown in the lower right portion of the figure.

but the relative and absolute amplitudes of these peaks are quite different from subject to subject (compare, e.g., subjects 2 and 3). Moreover, not all of the subjects have a distinct P2 peak, and the overall voltage from 200 to 300 ms is positive for three subjects, near zero for one subject, and negative for one subject. This is quite typical of the variability that one sees in an ERP experiment. I didn't fish around for unusual examples—this is truly a random selection.

What are the causes of this variability? To answer this question, we first need to determine how much of the variability reflects stable individual differences among the subjects and how much reflects random trial-to-trial noise that remains after averaging together a finite number of trials. You can get some idea of this by looking at the first three rows of the right side of figure 8.6, which show the waveforms from subject 6, who participated in three sessions of the same experiment. You can see a little bit of variability in subject 6's waveforms from session to session, but this variability is very small compared to the variability you can see across the five subjects shown on the left side of the figure. The waveforms were quite stable across sessions for subject 6 because we averaged together hundreds of trials to create the averaged waveforms for each session, resulting in a good SNR. You can tell that the SNR was quite good because there is very little noise in the prestimulus baseline period. Some of the differences in the waveforms from session to session reflect random trial-to-trial noise that has not been eliminated by averaging, but some of the differences may reflect real changes in the subject from session to session, ranging from global state factors (e.g., number of hours of sleep the previous night) to shifts in task strategy. John Polich has published an interesting series of studies showing that the P3 wave is sensitive to a variety of global factors, such as time since the last meal, body temperature, and even the time of year (see review by Polich & Kok, 1995).

Given the session-to-session reliability shown on the right side of figure 8.6, the differences in the waveforms across subjects on the left side of the figure mainly reflect stable differences between subjects. However, you cannot assume that this will always be the case. If we had averaged together fewer trials or if we hadn't minimized other sources of noise, we would have had greater session-to-session variability in subject 6, and more of the variance between subjects would have reflected noise rather than stable differences.

When noise is minimal, as in figure 8.6, there are several potential causes of the stable differences in waveforms across subjects. One major factor is the idiosyncratic folding pattern of the cortex. As was discussed in chapter 2, the location and orientation of the cortical generator source of an ERP component has a huge influence on the size of that component at a given scalp electrode site. Every individual has a unique pattern of cortical folding, and the relationship between functional areas and specific locations on a gyrus or in a sulcus may also vary. Although I've never seen a formal study of the relationship between cortical folding patterns and individual differences in ERP waveforms, I've always assumed that this is the most significant cause of waveform variation in healthy young adults (for additional discussion, see Kappenman & Luck, 2012). There are certainly other factors that can influence the shape of the waveforms, including drugs, age, psychopathology, and even personality. But in experiments that focus on homogeneous groups of healthy young adults, these factors probably play a relatively small role.

The differences in waveforms among subjects are usually ignored in ERP studies. In many cases, this is very reasonable. However, this variability can be problematic when you are trying to measure a component using the same time window and set of electrode sites in all subjects (see the discussion of figure 9.2 in chapter 9). Large and stable differences across subjects can actually be a good thing if you are trying to study individual differences in psychological or neural processes. However, if the differences reflect nonfunctional factors such as cortical folding patterns, much of the variance across individuals may be unrelated to psychological or neural processes. Nonetheless, some of the individual differences can be related to psychological or neural factors, especially if you isolate specific processes with difference waves, as described in chapter 4. My former graduate student, Ed Vogel, has used difference waves to show beautiful and robust correlations between ERP effects and behavioral performance in studies of working memory (see, e.g., Vogel, McCollough, & Machizawa, 2005; Drew, McCollough, Horowitz, & Vogel, 2009; Anderson, Vogel, & Awh, 2011, 2013; Tsubomi, Fukuda, Watanabe, & Vogel, 2013).

The waveforms in the bottom right portion of figure 8.6 show the grand average of the nine subjects in this experiment. An important attribute of the grand average waveforms is that the peaks are smaller than those in most of the single-subject waveforms. This might seem odd, but it is perfectly understandable. The time point at which the voltage reaches its peak values for one subject are not the same as for other subjects, and the peaks in the grand averages are not at the same time as the peaks for the individual subjects. Moreover, there are many time points at which the voltage is positive for some subjects and negative for others. Thus, the grand average is smaller overall than most of the individual-subject waveforms. This is an example of a principle that will be discussed later in the chapter; namely, that latency variability leads to reduced peak amplitudes in averaged waveforms.

The (Non-)Problem of Amplitude Variability

Signal averaging is based on several assumptions, the most obvious of which is that the neural activity related to the time-locking event is the same on every trial. This assumption is clearly an oversimplification because neural and cognitive processing will obviously vary from trial to trial. If the amplitude of a given component varies from trial to trial, we are violating this assumption, but this type of violation is not usually a problem. For example, if the amplitude of the N2 wave varies from trial to trial, then the N2 wave in the averaged ERP waveform will simply reflect the average amplitude of the N2 wave across trials. This is no different from the typical use of averaging in science, in which the mean of a set of values is used as a measure of central tendency. Thus, we can just treat the voltage values in an averaged ERP waveform as a series of measures of central tendency, one at each time point in the waveform.

The mean is not always a very good measure of central tendency, and this is true for almost any measure of psychological or neural activity including ERPs. For example, if N2 amplitude is very large on half the trials and very small on the other half, the N2 in the averaged waveform will be an intermediate value that was not present on any individual trial. Similarly, N2 amplitude could be small on most trials but very large on a few trials (i.e., a skewed distribution). ERP researchers are not usually very concerned about these possibilities, but they are occasionally important. For example, P3 amplitude is typically smaller in schizophrenia patients than in healthy control subjects when measured from averaged ERP waveforms, and this could reflect either a consistent reduction in P3 amplitude on every trial or a complete absence of the P3 on a subset of trials coupled with normal P3 amplitude on the majority of trials. Ford, White, Lim, and Pfefferbaum (1994) conducted a single-trial analysis to distinguish between these possibilities and found evidence that patients have both a reduced likelihood of having a P3 wave and a reduced amplitude on the trials that contained a P3.

In most cases, you can ignore the likely possibility that amplitudes vary across trials. This very rarely impacts the conclusions you will draw from your experiments.

The Problem of Latency Variability

Although trial-to-trial variability in amplitude is not usually problematic, trial-to-trial variability in latency (also called *latency jitter*) can be a big problem. The timing of neural processes always has some variability, and the variability tends to become greater for later, cognitive components. For example, the trial-to-trial variability of a relatively late component like N400 will tend to be greater than that of an earlier component like P1. P3 latency is especially variable because the onset of the P3 wave depends on the amount of time required for stimulus categorization (see chapter 3), and this can vary widely from trial to trial. The effects of latency variability on averaged ERPs are illustrated in figure 8.7A, which shows four individual trials in which a P3-like ERP component occurs at different latencies, along with the average of the four trials. The first thing you should notice is that the peak amplitude of the averaged waveform is much smaller than the peak amplitude of the individual trials. This is particularly problematic when the amount of latency variability differs across experimental conditions. You can see this by comparing panels A and B of figure 8.7: the amplitudes of the single trials are identical in these two panels, but panel B has less latency variability than panel A, and the averaged waveform in panel B therefore has a greater peak amplitude than the averaged waveform in panel A. Thus, if two experimental conditions or groups of subjects differ in the amount of latency variability for some ERP component, they may appear to differ in the amplitude of that component even if the single-trial amplitudes are identical. This could lead you to the incorrect conclusion that there was a difference in amplitude. For example, it is important to ask whether the reduced P3 amplitude observed in schizophrenia patients relative to healthy control subjects reflects greater variability in P3 timing rather than smaller single-trial amplitudes. A difference in latency variability seems likely because schizophrenia patients usually exhibit greater variability in reaction time than do control subjects. Ford et al. (1994) examined this possibility and found that patients did indeed have greater latency variability than controls but that this did not fully account for their reduced P3 amplitude.

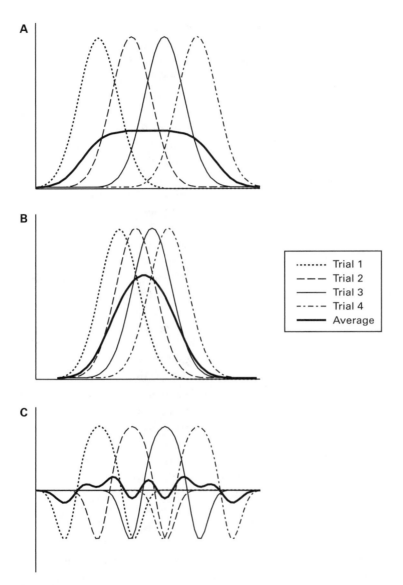

Figure 8.7
Example of the problem of latency variation. Each panel shows four single-trial waveforms, along with the average of the four waveforms. The same waveforms are present in panels A and B, but there is greater latency variability in panel A than in panel B, leading to a smaller peak amplitude and broader temporal extent in the averaged waveform for panel A than for panel B. Panel C shows that when the single-trial waveforms are not monophasic, but instead have both positive and negative subcomponents, latency variability may lead to cancellation in the averaged waveform.

When an ERP response contains both positive and negative portions, latency variability may cause the positive part of the response on one trial to be at the same time as the negative portion on another trial, leading to cancellation (see figure 8.7C). In extreme cases, the cancellation will be complete, and the ERP will be completely absent from the averaged waveform. For example, imagine that a sinusoidal oscillation is triggered by a stimulus but varies randomly in phase from trial to trial (which is not just a hypothetical problem—see Gray, König, Engel, & Singer, 1989). Such a response will average to zero and will be essentially invisible in an averaged response.

Example of Latency Jitter

A real example of latency jitter is shown in figure 8.8 (from Luck & Hillyard, 1990). In this experiment, we examined the P3 wave during two types of visual search tasks. In one condition (*parallel search*), subjects searched for a target with a distinctive visual feature that "popped out" from the display and could be detected immediately no matter how many distractor items were present in the stimulus array. In the other condition (*serial search*), the target was defined by the absence of a feature; in this condition, we expected that the subjects would search the array one item at a time until they found the target. Reaction time (RT) was expected to increase as the number of items in the array (the set size) was increased in the serial search condition, whereas no effect of set size was expected in the parallel search condition. This was the pattern of results that we obtained, replicating many previous visual search experiments (see, e.g., Treisman & Souther, 1985; Treisman & Gormican, 1988).

We also expected to see consistent differences across conditions in the trial-by-trial variability of RT. In the parallel search condition, the target pops out immediately, so the amount of time required to find the target should be relatively consistent from trial to trial. In the serial search condition, however, subjects shift attention randomly from one item to the next until they find the target, and this leads to variability in the amount of time that passes before the target is found. At set size 4, for example, the target could be the first, second, third, or fourth item searched, but at set size 12, the target might be found anywhere between the first item and the 12th item. We thus expected that RTs in the serial search condition would vary a great deal from trial to trial and that the variability would increase as the set size increased. This is exactly what we found.

Because the P3 is linked to the categorization of a stimulus as a target (see chapter 3), we expected that the timing of the P3 on each trial would be tightly linked to the timing of the behavioral response. Consequently, we expected that P3 latency would be relatively constant in the parallel search condition, because the amount of time required to find the target should not vary much from trial to trial in this condition. However, because the amount of time required to find the target varies greatly from trial to trial in the serial search condition, especially at the large set sizes, we expected that P3 latency would become progressively more variable as the set size increased in this condition. The increased latency variability would be expected to decrease the peak amplitude of the P3 wave in the averaged ERP waveforms.

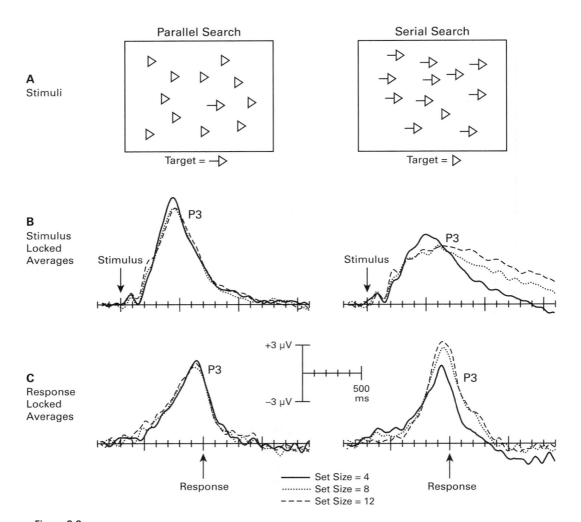

Figure 8.8
Example of an experiment in which significant latency variability was expected for the P3 wave (Luck & Hillyard, 1990). (A) Sample stimuli from the two conditions of the experiment. (B) Stimulus-locked averages from the Pz electrode site. (C) Response-locked averages from the Pz electrode site.

The averaged ERP waveforms from this experiment are shown in figure 8.8B, time-locked to the onset of the stimulus array. In the parallel search condition, the P3 wave was relatively large in amplitude and short in duration, and it did not vary much as a function of set size. In the serial search condition, the P3 had a smaller peak amplitude than in the parallel search condition but was very broad. In addition, the P3 peak in the serial search condition was greater at set size 4 than at set sizes 8 and 12. If you did not take into account the effects of latency variability, you might conclude that the brain generally produced a smaller P3 response when it performed the serial task than when it performed the parallel task, and that within the serial task the brain produced a smaller P3 at set sizes 8 and 12 compared to set size 4. However, given that the smaller peak amplitudes were observed in the conditions that were expected to have greater trial-by-trial latency variability, and that greater variability artificially reduces peak amplitudes in averaged waveforms (see figure 8.7), the peak amplitudes could be misleading. If we just look at the stimulus-locked P3 peaks, we can't tell whether the single-trial P3 amplitudes differed across conditions or if the peak differences are an artifact of differences in P3 latency variability.

How, then, can we factor out the effects of latency jitter to determine what really happened to P3 amplitude on single trials in each condition of this experiment? The following sections will describe some simple but effective methods for dealing with this type of latency jitter. Online chapter 11 will provide a more detailed analysis of latency jitter, introducing the mathematical concept of *convolution*, which will give you a deeper understanding of exactly how latency jitter influences the averaged ERP waveform.

Dealing with Latency Jitter

Area Measures   In many cases, you can eliminate the effects of latency jitter simply by measuring mean or area amplitude rather than peak amplitude. These measures are particularly effective for *monophasic* ERP components (components that are entirely positive or entirely negative at a given electrode site). When a component is monophasic, the area under the curve in an average of several trials is equal to finding the average of the area under the curve on the individual trials and then taking the average. Consequently, latency variability does not impact the area under the curve for monophasic components. For example, the area under the curve for the averaged waveform shown in figure 8.7A is the same as the area under the curve for the averaged waveform shown in figure 8.7B, even though the latency variability was greater in the former than in the latter. Mean amplitude is also unaffected by latency variability; the mean amplitude is equivalent for the averaged waveforms in figure 8.7A and B despite the difference in latency variability and peak amplitude.

We can therefore apply mean or area amplitude measures to the data from the experiment shown in figure 8.8 to determine whether the differences in P3 peak amplitude in the averaged waveforms reflect differences in the single-trial P3 amplitudes. When I measured mean amplitude from the waveforms shown in figure 8.8B, I found that P3 amplitude was actually larger in the serial search task than in the parallel search task (whereas the opposite was true for peak amplitude) and P3 amplitude increased as the set size increased in the serial search condition (whereas peak amplitude decreased). This implies that the single-trial P3 amplitudes were actually larger for the serial search task than for the parallel search task and that they increased as the set size increased in the serial search task. The next section will provide converging evidence for this conclusion using response-locked averages. Note that I would have erroneously reached exactly the opposite conclusion if I had only looked at the peak amplitudes.

In addition to being insensitive to latency jitter, area and mean amplitude measures have many additional advantages over peak amplitude measures, as will be discussed in detail in chapter 9 (which will also describe how to measure latency in a situation like this). However, these measures are ineffective if the jittered component is multiphasic (i.e., when it consists of both positive and negative periods). This is illustrated in figure 8.7C, which shows that the positive and negative deflections cancel each other when the onset time is variable. In this situation, a time–frequency analysis can be helpful, as will be described near the end of the chapter. In many cases, a single late component (e.g., the P3 wave) has substantial latency jitter, but this component is superimposed on the other positive and negative peaks in the waveform (e.g., P1, N1, P2, and N2). The overall waveform is multiphasic in this situation, but only a single monophasic component varies much in latency from trial to trial (P3). Area and mean amplitude measures will typically work quite well in this situation. The key is whether the jittered activity—not the overall waveform—is multiphasic. As long as the jittered brain activity is monophasic, area and mean amplitude measures are appropriate.

Response-Locked Averages   In some cases, the latency of an ERP component is tightly coupled with RT, and in these cases latency variability can be corrected by using response-locked averages rather than stimulus-locked averages. In a response-locked average, the response rather than the stimulus is used to align the single-trial EEG segments during the averaging process. As an example, figure 8.8C shows the response-locked waveforms from our visual search experiment, in which we expected the P3 to peak at approximately the same time as the behavioral response on each trial. In the serial search condition, the P3 was narrower and taller in the response-locked averages than in the stimulus-locked averages, which is exactly what would be expected if the P3 was more tightly time-locked to the response than to the stimulus. In addition, the response-locked P3 was larger in the serial search condition than in the parallel search condition and increased as set size increased in the serial search condition. Recall that this is exactly what we found when we measured mean amplitude in the stimulus-locked averages but is the opposite of the pattern found for peak amplitude in the stimulus-locked averages. Thus, when brain activity is likely to be more closely time-locked to the response than to the stimulus, response-locked averages can be very useful for figuring out how single-trial amplitudes change across conditions or groups (for an example in the context of group differences, see Luck et al., 2009).

The Woody Filter Technique   A third technique for mitigating the effects of latency variability is the *Woody filter* technique (Woody, 1967). The basic approach of this technique is to estimate the latency of the component of interest on individual trials and to use this latency as the time-locking point for averaging. The component is identified on single trials by finding the portion of the single-trial waveform that most closely matches a template of the ERP component. Of course, the success of this technique depends on how well the component of interest can be identified on individual trials, which in turn depends on the SNR of the individual trials and the similarity between the waveshape of the component and the waveshape of the noise.

The Woody filter technique begins with a best-guess template of the component of interest (such as a half cycle of a sine wave) and uses cross-correlations to find the segment of the EEG waveform on each trial that most closely matches the waveshape of the template.[3] The EEG epochs are then aligned with respect to the estimated peak of the component and averaged together. The resulting averaged ERP can then be used as the template for a second iteration of the technique, and additional iterations are performed until little change is observed from one iteration to the next.

The shortcoming of this technique is that the part of the waveform that most closely matches the template on a given trial may not always be the actual component of interest, resulting in an averaged waveform that does not accurately reflect the amplitude and latency of this component (Wastell, 1977). Moreover, this does not simply add random noise to the averages; instead, it tends to make the averages from each different experimental condition more similar to the template and therefore more similar to each other (this is basically just regression toward the mean). Thus, this technique is useful only when the component of interest is relatively large and dissimilar to the EEG noise. For example, the P1 wave is small and is similar in shape to spontaneous alpha waves in the EEG, and the template would be more closely matched by the noise than by the actual single-trial P1 wave on many trials. The P3 component, in contrast, is relatively large and differs in waveshape from common EEG patterns, and the template-matching procedure is therefore more likely to find the actual P3 wave on single trials.

Even when a large component such as the P3 wave is being examined, Woody filtering works best when the latency variability is only moderate; when the variability is great, a very wide window must be searched on the individual trials, leading to more opportunities for a noise deflection to match the template better than the component of interest. For example, I tried to apply the Woody filter technique to the visual search experiment shown in figure 8.8, but it didn't work very well. The P3 wave in this experiment could peak anywhere between 400 and 1400 ms poststimulus, and given this broad search window, the algorithm frequently located a portion of the waveform that matched the search template fairly well but did not correspond to the actual P3 peak. As a result, the averages looked very much like the search template and were highly similar across conditions.

It should be noted that the major difficulty with the Woody filter technique lies in identifying the component of interest on single trials, and any factors that improve this process will lead to a more accurate adjustment of the averages. For example, the scalp distribution of the component can be specified in addition to the component's waveshape, which makes it possible to reject spurious EEG deflections that may have the correct waveshape but have an incorrect scalp distribution (see Brandeis, Naylor, Halliday, Callaway, & Yano, 1992).

Basics of Time–Frequency Analysis

As shown in figure 8.7C, latency jitter can cause single-trial activity to disappear from the averaged ERPs if the single-trial activity consists of an oscillation (because an uppie on one trial will cancel a downie at the same time on another trial). The past decade has seen an explosion in research on these EEG oscillations, mainly because *time–frequency analysis* techniques have been developed to aggregate across trials in a manner that avoids the cancellation that occurs with conventional signal averaging. I will discuss these techniques in greater detail in online chapter 12, but I will also provide a brief overview here because these techniques are really just a fancy version of signal averaging. If you haven't already read the section on Fourier analysis in chapter 7, you should go back and read that section before proceeding. I would also recommend reading the excellent review paper by Roach and Mathalon (2008).

Time–Frequency Analysis via Moving Window Fourier Analysis

Time–frequency analysis is based on variants of the *Fourier transform*. As was discussed in chapter 7, the Fourier transform converts a waveform into a set of sine waves of different frequencies, phases, and amplitudes. For example, if you were to apply the Fourier transform to a 1-s EEG epoch, you would be able to determine the amount of activity at 10 Hz, at 15 Hz, at 20 Hz, or almost any frequency. Fourier analysis measures the amplitude of an oscillation independently of its phase, and this allows us to avoid the problems associated with trial-by-trial variations in latency (which is closely related to phase).

However, we can't use the standard Fourier transform here because it doesn't give us any information about the time course of an oscillation. That is, it would provide a single value for each frequency, representing the power (amplitude squared) of that frequency for the entire epoch. Instead, we want a method that gives us the power of a given time frequency at each time point in the waveform. Power isn't actually defined for a single time point, but we can provide an approximation by looking at the power of a given frequency over a short time window (e.g., 200 ms) and using that power to represent the value at the middle latency of the time window. There are two basic approaches to accomplishing this, a *moving window* version of Fourier analysis and a *wavelet* analysis.

In the moving window Fourier analysis, a Fourier transform is performed for each of several consecutive time windows (Makeig, 1993). This is illustrated in figure 8.9, which shows four single trials in which a 10-Hz oscillation occurred, but starting at different time points on each trial. The gray rectangle represents the time window that is being used in the analysis. A Fourier analysis is done in this time window, and the power at each frequency is then assigned to the midpoint of that window. The window is then slid over to the right by one sample period (e.g., 4 ms), and a new Fourier analysis is done in the new time window. Three windows are shown in the figure to demonstrate the concept of a window, but an actual analysis would have many overlapping windows, one centered on each sample point in the waveform. This gives us an estimate of the power at each frequency at each sample point in the waveform. Note, however,

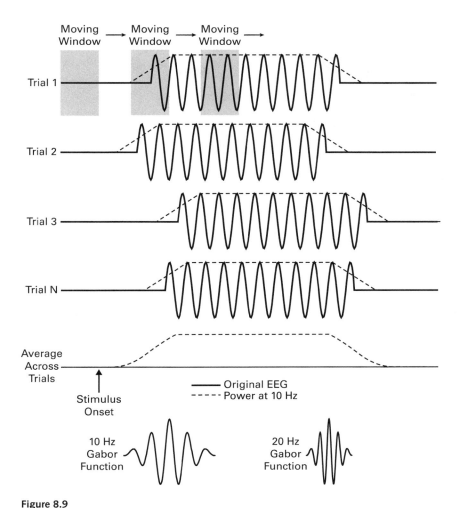

Figure 8.9

Basics of time–frequency analysis. On each trial of this example, a 10-Hz EEG oscillation is elicited at a variable time after stimulus onset (solid lines). If we were to compute a conventional average across trials, the oscillations would largely cancel, yielding a flat line in the average. To perform a time–frequency analysis, a window (200 ms wide in this example) is moved across the EEG data from each trial, and the power at a given frequency is estimated for each window position (dashed lines). Three different windows are shown here, but this is actually done for a very large number of overlapping windows, one centered on every sample point in the waveform. The power is measured at a given frequency for each sample point in the single-trial EEG waveforms, creating a single-trial power waveform for each trial. These single-trial power waveforms are then averaged together (shown in the bottom dashed waveform here). An alternative approach is to convolve the EEG with a set of Gabor functions, like the two shown at the bottom of the figure.

that we have lost some temporal resolution because the power at a given time point really reflects the entire time window centered at that time point.

Figure 8.9 illustrates how this works for a single frequency (10 Hz), using a 200-ms moving window. The EEG on each trial is shown with the solid lines, and the power at 10 Hz at the middle of each 200-ms window is shown with the dashed lines. The 10-Hz power is zero at the beginning of each EEG epoch because each epoch begins with a flat line. Once the window begins to reach the oscillation, the power at 10 Hz ramps up. The 10-Hz power levels off once the entire window is filled with the 10-Hz oscillation, and then it falls back to zero when the oscillation ends.

Once this has been done for every trial, the waveforms representing power at a given frequency on each trial can be averaged together, just as you would ordinarily average the original voltage waveforms across trials. This is shown near the bottom of figure 8.9. Whereas the oscillations will cancel out in a conventional average, leading to a flat line, the 10-Hz power in the average does a good job of representing the 10-Hz power on the single trials.

Time–Frequency Analysis via Wavelets

Although the moving window Fourier analysis method shown in figure 8.9 has some advantages, it has two disadvantages. First, it treats the power within the window as if it was the power at the center of the window, even though the entire window contributes equally to the power measurement.[4] Second, the same size window is used to calculate the power at each frequency, even though this yields lower precision for low frequencies than for high frequencies. Most people now use a different method based on *wavelets* that addresses both of these problems.

An example of a wavelet is shown at the bottom left of figure 8.9. There are many kinds of wavelets, and this particular wavelet is a Gabor function. It was created by taking a 10-Hz sine wave and multiplying it by a Gaussian (bell curve) function. Whereas the original sine wave was infinite in duration, the multiplication by the Gaussian function causes the oscillations to taper down over time. This solves the first of the two problems that arise from moving window Fourier transforms: rather than treating every point within a time period equally, a wavelet gives the greatest weight to the center of the time period. We have still lost some temporal resolution, because the power at a given time point is influenced by a range of surrounding time points, but this problem has been reduced somewhat because more distant time points receive lower weight.

The second problem—different precision for different frequencies—is solved by using Gabor functions with different widths for different frequencies. For example, the 20-Hz wavelet shown at the bottom right of figure 8.9 has the same number of cycles as the 10-Hz wavelet, thus yielding the same precision, but its duration is half as great. When multiple wavelets are created that are all identical but are squeezed or expanded horizontally to represent different frequencies, these wavelets are called a *wavelet family*. When each wavelet is a Gabor function, the family is called a *Morlet wavelet family*.

You may be wondering how a wavelet family is used to calculate the power at a given frequency and time point. The answer is a mathematical operation called convolution. The general idea of convolution is described in online chapter 11, and online chapter 12 explains exactly how time–frequency analysis is achieved with convolution, using simple math. In my view, you shouldn't use time–frequency analysis without understanding the basics of how it works, and you should definitely read online chapters 11 and 12 if you are planning to do time–frequency analysis in the near future. In addition, Mike Cohen has recently published a book that provides both the mathematical details and practical advice for performing time–frequency analyses (Cohen, 2014), and I strongly encourage you to read that book if you want to use this technique.

Using Time–Frequency Analysis to See Random-Phase Activity

Figure 8.9 shows what time–frequency analysis looks like for a single frequency, but people usually do it for many frequencies at the same time. The results are usually plotted with a *heat map*, which uses different colors (or different shades of gray) to indicate the power at a given time and frequency. An example is shown in figure 8.10A, which shows data from the study of Tallon-Baudry, Bertrand, Delpuech, and Pernier (1996). The *X* axis in this plot is time, just as in a traditional ERP average. The *Y* axis, however, is frequency (with frequencies from 20 to 100 Hz shown here). The gray-scale level indicates the power that was present at each frequency at each time point, with lighter shades indicating greater power. A band of activity centered at 40 Hz can be seen at approximately 90 ms poststimulus, and a somewhat weaker band of activity between 30 and 60 Hz can be seen at approximately 300 ms poststimulus.

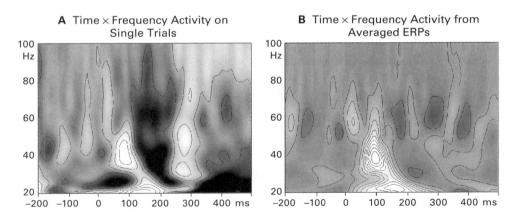

Figure 8.10
Example of time–frequency data from the study of Tallon-Baudry et al. (1996). The *X* axis is time; the *Y* axis is frequency; the intensity of the gray indicates the power at a particular time and frequency, with lighter shading for greater power. In panel A, the time–frequency transformation was applied to the individual trials, and the transformed data were then averaged. This plot therefore includes activity that was not phase-locked to stimulus onset as well as phase-locked activity. In panel B, the transformation was applied after the waveforms had been averaged together. This plot therefore includes only activity that was phase-locked to the stimulus, because random-phase activity is eliminated by the ERP averaging process. Adapted with permission from Tallon-Baudry et al. (1996). Copyright 1996 Society for Neuroscience.

The crucial aspect of this approach is that these bands of activity can be seen whether or not the activity varies in phase from trial to trial, whereas random-phase activity is completely lost in a traditional average. However, researchers often assume that they are seeing *only* random-phase oscillations in time–frequency analyses, but this is not usually a valid assumption. For example, figure 8.10B shows the results obtained by Tallon-Baudry et al. (1996) when the time–frequency transformation was applied to the averaged ERP waveform rather than to the single-trial EEG. The averaged ERP waveform should contain only activity that has a consistent phase from trial to trial, so any activity in the time–frequency transformation of the averaged ERP cannot be a result of random-phase oscillations (except under the conditions described in box 8.2). You can see that the 40-Hz activity at 90 ms is visible in the time–frequency transformation of the averaged ERP, so this activity was a part of the traditional averaged ERP waveform and was not a random-phase oscillation. However, the activity from 30 to 60 Hz at 300 ms in figure 8.10A is not visible in figure 8.10B, so this must have been random-phase activity.

Box 8.2
How Random-Phase Oscillations Can Survive in Conventional Averages

Although random-phase oscillations normally cancel out in conventional averages, Ali Mazaheri and Ole Jensen have shown that this is not always the case (Mazaheri & Jensen, 2008). Given the biophysics of neural circuits, oscillations may not always be symmetric around the baseline voltage. As shown in the illustration that follows, the voltage may instead rise up away from the baseline and then fall back to baseline on each cycle (or the opposite, depending on the orientation of the dipole). Because the voltage never dips below baseline, the downies are near zero rather than being negative, so they don't cancel the uppies (see, e.g., the time points indicated by the dashed lines in the illustration). As a result, the average across many trials with different phases may be a broad positivity (or a broad negativity).

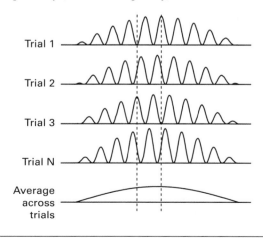

Figure 8.11 provides a closer look at how constant-phase and random-phase activity are influenced by averaging. Four trials of EEG are shown on the left of the figure, along with the conventional signal-averaged ERP. The EEG contains a 10-Hz burst at the beginning of each trial that has the same phase on every trial. It also contains a later 10-Hz burst that varies in phase from trial to trial. When the voltage waveforms are averaged together, the initial phase-locked 10-Hz burst is present in the average, but the later random-phase burst is absent. The right column contains the time–frequency transformations of the single trials, along with the average across the single-trial time–frequency transformations. You can see the two 10-Hz bursts on each single trial and also in the average. Thus, whereas only the first 10-Hz burst was present in the conventional average, both 10-Hz bursts can be seen when the averaging across trials is performed on the time–frequency transformations. Box 8.2 describes a special situation in which random-phase oscillations can be seen in conventional averages.

Although the groundbreaking study of Tallon-Baudry et al. (1996) showed the time–frequency transformation of the averaged ERP waveforms as well as the results of performing the time–frequency transformation before averaging, more recent studies often fail to show the results of applying the transformation to the averaged ERP waveforms. This makes it impossible to know whether the results reflect random-phase activity or simply reflect the traditional ERP waveform. When you read time–frequency studies, you should ask yourself whether the authors have provided the information needed to determine whether random-phase oscillations were actually present.

The Baseline in Time–Frequency Analyses

As discussed at the beginning of the chapter, the main purpose of baseline correction in conventional ERP studies is to remove large voltage offsets and slow drifts. These offsets and drifts occur at very low frequencies (mainly <1 Hz). Because time–frequency analyses provide estimates of power at specific frequencies that are usually much higher than 1 Hz, offsets and drifts have little or no impact on time–frequency results. Consequently, baseline correction is not always necessary in time–frequency analyses.

The main value of baseline correction in time–frequency analyses is to isolate stimulus-related brain activity from activity that was present prior to stimulus onset. If an experiment involves comparing activity elicited by stimuli that were presented in random order, the baseline activity should be the same for all trial types, and baseline correction is not strictly necessary (although it might be useful for minimizing random variations and therefore increasing statistical power).

However, some form of baseline correction will be necessary if the experiment involves comparing activity recorded in different trial blocks, from nonrandom stimulus orders, or from different groups of subjects. In these cases, any poststimulus differences in activity could simply be the continuation of prestimulus differences. If, for example, one condition is more boring than another, the more boring condition might yield greater alpha activity in both the prestimulus and poststimulus periods. In the absence of baseline correction, a finding of greater alpha from 200 to 300 ms poststimulus would not mean that there was greater stimulus-related alpha from

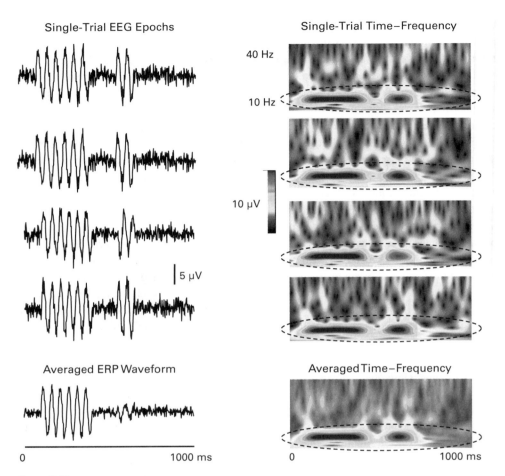

Figure 8.11
In-depth look at time–frequency analysis. The left column shows voltage waveforms for four single trials and the average. The single trials contain an initial 10-Hz oscillation that has the same phase on each trial, followed by a later 10-Hz oscillation that varies in phase from trial to trial. The constant-phase oscillation appears in the average, but the random-phase oscillation cancels out and is largely absent from the average. The right column shows the time–frequency transformation of the single trials, along with the average of the time–frequency transformations (note that this is not the time–frequency transformation of the averaged voltage waveform). You can see both the early and late 10-Hz oscillations on each trial and in the average (highlighted with the dashed ellipses). If the time–frequency transformation had been applied to the averaged ERP waveform, only the first 10-Hz oscillation would have been present. Adapted with permission from Bastiaansen et al. (2012). Copyright 2012 Oxford University Press.

200 to 300 ms, because it could just be a stimulus-independent difference in alpha power. The same is true of across-group comparisons: Without baseline correction, you cannot determine whether a difference between groups in the poststimulus period reflects a general difference that is unrelated to the stimuli or task and is also present prior to the stimulus.

However, when baseline correction is performed, you still need to be cautious about how you interpret the data. Recall from chapter 6 that alpha activity is typically suppressed shortly after the onset of a stimulus (see, e.g., the data from subject 1 in figure 6.6). Consequently, if a patient group has more alpha during the prestimulus period than a control group, but they both have the same amount of alpha from 300 to 400 ms poststimulus in the absence of baseline correction, then the patient group will appear to have less alpha than the control group after baseline correction is performed.

In my reading of the literature, studies using time–frequency analysis are quite variable in whether and how they use baseline correction. Some perform baseline correction and others don't. When baseline correction is performed, some studies subtract the average baseline power at a given frequency from the power at each poststimulus time point for that frequency. Other studies use division rather than subtraction (i.e., for each frequency, the power at each poststimulus time point is divided by the average prestimulus power). In other studies, the change between the prestimulus baseline and the poststimulus period is represented on a log scale (decibels) to take into account the fact that power typically falls off as the frequency increases. This variability across studies doesn't mean that the researchers are doing anything wrong, because different baseline correction procedures may be appropriate in different situations. However, when you read these studies, you need to look carefully at how the baseline is treated so that you understand exactly what information the time–frequency analyses are showing.

Is It Really an Oscillation?

Time–frequency analysis provides a very useful technique for making random-phase oscillations visible. However, it is very easy to draw an incorrect conclusion from time–frequency data; namely, that the brain activity actually *consists* of oscillations (i.e., a series of upward and downward deflections). As I discussed previously in chapter 7, a brief monophasic ERP deflection is mathematically equivalent to the sum of many sine waves at many different frequencies, just as a dollar bill has the same value as 100 pennies. However, the ERP deflection doesn't consist of the sine waves any more than a dollar bill consists of 100 pennies. Consequently, the presence of activity in a given frequency band does not entail the existence of a true oscillation.

As mentioned in chapter 7, a true oscillation will usually consist of a relatively narrow band of activity. In contrast, a non-oscillating transient brain response will typically lead to activity spread across a wide band of frequencies, beginning at the very lowest frequencies. For example, the alpha oscillations in figure 8.11 form a narrow band of activity at 10 Hz, with very little activity at lower frequencies. In contrast, figure 8.10 does not show any frequencies below 20 Hz, making it difficult to know if the observed effects are confined to a narrow frequency band or if they extend all the way down to 0 Hz. If a study does not show the data from relatively low frequencies, you should not accept any conclusions they draw about oscillations per se. This is discussed in more detail in online chapter 12.

Suggestions for Further Reading

Bastiaansen, M., Mazaheri, A., & Jensen, O. (2012). Beyond ERPs: Oscillatory neuronal dynamics. In S. J. Luck & E. S. Kappenman (Eds.), T*he Oxford Handbook of ERP Components* (pp. 31–49). New York: Oxford University Press.

Cohen, M. X. (2014). *Analyzing Neural Time Series Data: Theory and Practice*. Cambridge, MA: MIT Press.

Mazaheri, A., & Jensen, O. (2008). Asymmetric amplitude modulations of brain oscillations generate slow evoked responses. *Journal of Neuroscience, 28*, 7781–7787.

Roach, B. J., & Mathalon, D. H. (2008). Event-related EEG time-frequency analysis: An overview of measures and an analysis of early gamma band phase locking in schizophrenia. *Schizophrenia Bulletin, 34*, 907–926.

Urbach, T. P., & Kutas, M. (2002). The intractability of scaling scalp distributions to infer neuroelectric sources. *Psychophysiology, 39*, 791–808.

Urbach, T. P., & Kutas, M. (2006). Interpreting event-related brain potential (ERP) distributions: Implications of baseline potentials and variability with application to amplitude normalization by vector scaling. *Biological Psychology, 72*, 333–343.

9 Quantifying ERP Amplitudes and Latencies

Overview

This chapter describes methods for quantifying the amplitudes and latencies of ERP components. These are the values that go into your statistical analyses, so it is obviously important that they be both accurate and reliable. There are many very sophisticated approaches that can be used for quantifying ERP components (involving, for example, principal component analysis or dipole source analysis), but here I concentrate on approaches that are relatively simple and robust.

Some of the techniques I describe here are not available in all ERP analysis packages. In fact, this was one of my motivations for putting together my own package, ERPLAB Toolbox (which is freely available at http://erpinfo.org/erplab). These techniques are not complicated to implement, so if you don't want to switch to ERPLAB Toolbox, you should pressure the manufacturer of your analysis software to implement them. Although simple, they are often dramatically superior to the more widely available methods, and they are used by many leading laboratories around the world.

Note that all of the approaches described here require defining a *measurement window* during which the amplitude or latency is measured. Selecting an appropriate measurement window is one of the most difficult aspects of amplitude and latency measurement. If you choose the window on the basis of the time period that shows the largest differences between groups or conditions in your data set, you are biasing yourself to find a significant effect even if there is no true difference (because noise may influence the measurement window). Because the method used for choosing the measurement window has implications for statistical significance, I have postponed my discussion of this issue until the next chapter, which focuses on statistical analyses.

Before I get started, however, I want to give you a piece of very important advice: Make sure that you look at the measurements for every ERP waveform that is being measured. That is, don't just run some script that measures the values and then start doing statistical analyses. Compare each measured amplitude or latency value with a plot of the corresponding ERP waveform in every subject. The waveforms often differ markedly among subjects, and a measurement procedure that looks like it will work for the grand averages may fail miserably when applied to individual subjects. We spent a lot of time developing a visualization tool in ERPLAB Toolbox that allows you to plot each subject's waveforms and see exactly how the amplitude and latency values for those waveforms were computed, and other ERP analysis packages have similar tools. I strongly encourage you to use them!

Basic Measurement Algorithms: Peak Amplitude, Peak Latency, and Mean Amplitude

As illustrated in figure 9.1, the oldest method for measuring ERP amplitudes and latencies is to define a time window and find the maximum point in that time window (either the most positive or most negative point, depending on whether you are searching for a positive peak or a negative peak). The voltage at this point is called the *peak amplitude*, and the time of this point is called the *peak latency*. For example, the P3 wave in figure 9.1A reaches a peak voltage of approximately 17 μV at 404 ms after stimulus onset. As was described in chapter 8, amplitude is typically measured relative to the average voltage in the prestimulus period. Make sure you don't forget that any noise or systematic distortion in the baseline will have a big impact on your amplitude measurements (for peak amplitude or other measures).

The term *peak* can be ambiguous. For example, to measure the peak of the P2 wave in the data shown in figure 9.1, you might use a time window of 150–300 ms. The maximum voltage in this time window occurs at the edge of the time window (300 ms) due to the onset of the P3

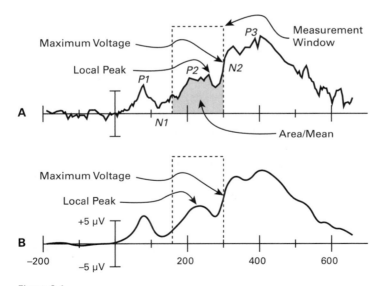

Figure 9.1
(A) Illustration of how the P2 wave can be measured, showing the measurement window (150–300 ms poststimulus), the maximum voltage (i.e., the simple peak), the local peak, and the region used to calculate area or mean amplitude. (B) The same waveform after a low-pass filter was applied to attenuate high-frequency noise.

wave. Consequently, the amplitude of the P2 wave would be measured at 300 ms, which is not very near the actual P2 peak. Clearly, this is not a good way to measure the P2 wave. This problem could be avoided by using a narrower measurement window, but a fairly wide window is usually necessary because of variations in peak latency across electrode sites, experimental conditions, and subjects. When we measure the true maximum value in this manner, this is called a *simple peak amplitude* or *simple peak latency*.

When I encountered this problem in graduate school, I developed a better measure of peak amplitude called *local peak amplitude*, which defines a local peak as the largest point in the measurement window that is surrounded on both sides by lower voltages. If we used this to quantify the P2 in figure 9.1, it would find the amplitude of the point labeled "Local Peak," which would reflect the P2 much better than the point labeled "Maximum Voltage." You can also measure *local peak latency*, which is simply the time of the local peak amplitude.[1]

Most commercial ERP analysis packages do not implement the local peak algorithm, but it is available in ERPLAB Toolbox and in some commercial packages (e.g., BrainVision Analyzer). When this algorithm is unavailable, people sometimes visually inspect the measurements and manually adjust them according to some informal rule. I recommend against this kind of manual approach because it may lead to bias and it is difficult to describe in a way that can easily be replicated by other researchers. Instead, you should use a package that implements the local peak algorithm or harass the manufacturer of your package so that they implement this simple but useful algorithm. In any case, you should be sure to specify exactly how peaks were defined when you publish your research. For example, you might write, "We measured local peak amplitude (as defined by Luck, 2014) between 150 and 300 ms" (which would make me happy because it will increase my citation count).

Mean amplitude is a common (and usually superior) alternative to peak amplitude. Mean amplitude is computed by simply taking the average voltage over a specified measurement window. That is, you take the voltage at each sample point in the time window and then compute the average of these voltages.

Strengths and Weaknesses of Peak and Mean Measures

In this section, I describe the relative advantages and disadvantages of peak and mean amplitude measures. In the vast majority of cases, mean amplitude is superior to peak amplitude, and it's important for you to know when and why this is true so that you can use the most appropriate measure in your own research.

Issue 1: Peaks and Components Are Not the Same Thing

The goal of any ERP quantification approach is to provide an accurate measurement of the size and timing of the underlying ERP component with minimal distortion from noise and from other spatially and temporally overlapping components. As discussed in chapter 2, measurements of amplitude and latency from a raw ERP waveform can easily be distorted by overlapping components. That chapter also described several strategies for avoiding this problem (e.g., using difference waves), but those strategies often rely on the use of an appropriate measurement approach. Chapter 2 also noted that the most important rule for interpreting ERP waveforms is this: *Peaks and components are not the same thing. There is nothing special about the point at which the voltage reaches a local maximum.*

Given that the point at which the voltage reaches a local maximum is not special, why should peak amplitude be used to measure the amplitude or latency of an ERP component? The main reason is largely historical. General-purpose computers were not yet available in the early days of ERP research, and the easiest way to assess amplitudes was to plot the waveforms and use a ruler to measure the peaks (see Donchin & Heffley, 1978). Once peak measurements became standard, they continued to be used even after researchers had access to computers that could perform more sophisticated measurements. Mean amplitude does a better job of treating an ERP component as something that is extended over time, and there has been a gradual trend away from peak amplitude and toward mean amplitude over the past few decades. However, peaks are occasionally useful to measure. The rest of this section will detail the relative advantages and disadvantages of peak versus mean measures.

Issue 2: Sensitivity to High-Frequency Noise

Peak measures are easily distorted by high-frequency noise. In figure 9.1A, for example, the local peak is not in the center of the P2 wave, but is shifted to the right because of a noise "blip." The effects of high-frequency noise can be mitigated somewhat by filtering out the high frequencies before measuring the peak, as illustrated in figure 9.1B. The local peak in the filtered waveform is closer to the "true" peak. However, this is still the peak in the overall waveform and may not be a particularly good measure of the amplitude of the underlying component.

Another common approach to dealing with high-frequency noise is to find the peak and then measure the average voltage over a window centered on that peak (e.g., a 50-ms window). This does filter out the high-frequency noise after the peak is found, but noise may still lead to the wrong point being chosen as the peak. For example, noise in the waveform causes the local peak for the P2 in the waveform shown in figure 9.1 to be shifted later in time than the true peak; if we found this local peak and then computed the average voltage over a 50-ms period centered on this peak, we would be getting a measure of a 50-ms period that is shifted to the right of the actual peak and is influenced by the falling edge of the N2 wave. I haven't seen any studies that have directly examined the accuracy of this measure, but I suspect you will get better results by simply applying a low-pass filter first (as in the waveform shown in figure 9.1B) and then measuring the amplitude of this peak.

Mean amplitude measures tend to be less sensitive to high-frequency noise than are peak amplitude measures. This is because all the little "uppies" and "downies" within the measurement window cancel each other out. For example, even though the data in figure 9.2 have been low-pass filtered, there is still quite a bit of noise that can influence the peak (especially in subject 3), but this noise has almost no effect on mean amplitude as long as the measurement

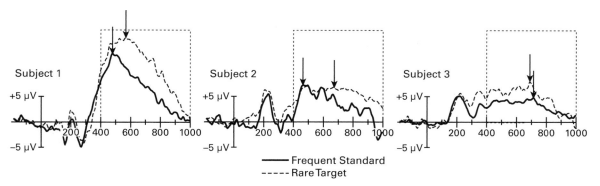

Figure 9.2
ERP waveforms elicited by frequent standards and rare targets, recorded from three individual subjects in an oddball paradigm. The dashed region shows a measurement window that could be used for measuring the P3 wave, and the arrows show the peaks in this window.

window is wide enough (e.g., >50 ms). In fact, there is no advantage to applying a low-pass filter prior to measuring mean amplitude, and box 9.1 explains why it is conceptually purer to measure mean amplitude without first applying a low-pass filter.

Issue 3: Measuring the Same Process at Different Times

Even when the high frequencies are filtered out, noise may still have a large impact on peak measures. As an example, consider the three individual subjects shown in figure 9.2, who participated in an oddball paradigm. The waveforms were low-pass filtered with a half-amplitude cutoff of 30 Hz to eliminate high-frequency noise, and then P3 amplitude was measured as the local peak between 400 and 1000 ms. Although subject 1 had a large and sharp P3 deflection, making the peak a reasonable estimate of the amplitude, subject 2 and subject 3 had broader P3 deflections without a very clear peak. Whereas the peak for the standard stimulus was a little before 500 ms in subject 1, the peak for this same stimulus was at approximately 700 ms for subject 3. For a simple oddball task, it is implausible that two neurologically normal people differ by more than 200 ms in the timing of the process that generates the P3 wave. And what would it mean to compare the voltage at 500 ms in one subject with the voltage at 700 ms in another subject? These huge differences in peak latency across subjects make it a little strange to use the P3 peak as a reflection of differences in a single underlying component.

A similar issue arises if we use peak amplitude to compare different conditions. For example, if we compare the amplitude of the P3 peak for the target and standard stimuli in subject 2, we will be comparing a voltage at 475 ms in one condition with a voltage at 675 ms in another condition. It seems unlikely that we are really measuring the same process at these two very different times in this simple oddball task.

Another related issue arises when peak amplitude is used to measure the amplitude of a component at multiple electrode sites. Because voltages propagate instantaneously from the brain to

Box 9.1
Filtering and Mean Amplitude

If you carefully examine the methods sections of papers published by my lab, you will see that we often measure mean amplitudes from data without any low-pass filtering (except for the very mild hardware anti-aliasing filter). However, we low-pass filter the data for figures showing ERP waveforms. This reflects our attempt at being very precise about timing. After all, one of the main virtues of the ERP technique is its high temporal resolution, so it makes sense to be precise about timing.

Why does our approach give us greater precision? As described in chapter 7, low-pass filters cause a temporal spread in the ERP waveforms (see especially figure 7.9B). If our method section said that we measured the mean amplitude between 150 and 250 ms, but we had applied a low-pass filter to the EEG or ERP prior to measurement, our measurements would be influenced by brain activity that happened prior to 150 ms and after 250 ms. Thus, the measurement window described in our methods section would not be a very precise way of describing the timing of our measurements if we filtered prior to measuring the data. Moreover, because high-frequency noise is canceled out in measurements of mean amplitude (assuming the measurement window is wide enough), filtering out the high frequencies prior to measuring mean amplitude doesn't help anything.

In reality, the filters we use are mild enough that we would get almost exactly the same results if we filtered prior to measuring mean amplitude. However, it seems worthwhile to be perfectly clear about the timing of the data that contribute to the analyses, just for the sake of conceptual purity.

However, we do apply a low-pass filter before plotting the data in our figures. This simply makes it easier for the reader to see the differences between the waveforms without being distracted by noise. You will see that we explicitly state that the waveforms have been filtered in our figure captions. Thus, we are being both precise and clear about what we are doing. I would encourage you to take this same approach. If nothing else, it will show the world that you understand what you are doing with your data processing and analysis procedures (although it might also indicate that you are a little bit obsessive-compulsive, just like I am).

all of the electrodes, brain activity from a given underlying component cannot have different latencies at different electrode sites. However, the latency of a peak may vary greatly across electrode sites because of differences in the amplitudes of overlapping components. Thus, if you compare peak amplitudes at different electrode sites, you will typically be measuring the amplitude at different times for the different sites, which makes no sense. One solution to this problem is to find the peak at one electrode site (typically the site where the peak is largest) and then measure the voltage at this time point at the other electrode sites. For example, you might find the P3 peak at the Pz electrode site and then measure the voltage at that time point at all electrode sites. This is a fine solution, as long as you have good reason to believe that the amplitude of the peak is an appropriate measure (which is not typically true).

All of these problems are a result of the fact that peak amplitude does not ordinarily use the same time point to measure a given component across subjects, across conditions, or across electrode sites. In contrast, the mean amplitude in a given time window measures—by definition—the voltages at the same time points in all subjects, in all conditions, and in all channels.

There are occasionally situations in which it would not be appropriate to use the same time window for all subjects or for all conditions. For example, if a component occurs later in one group than in another, it may be problematic to compare mean amplitude over the same time window in both groups. Instead, you may want to find the peak amplitude over a latency window that is broad enough to include all subjects in both groups. However, this sort of situation is usually very tricky even with peak amplitude. For example, if the latencies differ across groups, the latency variability probably also varies, and differences in latency variability invalidate the use of peak amplitude measures (as described earlier in the chapter). In addition, if other components overlap the component you are trying to measure, these components are likely to differ in amplitude between the two measurement windows you are using for the two groups, and this could distort your amplitude measurements. If you are in this situation, you should try to use difference waves to isolate the component of interest (as discussed in chapter 2), and you should consider using the *signed area* measures described later in this chapter (which are insensitive to differences in latency variability).

Issue 4: Biased Versus Unbiased Measures

Another problem with peak amplitude is that it is biased by the noise level (as was mentioned previously in chapter 4). The noisier the data, the larger the maximum amplitude will tend to be. Thus, it is not legitimate to use peak amplitude when comparing waveforms that are based on different numbers of trials or when comparing groups with different noise levels (which is common when comparing patients versus controls or younger versus older subjects).

In contrast, mean amplitude is not biased by the noise level. That is, mean amplitude does not become systematically larger as the number of trials decreases or as some other source of noise increases. Increasing the noise makes mean amplitude measures more variable, decreasing your statistical power, but it does not bias them to become larger. In contrast, peak amplitude becomes both more variable and systematically larger as the noise level increases (all else being equal). Consequently, it is not usually legitimate to compare peak amplitude values across groups or conditions with different noise levels, but it is perfectly fine to compare mean amplitude measurements from waveforms based on different numbers of trials.

Many people have difficulty understanding this distinction between bias and variability, and I see a lot of inappropriate analyses of experiments in which the number of trials differs across groups or conditions. I have therefore provided an online supplement to this chapter that provides a more complete description of what happens to peak and mean amplitude when the number of trials varies across groups or conditions (and what you should do in this situation).

Issue 5: Linear Versus Nonlinear Measures

Some ERP processing operations are linear and others are nonlinear. The measurement of mean amplitude is a linear operation, whereas the measurement of peak amplitude (or peak latency) is nonlinear. You can apply linear processes such as filtering, re-referencing, and averaging in any order, and the result will be the same. In contrast, the order of operations can make a big difference if one of the processes is nonlinear (see the appendix of this book for further discussion, including a definition of *linear* and *nonlinear* operations).

For example, imagine that you measure the mean amplitude from 150 to 250 ms in the ERP waveforms for 15 subjects, obtaining one value for each subject, and then you compute the average of these 15 values. Now imagine that you measure the mean amplitude from 150 to 250 ms from the grand average of the 15 single-subject ERP waveforms. This measurement from the grand average will be exactly the same as the average of the 15 mean amplitudes that you obtained from the single-subject waveforms. In other words, it doesn't matter whether you measure the mean amplitude from the single subjects and then average them together or whether you average the single-subject waveforms together and then measure the mean amplitude. Because measuring mean amplitude is a linear operation, and making a grand average is a linear operation, the order doesn't matter. The same thing is true when applied to single-trial EEG waveforms and averaged ERP waveforms: Measuring the mean amplitude from 150 to 250 ms on the single-trial EEG waveforms and then averaging these measurements together will give you the same result as averaging the single-trial EEG waveforms into an averaged ERP waveform and then measuring the mean amplitude from 150 to 250 ms. This is a very convenient feature of mean amplitude.

In contrast, peak amplitude is a nonlinear measure, and the order of operations matters a great deal. If you measure the P2 wave by finding the peak amplitude between 150 and 250 ms in each of 15 single-subject ERP waveforms, the average of these peak amplitudes will not be the same as the peak amplitude from a grand average of the 15 single-subject waveforms. This may cause a discrepancy between the grand-average waveforms that you present in your figures and the averaged peak amplitude values that you analyze statistically. Box 9.2 describes how I first noticed this issue as a graduate student. The same thing is true when applied to single-trial EEG waveforms and averaged ERP waveforms: Measuring the peak amplitude from 150 to 250 ms on the single-trial EEG waveforms and then averaging these measurements together will give you a very different result compared to averaging the single-trial EEG waveforms into an averaged ERP waveform and then measuring the peak amplitude from 150 to 250 ms.

Issue 6: Latency Jitter

Because peak amplitude is not a linear measure, peak amplitude in an averaged ERP waveform may radically misrepresent the amplitudes on the single trials. We encountered a special case of this when we discussed latency jitter in chapter 8. As you will recall, when the latency of a component varies from trial to trial in the raw EEG data, the peak amplitude in the averaged ERP waveform will be smaller than the single-trial amplitudes (see figure 8.7 in chapter 8; see also the extended discussion in online chapter 11). If latency variability is greater in one condition than in another, the peak amplitudes will differ between conditions even if there is no difference between conditions in the single-trial peak amplitudes. Thus, peak amplitude is not a valid measure if there are differences in latency variability across groups or conditions.

Box 9.2
Why Don't These Peaks Match?

When you are analyzing data from ERP experiments, you will typically have four, five, or even six factors in your analyses of variance. This is because you will have the same factors as in a behavioral experiment, plus one to two additional counterbalancing factors (if you implement the experimental design suggestions from chapter 4), plus one to two electrode factors (e.g., a left-to-right factor and a front-to-back factor). In most analysis of variance (ANOVA) programs, this requires a data file in which each row is a subject and each column is a single cell of your multifactor design. With four to six factors, and two to six levels of each factor, you will commonly have dozens or even hundreds of columns. To get the ANOVA program to work properly, you need to provide information about the ordering of all these columns. It's really easy to make a mistake, which will lead to completely bogus ANOVA results. Sometimes this is obvious (e.g., when you get an F value of 3432.12 for a counterbalancing factor that shouldn't have much of an effect). But sometimes it's not obvious. When I was in graduate school, I learned that I should always verify the ordering of the factors to avoid errors.

How can you verify this? The simplest way is to look at the table of means provided by the ANOVA program and compare these means with your grand average ERP waveforms. If the means match, then you almost certainly have the factors in the right order.

One day when I was a grad student, I was trying to match up the means from the ANOVA program with the grand averages, and I couldn't get them to match for my peak amplitude measures. However, they matched perfectly for the mean amplitude measures. I double-checked and triple-checked the data files, and everything was in exactly the same order for the peak and mean amplitude measures. Then I started looking at the individual subjects. The peak values for the single subjects in the ANOVA output matched the single-subject averaged ERP waveforms perfectly. Eventually I realized the problem: Measuring the peaks from the individual subjects and then taking the average is not equivalent to averaging the single-subject waveforms together and then measuring the peak from the grand average. In contrast, because mean amplitude and averaging are both linear operations, you get exactly the same result whether you measure the mean amplitudes from the single subjects and then average or measure the mean amplitude from the grand average.

Because mean amplitude is a linear measure, it doesn't matter whether you measure it from the single trials and then average the single-trial values together or whether you measure it from the averaged ERP waveform. Thus, mean amplitude is not influenced by latency jitter as long as your measurement window is sufficiently wide.

Issue 7: Comparing Peaks with Reaction Times

Another problem with peak measures is that peak latencies are difficult to relate to other measures of processing time, such as reaction time (RT). The difficulty arises because the peak of an ERP waveform is analogous to the mode of the RT distribution, not to the mean, and comparing modes and means is a little like comparing apples and oranges. This will be discussed in more detail in the section on "Comparing ERP Latencies with Reaction Times" later in the chapter.

Drawbacks of Mean Amplitude

Although mean amplitude has several advantages over peak amplitude, it is not a panacea. In particular, mean amplitude is still quite sensitive to the problem of overlapping components and can lead to spurious results if the latency of a component varies across conditions. There are some more sophisticated methods for isolating specific components (e.g., dipole source modeling, ICA, PCA, etc.), but these methods are based on a variety of difficult-to-assess assumptions and are beyond the scope of this book. Thus, I would generally recommend using mean amplitude measures in conjunction with the rules and strategies for avoiding the problem of overlapping components that were discussed in chapter 4.

Because the peaks vary in latency across subjects and across electrodes, you might think that measuring the mean amplitude over a fixed window would be problematic. However, these variations in peak latency are often inconsequential for mean amplitude measures. That is, these variations in peak latency are often driven by variations in overlap from other components, not by variations in the timing of the experimental effect. Thus, this is not usually a problem in practice.

Perhaps the most vexing problem that arises in using mean amplitude is defining the measurement window. Consider, for example, the Gedankenexperiment shown in figure 9.3, in which the N2 component is being compared between group A and group B. If we want to measure mean amplitude, what measurement window should we use? If we look at the grand averages, we see that the difference between groups is present from approximately 250 to 400 ms. We

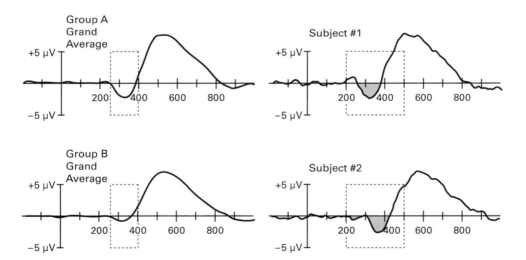

Figure 9.3
Data from a Gedankenexperiment in which the N2 component is compared between two groups of subjects, group A and group B. Grand averages are shown for the two groups, along with single-subject averages of two subjects from group A.

could use this as the measurement window, but because we are choosing the measurement on the basis of the data, we are biased to find a significant difference even if there is no true difference (this issue is described in more detail in chapter 10). A very good alternative is to use previous research to guide the selection of the measurement window. For example, if a previous study comparing these two groups found an N2 difference between 300 and 450 ms, we could use this interval as the measurement window for our new experiment. However, if the stimuli were brighter in the new experiment, the latencies would be a little shorter in the new experiment than in the previous experiment. Indeed, the P3 wave in figure 9.3 starts around 375 ms, and if we measured the mean amplitude from 300 to 450 ms, the P3 during this window would partially cancel the N2, canceling out the difference in N2 amplitude between conditions. We could measure the peak amplitude of the N2 wave, but that would have all the problems described earlier in this section.

In most cases, using previous experiments to determine the latency window for measuring mean amplitude is a good idea. However, it sometimes fails (as in the hypothetical example shown in figure 9.3). This doesn't mean that mean amplitude is a bad measure or that you shouldn't use previous experiments to select your measurement windows. The point of this example is that mean amplitude is very sensitive to the choice of latency window. This is the main shortcoming of using mean amplitude. Fortunately, area measures can sometimes overcome this limitation, as I will describe in the next section.

Area Amplitude

In the first edition of this book, I wrote that area amplitude is essentially the same as mean amplitude, except that mean involves an additional step of dividing by the duration of the measurement interval. However, when Javier Lopez-Calderon and I were developing the measurement procedures in ERPLAB Toolbox, Javier convinced me that this was not quite correct. Javier then implemented several different types of area measures that turned out to be extremely useful. In this section, I will first define what area amplitude means and how it is different from mean amplitude, and then I will describe how area measures can be advantageous.

Defining Area

Figure 9.4A illustrates the correct way to use the term *area*. To compute areas, geometric regions are defined using the ERP waveform, the baseline, and the edges of the measurement window as the boundaries. In this example, we have defined a measurement window of 200–300 ms to measure the N2 wave. This creates two distinct geometric regions, labeled A and B in the figure. Area is a geometric term, and it is always positive. For example, both region A and region B have a positive area, even though region A is below the baseline and region B is above. The area from 200 to 300 ms in this example is therefore the area of region A plus the area of region B. This is very different from the mean amplitude, in which negatives and positives cancel each other.

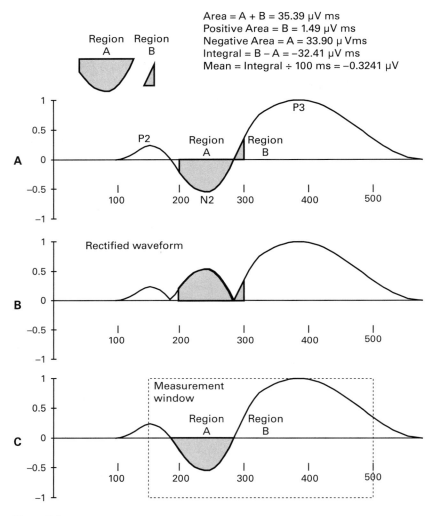

Figure 9.4
Example of how the term *area* is defined for ERP waveforms. (A) The ERP waveform, baseline, and measurement window (200–300 ms) are used to define two regions, A and B. Each region is a geometric shape and therefore has a positive area, expressed in units of μV·ms. (B) Rectified version of the waveform shown in panel A. Any negative voltages were multiplied by −1 to convert them into positive voltages. (C) Example of the *negative area* in the waveform shown in panel A. A much wider measurement window was possible because the positive regions are excluded when negative area is computed.

The units for area represent the multiplication of the X dimension and the Y dimension. The area of a room, for example, is expressed in square meters (m^2) if the X and Y dimensions are expressed in meters, or in square feet (ft^2) if the dimensions are expressed in feet. Similarly, the area of a region of an ERP waveform is expressed in units that multiply the X dimension (time) and the Y dimension (amplitude). If we measure amplitude in microvolts and time in milliseconds, the area would be given in units of $\mu V \cdot ms$. For example, a square region that is 1.5 μV high and 10 ms long would have an area of 15 $\mu V \cdot ms$.

When I used the term *area* in the first edition of this book, I was really thinking about the *integral*. The integral between 200 and 300 ms in figure 9.4A would subtract the area below the baseline (region A) from the area above the baseline (region B). The area of region A is 33.90 $\mu V \cdot ms$, and the area of region B is 1.49 $\mu V \cdot ms$, so the integral would be B – A = 1.49 – 33.90 = –32.41 $\mu V \cdot ms$. The mean amplitude is simply the integral divided by the duration of the interval. Thus, an integral of –32.41 $\mu V \cdot ms$ divided by a duration of 100 ms is equal to a mean amplitude of –0.3241 μV.

To clarify the distinction between the geometric area and the integral, ERPLAB Toolbox uses the term *rectified area* to refer to the geometric area. *Rectification* is just another term for taking the absolute value for each point in the waveform (multiplying each negative value by –1 to make it positive). Figure 9.4B shows what the waveform in figure 9.4A would look like if rectified. Once we rectify the waveform, all the areas are positive, so the integral is now equivalent to the geometric area. Note that rectified area is not usually used to measure ERP components, but rectification is often used in other contexts (e.g., measuring EMG activity), so it is worth knowing this term.

Once you start thinking about area, it becomes obvious that you might want to use only the area of the region below the baseline if you are measuring a negative component or only the area of the region above the baseline if you are measuring a positive component. In figure 9.4A, for example, we might want to quantify N2 amplitude with the area of the negative region (region A). We can therefore define *negative area* as the area of the region (or regions) below the baseline and *positive area* as the area of the region (or regions) above the baseline. More generally, negative area and positive area are cases of *signed area*.

If you use area or integral measures, I encourage you to adopt the specific terminology described in this section, which avoids the ambiguities that otherwise occur with the term *area*.

The Advantage of Signed Area Amplitude

As illustrated in figure 9.4C, an advantage of using negative area to measure the N2 wave is that you can use a very wide latency window without getting any cancellation from the preceding P2 wave or the subsequent P3 wave. In other words, as long as you use a fairly wide measurement window, it doesn't matter very much what measurement window you use.

This can be a huge advantage. Consider, for example, the Gedankenexperiment shown in figure 9.3. If we used the negative area measure, we wouldn't need to worry about the window being so wide that the P3 cancels the N2. We could select a fairly wide window (e.g., 200–500 ms), and it would capture the area of the N2 without any cancellation from the P2 or P3. We would also get approximately the same results with a window of 250–450 ms or 100–600 ms. This eliminates any bias that we might introduce by selecting a narrow window on the basis of the observed time course of the effect. It also gives us the one advantage of peak amplitude, which is the ability to measure something that might occur at different times in different conditions or subjects. But it eliminates several of the disadvantages of peak amplitude. For example, it is relatively insensitive to high-frequency noise. In addition, little noise blips won't cause it to use dramatically different time periods to measure different subjects. Instead, it tends to reflect the real differences across subjects in the timing of components.

The right side of figure 9.3 shows how this works with individual subjects who have somewhat different N2 latencies, using a window of 200–500 ms. Although the N2 is later in subject 2 than in subject 1, the N2 area is fully captured in both subjects. The only downside is that some negative-going noise deflections also contribute to the negative area in subject 2. This is a small price to pay given the advantages of using the negative area.

Risa Sawaki and I used this approach in a study in which a negative-going N2pc component was followed by a positive-going Pd component (Sawaki, Geng, & Luck, 2012). We had no previous experiments that we could use to determine an appropriate measurement window for the Pd component in this experiment. If we chose a narrow window on the basis of the data, this would have biased our results. If we chose a broad window, the N2pc would have partially canceled the Pd. We therefore decided to use the positive area, and it worked great.

The main disadvantage of signed area measures is that they are biased to be larger than the true value. For example, if you measure the positive area, the smallest possible value is zero, so the average across subjects will either be zero or greater than zero. In addition, noisy waveforms will tend to have larger values than clean waveforms (all else being equal). This is the same problem that arises with peak amplitude. Consequently, you cannot ordinarily use positive or negative area to compare groups or conditions with different noise levels. In addition, you can't determine if a component is significantly different from zero by just doing a one-sample t test, as you could with mean amplitude. To deal with this problem, Risa used nonparametric permutation statistics rather than conventional parametric statistics in her study (Sawaki et al., 2012).

Despite this one shortcoming, positive area and negative area have tremendous potential to minimize the problem of selecting the measurement window, and I encourage you to try this approach. But keep in mind that it is a relatively new approach, so you will want to think carefully about it.

Using Fractional Area Latency to Estimate the Midpoint Latency

The vast majority of ERP studies have quantified the timing of a component by measuring the latency of the peak (or local peak) within a given time window. Researchers don't usually provide an explicit justification for using the peak to quantify the latency, and it appears to be used mainly because of tradition (which is a result of the fact that peaks were the only thing that could be easily measured in the early days of ERPs, before powerful computers were available).

As is discussed in detail in online chapter 11, the shape of an averaged ERP waveform reflects the distribution of onset times of the single-trial ERP waveforms. This is why, for example, latency variability influences the shapes of the ERP waveforms. The peak of an ERP waveform is analogous to the mode of the underlying distribution of single-trial waveforms. The mode is not usually used as a measure of central tendency, and this makes it difficult to compare the timing of an ERP peak to other measures of processing time, such as RT. Moreover, the mode is typically less reliable than other measures of central tendency. In this section, I will describe a much better method for assessing the *midpoint* of a component, called *fractional area latency*. This approach involves using area amplitude, and it has many of the same advantages as using area to measure amplitude. In addition, it is much better than peak latency for making comparisons with reaction time.

The fractional area latency measure involves computing the area under the ERP waveform over a given latency range and then finding the time point that divides that area into a prespecified fraction (this approach was apparently first used by Hansen & Hillyard, 1980). Typically, the fraction will be one-half, in which case this would be called a *50% area latency* measure. An example of this is shown in figure 9.5A. The measurement window in this figure is 300–600 ms, and the area under the curve in this time window is divided at 432 ms into two regions of equal area. Thus, the 50% area latency is 432 ms. Just as amplitude can be measured using the geometric (rectified area), the positive area, the negative area, or the integral, you can find the 50% point using any of these definitions of area. Using just the positive area or just the negative area is usually the best approach, because it minimizes contributions from noise or overlapping components that do not have the same polarity as the component of interest.

The latency value that is estimated in this manner will depend quite a bit on the measurement window that is chosen. For example, if the measurement window for the waveform shown in figure 9.5A was shortened to 300–500 ms rather than 300–600 ms, an earlier 50% area latency value would have been computed. Consequently, this measure is appropriate primarily in two (fairly common) situations: (1) when the waveform is dominated by a single large component (e.g., the large P3 wave in the visual search experiment shown in figure 8.8); or (2) when a difference wave has been used to isolate a single component. It may also be useful under other conditions when combined with an automated procedure for determining the measurement window. Figure 9.5D shows an example of measuring the 50% area latency in a rare-minus-frequent difference wave that isolates the P3 wave. A wider measurement window can be used because there are no overlapping components to distort the measurement.

One advantage of the 50% area latency measure is that it is less sensitive to noise than is peak latency. To demonstrate this, I added random (Gaussian) noise to the waveform shown in figure 9.5A and then measured the 50% area latency and the local peak latency of the P3 wave. I did this 100 times for each of two noise levels, making it possible to estimate the variability of the measures. When the noise level was 0.5 µV (figure 9.5B), the standard deviation of the

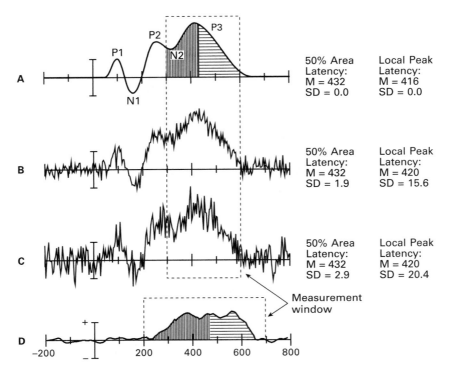

Figure 9.5
Application of 50% area latency and local peak latency measures to a noise-free ERP waveform (A), an ERP waveform with a moderate amount of noise (B), an ERP waveform with significant noise (C), and a rare-minus-frequent difference wave (D).

peak latency measure over the 100 measurements was 15.6 ms, whereas the standard deviation of the 50% area latency measure was only 1.9 ms. When the noise level was increased to 1.0 μV (figure 9.5C), the standard deviation of the peak latency measure was 20.4 ms, whereas the standard deviation of the 50% area latency measure was only 2.9 ms. The variability in peak latency measures can be greatly decreased by filtering the data, and a fair test of peak latency should be done with filtered data. When the waveforms were low-pass filtered quite severely with a half-amplitude cutoff at 6 Hz, the standard deviations of the peak latency measures dropped to 3.3 ms and 6.1 ms for noise levels of 0.5 and 1.0 μV, respectively, but this was still higher than the standard deviations observed with the 50% area latency measure without any filtering. Thus, when severe filtering was performed prior to measuring peak latency (well beyond the range I would ordinarily recommend), 50% area latency was still more reliable than local peak latency.

Kiesel et al. (2008) conducted a set of very rigorous and thoughtful simulations to test several different measures of latency. If your research involves looking for subtle latency differences, I strongly recommend that you read this paper. Their results were consistent with the simple simulation shown in figure 9.5, showing that fractional area latency tended to be the most reliable way of measuring changes in latency across conditions or groups, leading to the best statistical power (especially when combined with the jackknife approach to statistical analysis, which will be discussed in chapter 10).

Onset Latency

Because peaks were easily measured by early ERP researchers, peak latency became the standard way of measuring timing, and this tradition continues to this day. However, theories of neural, cognitive, and affective processes do not typically focus on the time at which a process reaches its peak. Instead, they typically focus on when a process begins, when it reliably discriminates between alternative input patterns, or its duration. Moreover, as discussed in chapters 2 and 4, the onset time of a difference between two conditions provides an excellent way of assessing the amount of time required for the brain to differentiate between these conditions. Thus, it often makes much more sense to measure the onset or offset of a component (or the onset or offset of a difference wave) rather than the peak. The use of such measures has been increasing in ERP research, which is a good thing.

Challenges in Measuring Onset Latency

Unfortunately, onset latency tends to be more difficult than midpoint latency to measure accurately and reliably. For example, the onset of the P3 wave in figure 9.5A is obscured by the P2 and N2 waves. I can't even make a reasonable guess about the onset time of the P3 wave in this waveform. The only component whose onset is reasonably clear in this waveform is the P1 wave, because it is the very first component. Thus, it is somewhere between difficult and impossible to assess the onset of an ERP component unless (a) it is much larger than the surrounding components or (b) it has been isolated by means of a difference wave.

When a difference wave is used, as in figure 9.5D, the onset is much easier to estimate, at least visually. However, there is still a conceptual problem that must be solved. Specifically, the onset of a difference between conditions is the point at which the difference is infinitesimally greater than zero (or infinitesimally greater than the noise level), which means that the signal-to-noise ratio at this point is essentially zero. People have attempted to address this problem in multiple ways. One intuitively attractive approach is to estimate the slope of the onset period and extrapolate to 0 μV. I have tried this, and it just doesn't work very well with single-subject ERP waveforms, which are highly variable. In many cases, the rising edge of the waveform is far from linear, making this approach difficult in practice.

Another approach is to find the time at which the waveform's amplitude exceeds the value expected by chance. The variation in prestimulus voltage can be used to assess the amplitude required to exceed chance, and the latency for a given waveform is the time at which this amplitude is first reached (for details, see Osman, Bashore, Coles, Donchin, & Meyer, 1992; Miller,

Patterson, & Ulrich, 1998). Unfortunately, this method is highly dependent on the noise level, which may vary considerably across subjects and conditions.

A related approach is to conduct a *t* test between two conditions or groups and find the time at which two conditions become significantly different from each other. To avoid spurious results arising from the use of a large number of *t* tests, you can find the first time point that meets two criteria: (1) the *p* value is less than 0.05, and (2) the *p* values for the subsequent *N* points are also less than 0.05 (where *N* is usually in the range 3–10). This approach doesn't really find the true onset time of the difference, but rather the point at which the difference is large enough to exceed the statistical threshold. Also, it is somewhat complicated to determine the number of consecutive significant points that should be required to lead to an overall Type I error rate of 5%. Moreover, because significance is tested for individual time points, which tend to be noisy, this approach tends not to have very good statistical power.

Although these techniques for measuring onset latency have significant disadvantages, there are two techniques that work quite well. Before I describe them, however, I would like to stress that all methods for assessing onset latency are easily distorted by high-frequency noise. I recommend using a low-pass filter with a half-amplitude cutoff of approximately 10 Hz and a slope of approximately 24 dB/octave when measuring onset latency. Keep in mind, however, that this can cause a shift in the onset times (see chapter 7). This shift will typically be equivalent across conditions, so it is not usually a problem, but you should be aware of it.

Measuring Onsets with Fractional Area Latency

One powerful approach for measuring onset latency is to use the fractional area latency measure shown in figure 9.5, but using a smaller fraction, such as the point that divides the first 25% of the area from the last 75% (which would be called the 25% area latency). The simulation study of Kiesel et al. found that a 30% area latency measure was highly reliable, although not quite as reliable as the 50% area latency in most cases. So why not just use the 50% area? Many ERP latency effects do not simply consist of a shift of the whole waveform, and the onset may change without much change in the midpoint, so it is sometimes important to measure the onset latency of an effect.

Measuring Onsets with Fractional Peak Latency

My favorite technique for measuring onset latency is called the *fractional peak latency* measure. As illustrated in figure 9.6, you first find the peak amplitude, and then you work backward in time until you find the point in the waveform at which the latency is a certain percentage of the peak value.[2] This time point is then considered to be the onset time. In most cases, the 50% point yields the highest reliability (Kiesel et al., 2008), in which case this can be called the *50% peak latency* measure.

The waveforms in figure 9.6 are from a visual search experiment in which subjects searched for a red target in some trial blocks and a green target in other trial blocks (Luck et al., 2006). Despite our efforts to equate the salience of the red and green targets, the red targets were quite

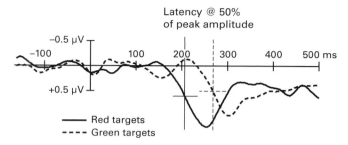

Figure 9.6
Grand average N2pc difference waves (contralateral minus ipsilateral) for red targets and green targets (from the study of Luck et al., 2006). The 50% peak latency is shown for both waveforms, defined as the latency at which the voltage reaches 50% of the peak (or local peak) voltage.

a bit more salient than the green targets. The figure shows the N2pc activity elicited by the red and green targets, isolated from the rest of the waveform with a contralateral-minus-ipsilateral subtraction (see the section on N2pc in chapter 3). The onset of this difference wave was clearly earlier for the red targets than for the green targets.

This example illustrates two virtues of the fractional peak latency measure. First, N2pc amplitude was a little larger for the red targets than for the green targets, and this is factored out by finding the time at which a percentage of the peak amplitude is reached. Second, the difference wave consisted of a rapid ramp-up from zero to the peak, then a decline, and then a long period of moderate amplitude. If we used fractional area latency instead of fractional peak latency, the long period of moderate amplitude late in the waveform would have influenced our measure of the onset time. In this situation (which is common), fractional peak latency provides a purer measure of onset time.

You might think that the 50% point is too late to provide a good measure of the onset time. However, it's actually a very good measure of the onset time in many cases. Recall from chapter 8 that when the latency of a component varies from trial to trial, the onset time of the average across trials is driven mainly by the trials with the earliest onset times (see figure 8.7). The point at which the average reaches 50% will often be close to the average onset time across trials.

The logic behind this assertion is illustrated in figure 9.7. In this example, I am assuming that the single-trial ERP is a 200-ms square wave (panel A). This is, of course, a little bit unrealistic, but it makes the example simpler. Panel B shows the probability distribution of the single-trial latencies. I am assuming that the most likely onset time is 200 ms, with a gradual fall-off in the probability of earlier and later onset times (with a normal distribution). Given this distribution, the average onset time is also at 200 ms. Panel C shows what the averaged ERP waveform would look like with the single-trial waveform shown in panel A and the distribution of onset times shown in panel B. The point at which the average reaches 50% of the peak amplitude is at 200 ms, which is exactly the average single-trial onset time. Perfect! This does not mean that the

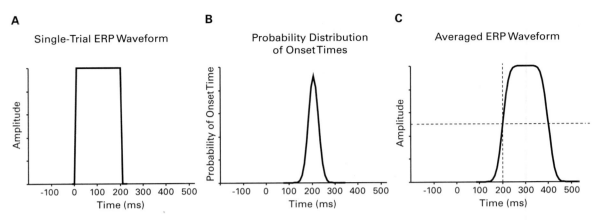

Figure 9.7
Example of how the 50% peak latency point may be a good estimate of the average onset time of a component. (A) Single-trial ERP waveform, which consists of a square wave in this simplified example. (B) Probability distribution of single-trial onset latencies. The average (and most common) onset latency is 200 ms, with a Gaussian distribution around this mean. (C) Averaged ERP waveform that results from combining the single-trial ERP waveform in panel A with the distribution of onset times in panel B.

50% peak latency measure will always capture the average single-trial onset time, but it will provide a reasonable approximation under many realistic conditions.[3]

Kiesel et al. (2008) found that the 50% peak latency measure was not quite as reliable as the 50% area latency measure, but both were quite good. The 50% area latency measure will typically be best when the onset and the midpoint of the component are both shifted across groups or conditions. However, with more complicated patterns such as that shown in figure 9.6, the 50% peak latency measure is likely to be best. In addition, when you want a measure that is likely to be close to the average of the single-trial onset times, the 50% peak latency measure is likely to be best. For examples of how I have used these two measures with the P3 wave, the N2pc component, and the lateralized readiness potential, see Luck and Hillyard (1990), Luck (1998b), Luck et al. (2006), and Luck et al. (2009).

Comparing ERP Latencies with Reaction Times

In many studies, it is useful to compare the size of an ERP latency effect to the size of an RT effect. This seems straightforward, but it is actually quite difficult.

To understand why this is true, you need to understand *frequency distributions* and *probability distributions*. Imagine, for example, that you have recorded RTs from a subject on 100 trials of a given condition. Ordinarily, you would just summarize the RTs with the mean of the 100 trials. However, this throws away a lot of information about the RTs. A frequency distribution provides much more information, showing how often RTs occurred within particular time bins. For example, figure 9.8A uses 50-ms time bins and shows that seven of the 100 RTs occurred in the

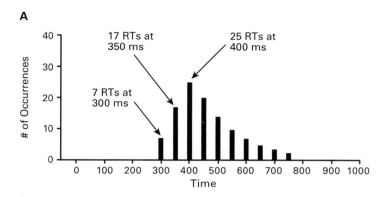

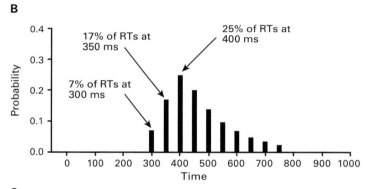

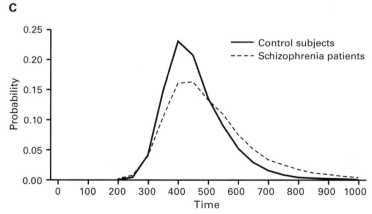

Figure 9.8

(A) Typical frequency distribution of RT. The height of each bar represents the number of responses that occurred within ±25 ms of the midpoint latency of the bar. (B) Same as panel A, but showing probability rather than frequency (by dividing each bar by the number of trials, which in this example is 100). (C) Actual RT probability distributions from a study of schizophrenia patients and control subjects (Luck et al., 2009). These distributions were aggregated across the subjects in each group, so they reflect both within-subject and across-subject variations in RT.

time bin centered at 300 ms (i.e., from 275 to 325 ms), 17 occurred in the time bin centered at 350 ms, 25 occurred in the time bin centered at 400 ms, and so forth. It is often useful to convert these frequencies into probabilities (i.e., the probability that a response occurred within a particular time bin), which you can do by simply dividing the value in each bin by the total number of trials. This gives us a probability distribution, as shown in figure 9.8B.

The probability distribution in figure 9.8B is skewed to the right (i.e., most of the RTs fall to the right of the peak). This is typical for RT, and many differences in RT between groups or conditions consist of an increase in the probability of relatively long RTs rather than a shift in the entire RT distribution. For example, figure 9.8C shows RT probability distributions for a group of schizophrenia patients and a group of control subjects (from the study of Luck et al., 2009). The fastest RTs were similar in the two groups, but patients had more long RTs than controls. The same pattern is often observed for comparisons between different experimental conditions in within-subject experiments.

As chapter 11 will describe in detail, finding the peak of an ERP waveform is like finding the mode (most frequent value) in a probability distribution. If you compare the latency of an ERP peak to a mean RT, this is like comparing a mode with a mean (which is like comparing apples with oranges). In many studies, such as the one shown in figure 9.8C, differences in the mean of the RT distribution will be substantially larger than differences in the mode of the RT distribution. Thus, differences in peak ERP latency will typically be smaller than differences in mean RT, even if the single-trial RT differences and single-trial ERP latency differences are exactly the same.

To make this clearer, figure 9.9A shows the probability distribution of RT in two conditions of a Gedankenexperiment, which we'll call the *easy* and *difficult* conditions. Each point represents the probability of an RT occurring within ±15 ms of that time. The RT distributions are right-skewed, as usual, and much of the RT difference between the conditions is due to a change in the probability of relatively long RTs rather than a pure shift in the RT distribution. Imagine that the P3 wave in this experiment is precisely time-locked to the response, always peaking 150 ms after the RT. The P3 wave will therefore occur at different times on different trials, with a probability distribution that is shaped just like the RT distribution from the same condition (but shifted rightward by 150 ms). Imagine further that the earlier components are time-locked to the stimulus rather than the response (which will typically be true). The resulting averaged ERP waveforms for these two conditions are shown in figure 9.9B (see chapter 11 for a more extensive discussion of how the shape of the averaged ERP waveform will reflect the probability distribution of single-trial component latencies).

Because most of the RTs occur within a fairly narrow time range in the easy condition, most of the single-trial P3s will also occur within a narrow range, causing the peak of the averaged ERP waveform to occur approximately 150 ms after the peak of the RT distribution (overlap from the other components will influence the precise latency of the peak). Some of the single-trial RTs occur at longer latencies, but they are sufficiently infrequent that they don't have much influence on the peak P3 latency in the averaged waveform.

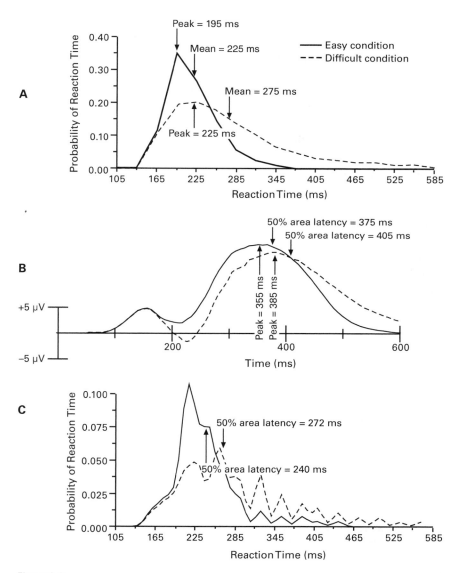

Figure 9.9
Gedankenexperiment comparing an easy condition with a difficult condition. (A) Probability distribution of reaction time, showing the probability of a response occurring in various time bins (bin width = 30 ms) in the easy and difficult conditions. (B) ERP waveforms that would be produced in this experiment if the early components were insensitive to reaction time and the P3 wave was perfectly time-locked to the responses. (C) Probability density waveforms, in which each single-trial reaction time was replaced by a Gaussian waveform (standard deviation = 8 ms), and then all the Gaussian waveforms were averaged together.

The mean RT is 50 ms later in the difficult condition than in the easy condition. However, because much of the RT effect consists of an increase in long RTs, the peak of the RT distribution is only 30 ms later in condition B than in condition A. Because the peak of the P3 wave in the averaged ERP waveform is tied closely to the peak of the RT distribution, P3 peak latency is also 30 ms later in the difficult condition than in the easy condition. Thus, the peak latency of the P3 wave changes in a manner that reflects changes in the peak (mode) of the RT distribution rather than its mean. Consequently, when RT effects consist largely of increases in the tail of the distribution rather than a shift of the whole distribution, changes in peak latency will usually be smaller than changes in mean RT, even if the component and the response are influenced by the experimental manipulation in exactly the same way. Consequently, you shouldn't compare ERP peak latency effects with mean RT effects. Box 9.3 describes my first encounter with this phenomenon as a college student.

How, then, can RT effects be compared to ERP latency effects? The answer is that they must be measured in the same way. One way to achieve this would be to use a peak latency measure for both the ERPs and the RTs (using the probability distribution to find the peak RT). However, the peak of the RT distribution is difficult to estimate reliably, and peak RT effects are likely to be smaller than mean RT effects (just as peak ERP latency effects tend to be smaller than mean RT effects).

An alternative is to use the 50% area latency measure for quantifying the ERP latencies and compare this with median RT. Median RT is the point that separates the fastest half of the RTs from the slowest half, which is almost the same thing as the point that divides the area into two equal halves. I have used this approach in several experiments, and the correspondence between

Box 9.3
Convolution, Peak Latency, and Mean Reaction Time

When I was a student at Reed College, I had a wonderful mentor named Dell Rhodes. Dell was trained as a physiological psychologist, which meant that she started her career in the 1970s studying the brains of rats. As the field of cognitive neuroscience started to emerge, Dell decided to take a new research path, and she spent a sabbatical year learning how to record and analyze ERPs. I started working with her the very next year (in my junior year of college), and I learned a tremendous amount from her as we muddled through the realities of setting up and running an ERP lab.

One day Dell remarked to me that she had been reading a lot of papers that looked at both P3 latency and mean RT, and the mean RT effects (i.e., differences in RT between groups or conditions) were always larger than the P3 latency effects. Both effects were on the same time scale (milliseconds relative to stimulus onset), so it was not obvious why the P3 effects should always be smaller than the RT effects. Dell's comment stuck with me for many years, and I finally figured out the answer to this riddle when I started to think about how peak latency was related to the mode of the RT distribution. The studies Dell had been reading were comparing peak latency with mean RT, and peak latency is analogous to the mode rather than the mean of the RT distribution. This explains why the P3 latency effects were smaller than the RT effects.

P3 latency and RT was excellent (Luck, 1998b; Luck et al., 2009). In most cases, this is the approach I would recommend for comparing ERP latencies with RTs.

You should note, however, that 50% area latency is not perfectly analogous to median RT, because median RT does not take into account the precise values of the RTs above and below the median. For example, a median RT of 300 ms would be obtained for an RT distribution in which half of the values were between 200 and 300 ms and the other half were between 300 and 400 ms, and the same median of 300 ms would be obtained if half of the RTs were between 290 and 300 ms and the other half were between 300 and 5000 ms. However, the correspondence between 50% area latency and median RT is close enough for most purposes, and median RT has the advantage of being a familiar measure.

If you want a measure of RT that is perfectly analogous to the 50% area latency measure, you need to find the point that bisects the *area* of the RT distribution. However, RTs are discrete, instantaneous events with zero area, making it difficult to measure the area of the RT distribution. Figure 9.9C shows how you can convert the data into an ERP-like waveform and then measure the area, using a technique borrowed from single-unit recording studies (see, e.g., Szücs, 1998). Each individual RT is replaced by a Gaussian function to turn it into a continuous function, and then the average of all the RTs is computed. The result is a waveform that looks a little bit like an ERP waveform, making it possible to calculate the 50% area latency for RT. In the example shown in figure 9.9C, the resulting latencies were 240 ms for the easy condition and 272 ms for the difficult condition, and the 32-ms difference between these latencies was nearly identical to the 30-ms effect that was obtained by measuring the 50% area latency in the ERPs. I don't know of anyone who has tried this approach, but it seems like the optimal way of comparing ERP latencies and RTs.

Suggestions for Further Reading

Dien, J., Spencer, K. M., & Donchin, E. (2004). Parsing the late positive complex: Mental chronometry and the ERP components that inhabit the neighborhood of the P300. *Psychophysiology, 41*, 665–678.

Donchin, E., & Heffley, E. F., III. (1978). Multivariate analysis of event-related potential data: A tutorial review. In D. Otto (Ed.), *Multidisciplinary Perspectives in Event-Related Brain Potential Research* (pp. 555–572). Washington, DC: U.S. Government Printing Office.

Kiesel, A., Miller, J., Jolicoeur, P., & Brisson, B. (2008). Measurement of ERP latency differences: A comparison of single-participant and jackknife-based scoring methods. *Psychophysiology, 45*, 250–274.

Sawaki, R., Geng, J. J., & Luck, S. J. (2012). A common neural mechanism for preventing and terminating attention. *Journal of Neuroscience, 32*, 10725–10736.

Spencer, K. M., Dien, J., & Donchin, E. (2001). Spatiotemporal analysis of the late ERP responses to deviant stimuli. *Psychophysiology, 38*, 343–358.

10 Statistical Analysis

Overview

Once you have recorded ERP waveforms from a sample of subjects and measured the amplitude and latencies of the components of interest, it will be time to perform statistical analyses to see whether your effects are significant. In the large majority of cognitive and affective ERP experiments, the investigators are looking for a main effect or an interaction in a completely crossed factorial design, and ANOVA-based statistical analyses are therefore the dominant approach. Consequently, this is the only approach I will describe, although other approaches can be useful in some cases.

Before I begin describing how statistical analyses are applied to ERP data, I would like to make it clear that I consider statistics to be a necessary evil. We often treat the 0.05 alpha level as being somehow magical, with experimental effects that fall below $p < 0.05$ as being "real" and effects that fall above $p < 0.05$ as being nonexistent. This is, of course, quite ridiculous. The 0.05 cutoff is purely arbitrary, and if the field had chosen a standard criterion of $p = 0.06$, we would have only a slightly higher rate of false positives (accompanied by a slightly lower rate of false negatives). Moreover, the assumptions of ANOVA are violated by almost every ERP experiment, so the p values that we get are only approximations of the actual probability of a Type I error. However, it is difficult to imagine how the publication process would work if we didn't have a commonly accepted criterion for deciding which effects to treat as real (although that may simply be a lack of imagination on my part). Unless Bayesian statistics completely take over, we are stuck with the need to evaluate statistical significance. This chapter therefore describes the common practices for dealing with this necessary evil in the context of ERPs. I'm assuming that you have already had an introductory statistics course, so this chapter focuses on the specific issues that arise when analyzing ERP data.

Before you read any more, take a look at box 10.1. It describes the most important principle for assessing statistical significance, and it supersedes everything else I will say in this chapter.

This chapter begins by describing the conventional approach to analyzing ERP data, in which ERP amplitudes and latencies are treated just like behavioral variables such as reaction time and accuracy. This approach was initially developed when the technology for ERP research was

Box 10.1
The Best Statistic

I first met Steve Hillyard when I visited UCSD during my senior year of college. When I met with Steve, I proudly told him about all the fancy multivariate statistics I had been using to analyze ERP data in my undergraduate senior thesis. He looked at me and said, "Around here, we think that replication is the best statistic." My initial thought, of course, was that this guy was a technically unsophisticated Luddite. By the time I finished my first year of grad school, however, I realized that he was absolutely correct (and wasn't a Luddite at all). Replication does not depend on assumptions about normality, sphericity, or independence. Replication is not distorted by outliers. Replication is a cornerstone of science. Replication is the best statistic.

A corollary principle—which I also learned from Steve—is that the more important a result is, the more important it is for you to replicate it before you publish it. An obvious reason for this is that you don't want to make a fool of yourself by making a bold new claim and being wrong. A less obvious reason is that if you want people to give this important new result the attention it deserves, you should make sure that they have no reason to doubt it. Of course, it's rarely worthwhile to run exactly the same experiment twice. But it's often a good idea to run a follow-up experiment that replicates the result of the first experiment and also extends it (e.g., by assessing its generality or ruling out an alternative explanation).

There are some areas of science in which the cost of running an experiment—in money or time—is so great that it is unrealistic to replicate a result before publishing it. If you are doing research of this nature, you will need to work extra hard to make sure that your results are real and not the result of a biased analysis approach. In these areas, replication is still important but usually occurs via meta-analyses across studies rather than via within-study replications.

primitive, and the data consisted of peak amplitudes or latencies measured at a few electrode sites. This approach evolved gradually as researchers began measuring other features of the waveform (e.g., mean amplitude) and were able to record from a couple dozen electrode sites. This approach is still valuable in many situations, especially when the researcher is testing a very specific hypothesis about the amplitude or latency of a component that is measured from a well-justified latency range at a reasonably small number of electrode sites (or if the data are averaged over clusters of nearby sites). A new variant of this approach—the *jackknife* approach—can dramatically improve statistical power under some conditions.

This chapter will also describe a newer and very different approach that is needed when large numbers of time points and/or electrode sites are analyzed, which leads to the *problem of multiple comparisons*. This newer approach is based on methods that were originally developed for the analysis of neuroimaging data, in which thousands of voxels must be tested and the problem of multiple comparisons is very obvious. These methods are just starting to hit the mainstream of ERP research, but I suspect they will become very common over the coming years.

Across the fields of psychology and neuroscience, the past few years have seen growing sensitivity to a variety of data analysis practices that dramatically increase the likelihood of Type I errors (i.e., significant effects that are actually bogus) (Vul, Harris, Winkielman, & Pashler,

2009; Simmons, Nelson, & Simonsohn, 2011; John, Loewenstein, & Prelec, 2012; Pashler & Wagenmakers, 2012; Button et al., 2013). This leads to a proliferation of incorrect conclusions in the literature, which is a very bad thing for scientific progress. The pressure to publish is partially responsible for leading people toward questionable data analysis practices. But there are many common practices people use that unintentionally inflate the Type I error rate. One goal of this chapter is to explain how these seemingly innocuous practices are problematic and to provide you with simple strategies for avoiding bogus significant results when analyzing ERP data.

This chapter has four main sections. First, I will review a little bit of statistical terminology that is used throughout the chapter. Second, I will describe the conventional approach to statistics. Third, I will describe the jackknife approach, which is a slight variation on the conventional approach that can dramatically improve your statistical power under certain conditions. Fourth, I will describe how the richness of an ERP data set often leads to a large number of (implicit or explicit) comparisons, which in turn complicate the analyses. In particular, the need to choose specific time windows and electrode sites can inflate the Type I error rate. This section provides several suggestions for avoiding this problem. One of them—called the *mass univariate approach*—is described in detail in online chapter 13. This chapter also describes a completely different general approach to statistics—called the *permutation* approach—which is becoming increasing popular in ERP research.

Terminology

This chapter uses a variety of basic statistical terms that should be familiar to most readers, but you may want to review some of them in the glossary before reading further. Here are the key terms: *null hypothesis; alternative hypothesis; alpha; p value; Type I error; Type II error; statistical power.*

In addition, this chapter uses the terms *experimentwise error rate* and *familywise error rate*, which may be less familiar. Imagine that you are analyzing the data from an oddball experiment with a three-way ANOVA for P3 amplitude and another three-way ANOVA for P3 latency. Each of these two ANOVAs will produce seven different p values (three main effects, three two-way interactions, and one three-way interaction), yielding 14 total p values across the two ANOVAs. If these were the only analyses in your experiment, the *experimentwise error rate* would be the probability that even one of these 14 p values was a false positive (a Type I error). With 14 p values and a standard alpha of 0.05 for each individual p value, the experimentwise error rate would be substantially higher than 5%. In other words, if the null hypothesis is actually true for all 14 of these effects, the chance of getting one or more significant p values ($p < 0.05$, uncorrected) would be greater than 5%. More generally, the experimentwise error rate is the probability that at least one p value among all the p values for a given experiment will be a false positive. The *familywise error rate* is the same concept, but refers to a subset of related p values from a given experiment (a "family" of related statistical tests, such as the seven p values from the P3 amplitude ANOVA). All else being equal, the more p values that are calculated in a given experiment or in a given family of analyses, the higher the experimentwise or familywise error rate will be.

The Conventional Approach

The conventional approach to ERP statistical analysis treats amplitude and latency measurements just like any other dependent variable. You obtain these values from each subject and enter them into a t test or ANOVA, just like you would for each subject's mean reaction time. The main difference from a behavioral analysis is that an ERP analysis will typically involve measurements from multiple electrode sites in each subject. You may also have both amplitude and latency measurements in an ERP study, and you may be measuring multiple components. However, amplitude and latency measurements are virtually always analyzed separately, just as you would analyze accuracy and RT separately in a behavioral experiment, and measurements of different components are almost always analyzed separately. Thus, a conventional ERP analysis is usually just like a behavioral analysis, except with measurements from each of several electrode sites. The inclusion of measurements from multiple electrode sites leads to two complications that you need to know about, and I will describe them in the next two subsections. First, however, I will give you a simple example of the conventional approach.

An Example of the Conventional Approach

This example is based on an unpublished oddball experiment described briefly near the beginning of chapter 1 (see figure 1.1). In this experiment, recordings were obtained from nine electrode sites (F3, Fz, and F4; C3, Cz, and C4; P3, Pz, and P4). Subjects saw a sequence of Xs and Os, and they were required to press one button for the Xs and another for the Os. In some trial blocks, X occurred frequently ($p = 0.75$) and O occurred infrequently ($p = 0.25$), and in other blocks this was reversed. We also manipulated the difficulty of the X/O discrimination by varying the brightness of the stimuli.[1]

In most cases, I recommend focusing your analyses on a single component (see strategy 1 in chapter 4). However, you may sometimes have a good reason to analyze multiple components. In addition, you may see effects for other components that are not the main focus, and you may want to analyze the data from these other components for the sake of completeness. In general, your experimentwise error rate will be lower if you analyze fewer components. If, for example, you conduct three-way ANOVAs on the amplitude and latency of five different components, you will have 70 total p values, and you are very likely to have several spurious significant effects. Practically speaking, the best approach is often to analyze multiple components but take the results seriously only for a few *a priori* comparisons (and rely on replication for assessing the reliability of the other significant effects).

In our example experiment, we focused on P3 amplitude, but we also measured the amplitudes and latencies of the P2 and N2 components. We first combined the data from the Xs and Os so that we had one waveform for the improbable stimuli and one waveform for the probable stimuli. We did this for the simple reason that we didn't care if there were any differences between Xs and Os per se, and collapsing across them reduced the number of factors in the ANOVAs. The more factors are used in an ANOVA, the more individual p values will be calculated, and the greater is the chance that one of them will be less than 0.05 due to chance (i.e., this increases the familywise error).[2] By collapsing the data across irrelevant factors, you can avoid this problem (and avoid having to come up with an explanation for a weird five-way interaction that is probably spurious). Of course, it's a good idea to look at the waveforms separately first, just to make sure that they aren't radically different. But if you see an interaction with one of your counterbalancing factors, there is a good chance that it is spurious, so don't take it seriously until you see whether it is replicable.

The results of this experiment are illustrated in figure 10.1, which shows the ERP waveforms recorded at Fz, Cz, and Pz. From this figure, it is clear that the P2, N2, and P3 waves were larger

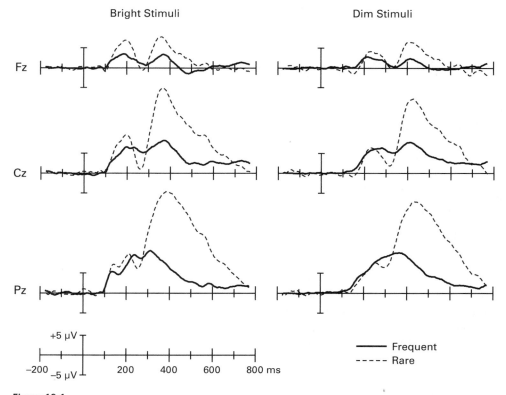

Figure 10.1
Grand average ERP waveforms from an unpublished oddball experiment in which the stimuli were either bright or dim. The data are referenced to the average of the mastoids and have been low-pass filtered (half-amplitude cutoff = 30 Hz, slope = 12 dB/octave).

for the rare stimuli than for the frequent stimuli, especially when the stimuli were bright. Thus, for the amplitude of each of these components, we would expect to see a significant main effect of stimulus probability and a significant probability × brightness interaction.

To quantify P3 amplitude, I measured the mean amplitude between 300 and 800 ms at each of the nine electrode sites in each subject, and I entered these data into a within-subjects ANOVA with four factors: stimulus probability (frequent vs. rare), stimulus brightness (bright vs. dim), anterior-to-posterior electrode position (frontal, central, or parietal), and left-to-right electrode position (left hemisphere, midline, or right hemisphere). Consistent with the waveforms shown in figure 10.1, this ANOVA yielded a highly significant main effect of stimulus probability, $F(1, 9) = 95.48$, $p < 0.001$. It also yielded a significant interaction between probability and brightness, $F(1, 9) = 11.66$, $p < 0.01$, because the difference between rare and frequent stimuli was larger for the bright stimuli than for the dim stimuli.

Using Electrode Site as an ANOVA Factor

I could have used a single factor for the electrode sites, with nine levels, but it is usually more informative to divide the electrodes into separate factors representing different spatial dimensions. That way, you can more readily determine if an electrode effect reflects a difference across hemispheres or a difference across regions within hemispheres. If you have large numbers of electrodes, you might want to average across clusters of nearby electrodes, yielding a $3 \times N$ set of values (left/middle/right × N anterior-to-posterior clusters). It is typically simplest to average across the waveforms in a cluster and then measure the amplitudes rather than measuring the amplitudes from each electrode and then averaging across the cluster (see the appendix of this book for a discussion of when the order of operations will impact your results). The downside of using separate anterior-to-posterior and left-to-right factors is that it increases the number of factors in the ANOVA and therefore increases the familywise error rate. However, this isn't a big problem if you are mainly using electrode site in your ANOVA to increase power (by including multiple sites at which the effect of interest is present) and to assist in your description of the scalp distribution of your effects.

You could, in principle, perform a separate ANOVA for each electrode site (or each left-midline-right set) rather than performing a single ANOVA with electrode site as a factor. Although this approach is occasionally appropriate, it is likely to increase the probability of both Type I and Type II errors. Type I errors will be increased because more p values must be computed when a separate ANOVA is performed for each electrode, leading to a greater probability of a spurious effect with a p value of less than .05. Type II errors may be increased because a small effect may fail to reach significance at any individual site even though the same effect would be significant in an analysis that includes multiple sites.

Even when multiple electrode sites are included in a single ANOVA, you may want to include only the sites where the component is actually present rather than including electrodes from the entire scalp. A component cannot be measured very precisely when it is small, so including these sites may add noise to the analysis, decreasing your statistical power. In addition, it is sometimes useful to analyze only the sites at which the component of interest is large *and* other components are relatively small so that the measurements of the component of interest are not distorted as much by the other components. In the current study, for example, I used all nine sites for analyzing the P3 wave, which was much larger than the other components, but I restricted the P2 analyses to the frontal sites, where the P2 effects were large but the N2 and P3 waves were relatively small. However, when you are trying to draw conclusions about the scalp distribution of a component, it may be necessary to include measurements from all electrodes. The issue of how to select the electrode sites for a given analysis is discussed in detail later in this chapter in the section on "Choosing Time Windows and Electrode Sites: The Problem of Multiple Implicit Comparisons."

An increasingly common approach is to average across all electrode sites within the relevant region of the scalp, measure the component of interest from this average, and then conduct the statistical analysis on this single value. This approach has several virtues. First, nonlinear measures such as fractional peak latency are more robust when measured from cleaner waveforms, and averaging across multiple sites will tend to reduce the noise level. Second, it completely avoids the use of electrode factors in the ANOVA, reducing the number of p values calculated in the ANOVA and thereby reducing the familywise error rate. Third, it makes the analysis easier to explain, which is especially important if you are trying to reach a broad audience. I would encourage you to use this approach whenever appropriate. (If reviewers hassle you about it, you can explain to them that it reduces the familywise error rate, and then you will have educated them about a very important issue! You can also cite this chapter, thereby increasing my citation count.)

Analyzing Difference Scores

In the example experiment shown in figure 10.1, it is problematic to compare the P3 elicited by bright and dim stimuli because of the sensory differences between these stimuli. This issue was discussed in the context of a very similar Gedankenexperiment in chapter 4. One solution described in that chapter is to make rare-minus-frequent difference waves for the bright stimuli and for the dim stimuli and then compare these difference waves (see figure 4.2). Any pure effects of brightness on the ERP waveform should be the same for rare and frequent stimuli, so the rare-minus-frequent difference wave eliminates any pure effects of brightness on the waveform.[3] This is one of many ways in which difference waves can be used to isolate a specific effect.

It is often useful to measure amplitudes and latencies from difference waves and use these measures as the dependent variables in your statistical analyses. First, difference waves can help you isolate a process of interest, as described in chapters 2 and 4. Second, using difference waves will reduce the number of factors in the ANOVA, decreasing the number of p values being calculated and thereby decreasing the familywise error rate. Again, I would encourage you to use this approach whenever it seems appropriate.

Interactions with Electrode Site

A well-known complication in interpreting ANOVAs in ERP experiments occurs when you find an interaction between condition and electrode site. For example, it is clear from figure 10.1 that the difference in P3 amplitude between the rare and frequent stimuli was larger at posterior sites than at anterior sites. This led to a significant interaction between stimulus probability and anterior-to-posterior electrode position, $F(2, 18) = 63.92$, $p < 0.001$. In addition, the probability effect for the bright stimuli was somewhat larger than the probability effect for the dim stimuli at the parietal electrodes, but there wasn't much difference at the frontal electrodes. This led to a significant three-way interaction between probability, brightness, and anterior-to-posterior electrode position, $F(2, 18) = 35.17$, $p < 0.001$.

From this interaction, you might be tempted to conclude that different neural generators were involved in the neural responses to the bright and dim stimuli. In other words, if the scalp distribution changes, this seems like it implies a change in the underlying generators. However, as McCarthy and Wood (1985) pointed out, ANOVA interactions involving an electrode position factor are ambiguous when two conditions have different overall amplitudes. This is illustrated in figure 10.2A, which shows the ERP amplitudes that would be expected at the Fz, Cz, and Pz electrode sites from a single generator source in two different conditions, A and B. If the magnitude of the generator's activation is 50% larger in condition B than in condition A, the amplitude at each electrode will be 50% larger in condition B than in condition A. This is a multiplicative effect, and not an additive effect. That is, the voltage increased by 50% at each site, leading to an increase from 1 μV to 1.5 μV at Fz (a 0.5-μV increase) and an increase from 2 μV to 3 μV at Pz (a 1-μV increase). This shows up as an interaction in an ANOVA, even though it arises from a change in the magnitude of a single generator source. An additive effect is shown in condition C in figure 10.2A. In this condition, the absolute voltage increases by 1 μV at each site relative to condition A, which is not the pattern that would result from a change in the amplitude of a single generator source. Thus, when a single generator source has a larger magnitude in one condition than in another condition, an interaction between condition and electrode site will be obtained (as in condition A vs. condition B). In contrast, a change involving multiple generator sites may sometimes produce a purely additive effect (as in condition A vs. condition C).

To determine whether an interaction between an experimental condition and electrode site really reflects a difference in the internal generator sources, McCarthy and Wood (1985) proposed *normalizing* the data to remove any differences in the overall amplitudes of the conditions. An example of this is shown in figure 10.2B (for details of the normalization procedure, see the online supplement to chapter 10). Once the data have been normalized, the scalp distribution is the same for conditions A and B and different for condition C, which tells us that the generator has simply changed in magnitude between conditions A and B whereas the generator has changed in some way for condition C.

However, Urbach and Kutas (2002) convincingly demonstrated that this normalization procedure doesn't actually work under most realistic conditions. In most cases, I would therefore

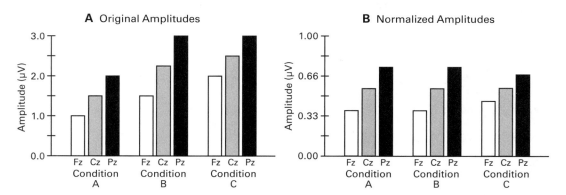

Figure 10.2
(A) Examples of additive and multiplicative effects on ERP scalp distributions. Condition B is the same as condition A, except that the magnitude of the neural generator site has increased by 50%, thus increasing the voltage at each site by 50%. This is a multiplicative change. Condition C is the same as condition A, except that the voltage at each site has been increased by 1 μV. This is an additive change, and it is not what would typically be obtained by increasing the magnitude of a single neural generator source. (B) Same as panel A, except that the amplitudes from each condition have been normalized by dividing by the vector length from that condition (see the online supplement to chapter 10 for a detailed description of this normalization procedure). Now we can see that condition A and condition B have the same scalp distribution, but condition C has a different distribution.

recommend simply reporting that you found an interaction with electrode site and then saying very little about it. If the difference between conditions is largest at the scalp sites where the component is largest (e.g., if the difference between the difference waves in the bright and dim conditions was largest at the Pz electrode site), you can simply state that the pattern of results is approximately what would be expected if a single component varied in amplitude across conditions (without even performing a formal analysis on the normalized data). This issue is discussed in more detail in the chapter 10 supplement.

Heterogeneity of Covariance and the Epsilon Adjustment

A second complication with including electrode site as an ANOVA factor is that this often leads to a violation of the assumption of *homogeneity of covariance*. You probably already know that ANOVA assumes *normality* (Gaussian distributions) and *homogeneity of variance* (equal variances across the different conditions). These assumptions are often violated, but ANOVA is fairly robust when the violations are mild to moderate, with very little change in the actual probability of a Type I error (Keppel, 1982). Unless the violations of these assumptions are fairly extreme (e.g., greater than a factor of 2), you don't need to worry about them.

However, when an ANOVA includes within-subject factors, such as electrode site, we must also assume homogeneity of covariance. The basic idea is illustrated in figure 10.3, which shows data from three subjects who were each tested in three conditions (A, B, and C). For the sake of this example, let's assume that weight is being measured. Although there is quite a bit of variance among subjects in each condition, all three subjects show exactly the same pattern of

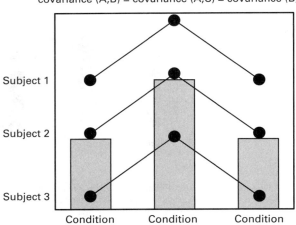

Within-Subjects ANOVA assumes that
covariance (A,B) = covariance (A,C) = covariance (B,C)

Figure 10.3
Example of a within-subjects design and the concept of homogeneity of covariance. Data are shown from three subjects in three conditions. It doesn't matter what is being measured, but you could think of the Y axis as representing weight.

differences among conditions. A within-subjects ANOVA factors out the overall differences among subjects, focusing on the consistency of the effect across conditions (i.e., the fact that all three subjects go up by the same amount in condition B compared to condition A and then go down by the same amount in condition C). This can dramatically increase statistical power. However, power is increased only to the extent that subjects who have high values in one condition tend to have high values across all conditions, and subjects who have low values in one condition tend to have low values across all conditions. This is equivalent to saying that a subject's score in one condition covaries with that subject's scores in the other conditions. Sometimes this covariance[4] is very strong (as in the example shown in figure 10.3). Sometimes it is weak. The assumption of homogeneity of covariance is simply the assumption that the degree of covariance between conditions A and B is equal to the degree of covariance between conditions A and C and between conditions B and C. This assumption does not apply if there are only two levels of a factor, because there is only one covariance in this case.

To see how this assumption might be violated, imagine that each subject's weight was measured three times, once at age 3, once at age 21, and once at age 22. A person's weight at age 21 will be much more strongly related to his or her weight at age 22 than to his or her weight at age 3. Thus, the covariance between ages 21 and 22 would be higher than the covariance between ages 3 and 21. This would violate the assumption of homogeneity of covariance.

Violations of the assumption of homogeneity of covariance are very common in ERP experiments that include multiple electrode sites as a factor in the analysis, because data from nearby electrodes tend to covary more than data from distant electrodes. For example, random EEG noise at the Fz electrode will spread to Cz more than to Pz, and the correlation between the data at Fz and the data at Cz will be greater than the correlation between Fz and Pz. In addition, a real ERP signal will tend to impact nearby sites to similar degrees but will not impact distant sites to different degrees, which also creates more covariance between nearby sites.

Unfortunately, ANOVA results become very inaccurate when the covariance is heterogeneous. Violating the assumption of homogeneity of covariance leads to artificially low p values, such that you might get a p value of less than 0.05 even when the actual probability of a Type I error is 0.25. This was brought to the attention of ERP researchers very forcefully in a paper published in the journal *Psychophysiology* by Jennings and Wood (1976). The journal subsequently developed an explicit policy stating that all papers published in the journal must address this problem.

The most common solution is to use the Greenhouse–Geisser epsilon adjustment, which counteracts the inflation of Type I errors produced by heterogeneity of covariance. For each factor or interaction that has more than two within-subjects levels, a value called *epsilon* is computed along with the F value. The epsilon value for a given F value is then multiplied by the degrees of freedom for that F value, and the adjusted degrees of freedom are used to compute the p value. Epsilon varies between 0 and 1, with small values corresponding to a large heterogeneity of covariance. If the covariances are homogeneous, epsilon is near 1, and multiplying the degrees of freedom by a value near 1 doesn't change them much. Thus, little or no change in the degrees of freedom occurs if the assumption of homogeneity of covariance is met, but the degrees of freedom move downward—and the p value therefore gets worse—as the heterogeneity of covariance increases. This adjustment is provided by most major statistics packages and is therefore easy to use. For example, the SPSS ANOVA output contains the adjusted p values along with the unadjusted p values.

I used the Greenhouse–Geisser adjustment in the statistical analysis of the P3 amplitude data shown in figure 10.1. It influenced only the main effects and interactions involving the electrode factors, because the other factors had only two levels (i.e., frequent vs. rare and bright vs. dim). For most of these F tests, the adjustment didn't matter very much because the unadjusted effects were either not significant to begin with or so highly significant that a moderate adjustment wasn't a problem (e.g., an unadjusted p value of 0.00005 turned into an adjusted p value of 0.0003). However, there were a few cases in which a previously significant p value was no longer significant. For example, when I normalized the data before conducting the ANOVA, the main effect of anterior-to-posterior electrode site was significant before the adjustment was applied ($F[2,18] = 4.37$, $p = 0.0284$) but was no longer significant after the adjustment ($p = 0.0586$). This may seem like a bad thing, because a significant effect was made non-significant by the adjustment. However, the original p value was not accurate, and the adjusted p value is closer to the actual probability of a Type I error. In addition, when very large numbers of electrodes are used, the adjustments are usually much larger, and spurious results are quite likely to yield significant p values without the Greenhouse–Geisser adjustment.

It is absolutely necessary to use the Greenhouse–Geisser adjustment—or something comparable[5]—whenever there are more than two levels of a factor in an ANOVA, especially when one of the factors is electrode site. Of course, you should use this adjustment for other within-subjects factors that include more than two levels, and not just for the electrode site factor. And you should use it for analyses of behavioral data as well. If you don't, your p values will not be correct, and you will be likely to draw conclusions on the basis of false positives.

The Jackknife Approach

Error Variance

In traditional statistical approaches, the p value depends on both the size of the difference between conditions and the amount of variance within each condition. As the variance within a condition increases, it becomes less likely that the difference between conditions is real (all else being equal). The variance within a condition is called the *error variance*. As described in the supplement to chapter 8, the error variance contains both *measurement error* and *true score variance*. Measurement error occurs when we do not obtain an accurate measurement of the subject's true value. True score variance reflects the real differences between subjects. The most obvious cause of measurement error in ERP experiments is noise in the single-subject averaged ERP waveforms. If we could average together an infinite number of trials for each subject, we would dramatically reduce measurement error. And if we reduced the measurement error, we would reduce the overall error variance, which would in turn give us more statistical power.

Another common problem is that outliers can have a huge effect on the error variance. If we had a principled way to reduce the impact of outliers, this would also give us greater power.

The jackknife technique helps us reduce measurement error and the effects of outliers by measuring amplitudes or latencies from grand averages rather than from single-subject averages. Several times I have had the experience of looking at grand average waveforms, seeing a beautiful effect that was far larger than the noise in the prestimulus period, and then being very surprised to find a p value that was far from significant when I conducted the statistical analyses. When I then took a closer look at the individual subjects, I realized that this was because the measured values were "crazy" for some of the subjects. These crazy values dramatically increased the error variance, resulting in a p value that was not significant. This is one reason why I recommended—at the beginning of chapter 9—that you always compare the measurements for each subject with that subject's ERP waveforms. This might make you realize that something is wrong with your measurement procedure. Sometimes, however, there is nothing wrong with the measurement procedure, and the problem is that the waveforms from some of the subjects are noisy or unusual.

As an example, the left column of figure 10.4 shows the lateralized readiness potential (LRP) from three individual subjects, isolated with a contralateral-minus-ipsilateral difference wave (see chapter 3 for details about the LRP). The data from each subject are pretty noisy, and the

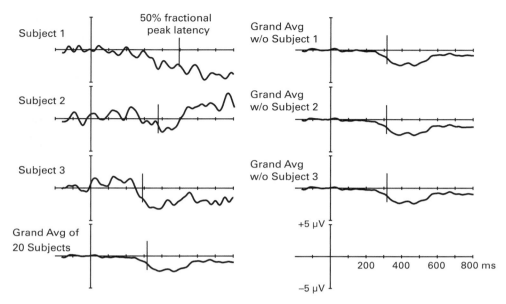

Figure 10.4
Lateralized readiness potential data collapsed across the C3 and C4 electrode sites, formed by subtracting the ERP over the hemisphere ipsilateral to the response hand from the ERP over the hemisphere contralateral to the response hand (i.e., contralateral minus ipsilateral). The left column shows averaged ERP waveforms from three individual subjects and the grand average of all 20 subjects. The right column shows leave-one-out grand averages, each of which was formed by averaging together 19 of the 20 subjects and leaving out one subject.

LRP develops much more gradually in subject 1 than in subjects 2 and 3. Consequently, the measured onset latency (measured as the 50% peak latency) of the LRP varies a great deal among these subjects. This variance gives us low statistical power.

The grand average across 20 subjects is shown in the bottom left of figure 10.4. You can see that it is very clean, and we could quantify the onset latency from this grand average with very little measurement error. However, measuring the onset latency from the grand average doesn't do us much good, because we can't do statistical analyses if we just have this one measure. This is where the jackknife technique comes in.

The jackknife technique is basically a method that makes it possible for you to measure amplitudes and latencies from your beautifully clean grand averages and still perform conventional statistical analyses.[6] This technique has been used for decades in some areas of statistics, and it was imported into ERP research by Jeff Miller and his colleagues in the late 1990s (Miller, Patterson, & Ulrich, 1998). It's easy to use, and it can dramatically improve your statistical power in some conditions, decreasing the probability of a Type II error without increasing the probability of a Type I error. When I explain how it works, you may not believe that it could possibly be a legitimate technique. However, it has been demonstrated to work both by mathematical proofs and by rigorous simulation studies (see, e.g., Miller et al., 1998; Ulrich & Miller, 2001; Kiesel, Miller, Jolicoeur, & Brisson, 2008). It does have some limitations, however, so make sure you read this entire section before using it.

Essence of the Jackknife Approach

Imagine that you have 10 subjects who have been tested in two conditions, A and B, and you want to know if the mean onset latency of the LRP is earlier in condition A than in condition B. If you measure the onset latency from the individual subjects, as in the left column of figure 10.4, you might find that you have so much error variance that the difference between groups is far from significant. If you measure the onset latency from the grand average of each condition, you could avoid much of this measurement error. However, you would have only one measurement from each condition, so you wouldn't have a measure of the error variance, and this would make it impossible to determine if the difference between conditions was larger than would be expected by chance. In other words, you couldn't get a p value if you just had one measure from the condition A grand average and one measure from the condition B grand average.

To be able to assess the variance within each condition without giving up the advantages of measuring from a grand average, you can create a series of grand averages, each of which is missing one of the subjects. In our example, you would make 10 grand averages for each of the two conditions, each created by averaging together the waveforms from nine of the 10 subjects for that condition. The first grand average would include everyone except subject 1; the second would include everyone except subject 2; and so forth. Examples are shown in the right column of figure 10.4. These are called *leave-one-out* grand averages. All of the leave-one-out grand averages for a given condition will be highly similar to the others, because most of the single-subject waveforms in one leave-one-out grand average are also in the other leave-one-out grand averages. However, the individual leave-one-out grand averages will be slightly different from each other, reflecting the subject who was left out of each average.

We can measure the onset latency from each of these leave-one-out grand averages, which will work very well because these waveforms are extremely clean. We can also assess the variance across the leave-one-out grand averages for a given condition. This will not be the same as the variance across individual subjects, but there is a mathematically principled way of using the variance of the leave-one-out grand averages to estimate the variance of the single subjects. An extension of this allows us to conduct a t test with the values measured from the leave-one-out grand averages. Specifically, we can take the 20 onset latencies that we've measured (from 10 leave-one-out grand averages for condition A and 10 leave-one-out grand averages from condition B) and enter these values into a t test (just like you would ordinarily enter the 10 values from each subject in each condition into a t test). The resulting t value will be unnaturally large, because the error variance has been artificially reduced by measuring from the leave-one-out grand averages. However, we can adjust for this artificial reduction in error variance by simply dividing the t value by $N - 1$ (which would be 9 in this example), giving us an adjusted

t value. We can then look up the p value for this adjusted t value, and the difference between conditions will be considered significant if the p value is less than 0.05 (or whatever alpha level you want to use). This is the essence of the jackknife approach.

This is just as simple as it sounds: You simply make the leave-one-out grand averages and treat them as if they were single-subject ERP waveforms. The only trick is to divide the t value by $N - 1$ before looking up the p value.

You can do the same thing with a more complex experimental design, using an ANOVA instead of a t test. You just make the leave-one-out grand averages for each condition, measure the latencies from these leave-one-out grand averages, enter the latencies into an ANOVA just as if they were measured from single-subject averages, and compute the F ratios. You then need to divide the F ratios by $(N - 1)^2$. This works for the interactions as well as the main effects. It also works for between-subject factors. If you have N subjects in each group, you will again divide each F ratio by $(N - 1)^2$. If you have different numbers of subjects in each group, then it's a little more complicated (for the details, see Ulrich & Miller, 2001). But keep in mind that you divide by $N - 1$ for a t test and $(N - 1)^2$ for an F test.

In many cases, you might find that a p value of 0.20 in conventional analysis (in which you measured the latencies from the single-subject ERPs) becomes a p value of 0.005 when you do the same analysis with the jackknife approach, even after you've adjusted the t value by dividing by $N - 1$. This may seem too good to be true. However, many simulation studies have shown that the jackknife approach does not lead to an increase in the Type I error rate. If you have a real effect, the jackknife technique often helps you get a much better p value. However, if the null hypothesis is true, you will get a significant p value only 5% of the time (assuming an alpha of 0.05). I have used the jackknife approach many times. I often find a radically better p value with the jackknife analysis than with the conventional analysis. But not always, because sometimes the null hypothesis is true. You are no more likely to falsely reject the null hypothesis with the jackknife technique than with conventional statistics. It mainly operates by reducing the error variance, which increases your statistical power. And increased statistical power means that you can find real effects with less effort, publish more papers in better journals, and become rich and famous. What could be better than that?

The jackknife technique can also be used for the Pearson r correlation coefficient (Stahl & Gibbons, 2004). Imagine that you want to look at the correlation between P3 onset latency and RT. You would create pairs of latency and RT values, where the latency came from a leave-one-out grand average that left out a given subject and the RT came from that subject. If P3 onset latency tends to be later in subjects with long RTs, but we pair the latency from a grand average that is missing a subject with the RT from that subject, this will reverse the direction of correlation. Imagine, for example, that subject 4 has a very late P3 onset latency and a very long RT. The leave-one-out grand average that excludes subject 4 will have a somewhat earlier-than-usual P3 onset latency (because a subject with a late onset latency has been left out), and this will be paired with the long RT from this subject. This will turn a positive correlation into a negative correlation (or vice versa). It turns out that the adjustment for the Pearson r correlation coefficient is simply to multiply the jackknife r value by –1. That is, you compute the *r* value from the pairs of values and then multiply this value by –1. I've used it, and it works.

Limitations of the Jackknife Approach

The jackknife approach can be truly amazing, but it does have some limitations. Most of these limitations are minor, but the last one I will describe is significant.

Linear Measures   One limitation of the jackknife technique is that it gives you the same result as the conventional analysis if you apply it to a linear measure, such as mean amplitude. This is because the jackknife technique is really just a way of changing the order of operations in your analysis pipeline. Rather than measuring the amplitudes or latencies from the single-subject averages and then doing your statistical analysis (which involves averaging across the single-subject values), you measure from waveforms that have already been averaged across the single subjects. As described in detail in the appendix, the order of operations does not matter if all the operations are linear. Thus, the jackknife technique neither helps nor hurts if you are analyzing a linear measure such as mean amplitude.

You can actually use this to your advantage, because it gives you a way of determining whether you are using the jackknife technique correctly. You can simply measure mean amplitude in an experiment, do both the conventional analysis and the jackknife analysis, and then compare the results. The *p* values should be the same in both analyses (assuming that you divided by the appropriate adjustment factor before computing the *p* value in the jackknife analysis). The values might be slightly different due to rounding errors, but they should be very close. If you verify that this is true, then you will trust the results you get by jackknifing a nonlinear measure, such as onset latency.

Note that my examples so far have focused on onset latency. This is because the jackknife technique was originally developed to analyze onset latency, and most of the simulations have focused on onset latency. However, the same principles apply to any nonlinear measure (although you would need to be very careful when using it with peak amplitude, for reasons that will be discussed at the end of this section). Simulations examining the effectiveness of the jackknife technique using a variety of measures from several components can be found in Kiesel et al. (2008), which is on my list of the Top Ten Papers Every New ERP Researcher Should Read (see chapter 1).

Equal Sample Sizes   If you have any between-groups factors, the simple adjustment procedures that I have described require that you have the same number of subjects in every group. However, it is still possible to do the adjustment if you have unequal *N*s. It's just more complicated (for a description of the adjustment procedure, see Ulrich & Miller, 2001). This is a very minor limitation.

The Jackknife *p* Value Is Sometimes Not Significant   This is not really a limitation, because the null hypothesis may actually be true. The jackknife technique is no more likely than a conventional analysis to give you a significant p value when the null hypothesis is true. In addition, even if your effect is real, you may not have enough statistical power to get a significant p value with any statistical technique. But you will usually have more power with the jackknife technique than with conventional statistics.

The Jackknife p Value Is Sometimes Worse   The p value from the jackknife technique is sometimes worse than the p value from the conventional analysis. This typically happens when you have one outlier subject who has a really big effect on the leave-one-out grand averages (for a description of other conditions that may lead to poor performance by the jackknife technique, see Miller, Ulrich, & Schwarz, 2009). What should you do in this situation?

I asked Jeff Miller about this, and he told me two things. First, if the null hypothesis is true, random noise will determine whether the jackknife p value is better or worse than the conventional p value, so you can't just pick whichever one is significant. That would allow you to capitalize on random variation, which would inflate your Type I error rate. Second, it would be valid to use the jackknife technique only when the standard error of the mean is much better for the jackknifed data than for the conventional data (because this means that jackknifing is helping reduce error). In other words, although you can't just see which one gives you the better p value, you can decide which one to use by determining whether the jackknife technique produces a substantial improvement in the standard error. Miller et al. (1998) explain how to compute the jackknifed standard error (which is quite simple).

The One Major Limitation: Testing a Different Null Hypothesis   You are testing a slightly different null hypothesis with the jackknife technique than with conventional statistics, and this is the one significant limitation of the jackknife technique. For several years, I couldn't imagine a situation in which this small difference would matter. But then my imagination improved, and I realized that this different null hypothesis could lead to misinterpretations of the results under certain conditions (see box 10.2 for the story of how this transpired). I will first explain the different null hypotheses, and then I will explain why it sometimes matters which null hypothesis you are testing.

In informal terms, the conventional and jackknife null hypotheses are as follows:

Conventional null hypothesis   If we could measure an amplitude (or latency) value from the single-subject ERP waveform in every individual in the infinitely large population, the average of these measures would not differ across conditions.

Jackknife null hypothesis   If we could make grand averages that included every individual in the infinitely large population and then measure an amplitude (or latency) value from these grand averages, the values from these grand averages would not differ across conditions.

These null hypotheses are equivalent except for the order of operations. When you are using a linear measure, such as mean amplitude, these two null hypotheses end up being exactly the same. When you are using a nonlinear measure, such as peak amplitude or fractional peak

Box 10.2
Imagining How the Jackknife Might Fail

Jeff Miller—who introduced the jackknife technique to ERP research—was on the faculty at UCSD when I was a graduate student there. I took his graduate ANOVA course, and he served on my dissertation committee. Jeff is an incredibly thoughtful and careful scientist, and he's particularly adept at developing quantitative techniques that can be applied to answering important questions about cognition.

I first encountered the jackknife technique when I saw the initial paper that Jeff and his colleagues wrote about it (Miller et al., 1998). I was intrigued, and I mentioned it briefly in the first edition of this book. A few years later, I did my first LRP study, and some of the latency effects that looked real were not significant in the conventional statistical analyses. I decided to try the jackknife technique, and the results were amazing! Several key effects that were far from significant in the conventional analysis were highly significant in the jackknife analysis. I was hooked.

Shortly after that, I gave a mini ERP Boot Camp at Merck Pharmaceuticals. The "audience" consisted of four extremely smart researchers who had each earned multiple advanced degrees in fields like biostatistics, mathematics, and biomedical engineering. I started talking about the jackknife technique, and the leader of the group told me that they couldn't use it, because the FDA would never allow it in a clinical trial. The reason, he explained, was that the jackknife technique tested a different null hypothesis (as explained in the main text).

I talked about this with Jeff Miller (via e-mail), and neither of us could figure out a situation in which it would matter which null hypothesis was being tested. But I am always concerned about "proof by lack of imagination" (see box 4.3 in chapter 4). So I kept thinking about it. A few years later, I was looking through the slides that I use to discuss the problem of latency variability in ERP averages, and it suddenly occurred to me that this problem also had implications for the jackknife technique. That is, the same problems that arise in creating a single-subject averaged ERP waveform from single-trial EEG epochs can also be a problem when you average together the ERPs from multiple subjects to create a grand average ERP waveform. For example, increased latency variability across subjects will lead to decreased peak amplitude in the grand average. This does not invalidate the jackknife technique, but it places some important limits on how it is applied and interpreted.

latency, they are not the same. But does it really matter which null hypothesis you're testing? After all, if two infinitely large populations are truly equivalent, then they should be the same for both null hypotheses.

It's not quite this simple. As described in box 10.2, the same problems that arise when you average multiple single trials together to create an averaged ERP waveform for a single subject can also arise when you average multiple single-subject ERPs together to create a grand average ERP waveform (or a leave-one-out grand average). Recall from figure 8.7 in chapter 8 that the peak amplitude in an averaged ERP waveform is smaller if the latency of the component varies from trial to trial. The more trial-to-trial latency variability is present, the smaller the peak amplitude will be. The same principle applies to grand averages. If you have a lot of subject-to-subject variability in the latency of a component, the peak amplitude of the grand average will be reduced. Moreover, you may recall that the onset latency of a single-subject averaged ERP waveform reflects the trials with the earliest onset latencies, not the average of the single-trial onset latencies. Similarly, the onset of a difference between conditions in a grand average will reflect the subjects with the earliest onset of the effect, not the average of the single-subject onset times.

Imagine that you were comparing the peak amplitude of the P3 wave in a patient group and a control group using the jackknife technique. Imagine also that the latency of the P3 wave was more variable across subjects in the patient group than in the control group (which is very likely), but the amplitude of the P3 was the same in the individual patients as in the individual control subjects. In this scenario, the amplitude of the P3 peak in the patient grand average would be smaller than the amplitude of the P3 peak in the control grand average, even though there was no amplitude difference between the individual subjects in these two groups. In a conventional analysis with peak amplitude, you would see no difference between groups. This would be the correct result (although it might be confusing if you were looking at the grand averages, which do have different peak amplitudes). In a jackknife analysis, you would likely see a significant difference between the two groups. If you interpreted this result in the same way that we would interpret the result of a conventional analysis, you would conclude that the subjects in the patient group had smaller P3 amplitudes than the subjects in the control group. This would be an incorrect interpretation. The two groups do differ, but the difference is in latency variability and not in peak amplitude per se. Thus, the jackknife technique will not lead you to conclude that two groups or conditions are different if the entire waveforms are exactly the same in the two groups or conditions. However, a difference that appears to be in one aspect of the waveforms can be a result of a difference in another aspect of the waveforms (e.g., apparent differences in peak amplitude may actually reflect differences in latency variability).

This is a significant issue, but it is not insurmountable. If the jackknife effect is real, then you should see the same basic pattern of means in the conventional analysis that you see in the jackknife analysis. That is, if you look at the table of means produced by your statistics software for the conventional analysis and for the jackknife analysis, you should see the same basic pattern of differences across groups or conditions in both tables. The differences may be somewhat larger in one table than the other, and the effects might not be significant in the conventional analysis, but the patterns of means should be similar. If they are not similar, you may be seeing an unintended side effect of the process of making grand averages.

It is also worth considering the fact that the onset time of the grand average is not the same as the average of the single-subject onset times. In theory, the average of the single-subject onset times could be the same for two conditions, but the onset time of one grand average could be earlier than the onset time of the other grand average. For this to happen, however, the condition with the earlier onset time in the grand average must have some subjects with unusually early onset times and other subjects with unusually late onset times. This would lead to the same average of single-subject onset times across groups but greater variance in the group that had some subjects with very early onset times and other subjects with very late onset times. This would be a very unusual situation. If one group has greater variance in onset times, this almost always arises from having more subjects who have longer-than-average onset times, without being balanced out by other subjects who have shorter-than-average onset times. In other words, the group with greater variability almost always has a greater mean onset latency. In addition, if you use the 50% peak latency measure, you are not attempting to measure the absolute onset time (the time at which the waveform deviates just slightly from zero), and it's likely that the value measured from the average will be a good approximation of the average onset time. This was discussed in chapter 9 (see figure 9.7), and several simulation studies have shown that the jackknife technique works very well at accurately estimating onset times, increasing statistical power without increasing the Type I error rate (Miller et al., 1998; Ulrich & Miller, 2001; Kiesel et al., 2008; Miller et al., 2009).

The bottom line is that the jackknife technique can be extremely helpful, but you need to be thoughtful and careful when using it (just like anything else in statistics). Using it to analyze peak amplitude can easily lead you astray (especially if there are differences in latency variability between groups or conditions), but you can assess this by making sure that the same pattern is present in the means from the conventional analysis. Using it to measure onset latencies requires some thought, but it is likely to work very well (both in terms of statistical power and the accuracy of the results). It's definitely a great tool to have in your data analysis "toolbox," but like all tools, it needs to be used properly.

Choosing Time Windows and Electrode Sites: The Problem of Multiple Implicit Comparisons

We're now going to switch to a very different issue that arises because of the richness of an ERP data set. If you computed a t test comparing two conditions at every time point and every electrode site in a typical ERP experiment, random fluctuations would be expected to lead to many large t values that would exceed the usual criterion for statistical significance. This is the widely known *problem of multiple comparisons*, and you would never be allowed to publish results from this approach without some kind of *correction for multiple comparisons*. However, an implicit variation on this problem arises all the time in ERP research when researchers first look at their waveforms, then find the combination of time window and electrode site where a big difference between conditions is present, then measure the amplitude or latency at that combination of time and electrode, and finally enter the resulting values into a statistical analysis. This will often lead to significant p values that are a result of noise rather than real effects. If you do this, you are implicitly performing multiple comparisons (by visually comparing the waveforms from multiple time points and multiple electrode sites), so you are biased to obtain a significant p value even if no real effect is present. I call this the *problem of multiple implicit comparisons*. It is particularly important when *a priori* information is not used to choose the time windows and electrode sites used in the statistical analyses.

Performing multiple comparisons leads to inflation of the Type I error rate, which means that the true rate of Type I errors (rejecting the null hypothesis when it is true) is greater than 5%

(or whatever alpha you are using). This can be an enormous problem in ERP research. Indeed, it is a problem in many areas of psychology and neuroscience, and it is receiving increasing attention (Simmons et al., 2011; Pashler & Wagenmakers, 2012). Over the coming years, I expect that journals, editors, and reviewers will become increasingly strict about factors that inflate the Type I error rate. For example, *Nature Neuroscience* and *Psychophysiology* have both recently instituted a methods checklist, which is designed in part to help editors and reviewers determine whether the statistical methods of a paper may have led to an inflation of the Type I error rate (Editorial, 2013; Keil et al., in press).

To make this more concrete, the following discussion will refer to an actual set of data from a published oddball experiment. The actual experiment, which was described in some detail in chapter 1 (see figure 1.4), involved rare and frequent categories of visual stimuli that were presented to schizophrenia patients and healthy control subjects (Luck et al., 2009). To keep things simple, I will discuss only the data from the control subjects, focusing on the comparison between the rare and frequent trials. In addition, I will include the data from only 12 of the subjects. Note that there wasn't anything wrong with the way we analyzed the data in the published paper, but it would be possible for someone to analyze these data in a way that capitalized on noise and inflated the Type I error rate.

Figure 10.5A shows the grand average waveforms for the rare (target) and frequent (standard) stimuli at a set of left- and right-hemisphere electrode sites (midline sites were recorded in the original experiment, but they will be ignored here). The data were recorded with a sampling rate of 500 Hz and an online band-pass of 0.05–100 Hz (half-amplitude cutoffs) using a right earlobe reference. For the analyses presented here, the data were re-referenced offline to the average of the left and right earlobes and filtered again with a low-pass filter (half-amplitude cutoff = 30 Hz, slope = 12 dB/octave). You can see a much larger P3 for the rare stimuli than for the frequent stimuli, especially at central and parietal electrode sites. You can also see a more negative voltage for the rare stimuli than for the frequent stimuli in the N2 latency range (250–350 ms) at the central and parietal electrode sites.

An Example of the Problem of Multiple Implicit Comparisons

To make the problem of multiple implicit comparisons clear in the context of ERPs, I performed a simple simulation using the data from this example experiment. The goal of the simulation was to create realistic data from two conditions in which the null hypothesis was clearly true (i.e., no real differences between conditions). To do this, I started by extracting the EEG epochs from the standard (frequent) trials in the oddball experiment. The target (rare) stimuli were left out of this simulation. For each subject, I randomly divided the 640 standard trials into two sets, 512 that that were averaged together to form condition A and 128 that were averaged together to form condition B (80% and 20%, respectively). All 640 trials were actually from the standard stimuli, and the division into conditions A and B was purely random. Thus, this simulates an oddball experiment in which there is true difference between the standards (condition A) and the targets (condition B). Any differences between them in this simulation are purely due to noise. If each

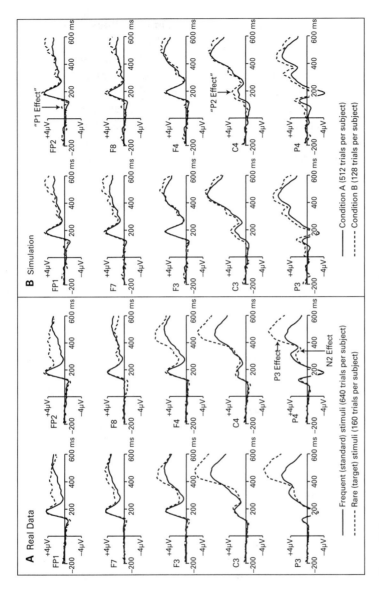

Figure 10.5

(A) Grand average ERP waveforms from an example oddball experiment in which 20% of the stimuli were letters and 80% were digits (or vice versa, counterbalanced across trial blocks). The data are from a subset of 12 of the control subjects who participated in a published study comparing schizophrenia patients with control subjects (Luck et al., 2009). (B) Grand average ERP waveforms from a simulated experiment in which the null hypothesis is guaranteed to be true. This simulation was based on the data collected on the standard (frequent) trials in panel A. The event codes from the 640 standard trials for each subject were randomly sorted into a set of 512 trials for condition A and a set of 128 trials for condition B. Averaged ERP waveforms were then created from these sets of trials. Note that the two conditions differ in P1 amplitude over the entire right hemisphere and in P2 amplitude over the central and parietal regions, but these differences are the result of false positives. The data shown in this figure are referenced to the average of the left and right earlobes and were low-pass filtered offline (half-amplitude cutoff = 30 Hz, slope = 12 dB/octave).

subject had 1 million trials instead of 640 trials, the averaged waveforms for the 800,000 trials in condition A would have been virtually identical to the 200,000 trials in condition B. However, with only 512 trials in condition A and 128 trials in condition B, some random noise was present in the averaged ERPs from these conditions. Thus, the null hypothesis was true (i.e., the data were sampled at random from a single population), but the actual waveforms were not exactly identical (as will always be the case when you are averaging together a finite number of trials). It's important that you understand this, because the idea of randomly combining the observed data in this manner to simulate the null hypothesis will come up again later in the chapter.

Figure 10.5B shows the grand average waveforms from conditions A and B in this simulated experiment. The waveforms from these conditions are fairly similar, and in this simulation we know that the differences must reflect random noise. However, if you ran an experiment with two different conditions and you saw these waveforms, you wouldn't know if the small differences between the waveforms were a result of random noise or true differences in brain activity produced by the two different conditions. You might notice that there are two interesting "effects" in the data. First, the waveforms exhibit a more positive potential in condition A than in condition B in the P1 latency range (50–150 ms), especially in the right hemisphere electrode sites. Second, the waveforms exhibit a more positive potential in condition A than in condition B in the P2 latency range (150–250 ms) at the central and parietal electrode sites.

To quantify the "P1 effect," you might measure the mean amplitude between 50 and 150 ms at all of the electrode sites and look for a condition × hemisphere interaction. I did this with the simulation data, and I found a marginally significant main effect of condition ($p = 0.051$) and a significant condition × hemisphere interaction ($p = 0.011$). I did a follow-up comparison, and I found a significant difference between conditions A and B at the right hemisphere sites ($p = 0.031$). To quantify the "P2 effect," you might measure the mean amplitude between 150 and 250 ms at the C3, C4, P3, and P4 electrode sites. I did this with the simulation data, and I found a significant difference between conditions A and B ($p = 0.026$). This seems like a real difference, but in this simulation we know with 100% certainty that this "effect" is completely bogus. If you saw these data and analyses in a journal article, they might seem like convincing evidence that these two conditions differed in P1 amplitude over the right hemisphere and in P2 amplitude over the central and parietal regions. But in fact these differences were false positives, because the two conditions were the same except for random noise.

Why are we seeing these effects that look real and are statistically significant even though the two conditions are actually just random samples from a single condition? The answer is that we have enough data points and channels that random fluctuations in the data are likely to create substantial differences at some channels and some sites. And these effects are likely to form realistic-looking clusters across nearby time points and nearby electrode sites, because EEG noise changes gradually over periods of tens of milliseconds and is blurred by the high resistance of the skull so that it is present at multiple nearby electrode sites. These data were also low-pass filtered (half-amplitude cutoff at 30 Hz, 12 dB/octave), which also causes a temporal spreading of noise in the data (see chapter 7).

Box 10.3
I Plead Guilty!

Many decades ago, when ERP researchers recorded from only a few electrodes and used peak measures of amplitude and latency based on relatively wide measurement windows, the problem of multiple implicit comparisons wasn't a big deal. There just weren't many choices that researchers could make when analyzing their data. Consequently, no one worried much about this problem. As the number of electrode sites gradually increased, and the use of measures that are more dependent on the time window became more prevalent, the problem of multiple implicit comparisons slowly increased. Because this problem increased so gradually—and the practice of looking at the waveforms to choose the analysis parameters was ingrained in the culture—it took a while before people started taking this problem seriously. By comparison, the problem of multiple comparisons was obvious from the very first days of functional neuroimaging, because researchers were measuring activity from thousands of voxels even in the first studies. Consequently, this problem has been addressed by neuroimaging analysis methods from the beginning, but most ERP researchers didn't take it very seriously until about 10 years ago.

When I think back to my own published ERP studies from the 1990s, I realize that I often used the data to guide my choices of time windows and electrode sites, and I'm certain that this inflated my Type I error rate. That is, it's very likely that more than 5% of the results that I reported as being significant were bogus. This is not an easy thing to admit! However, I have always taken seriously the idea that replication is the best statistic (see box 10.1), and most of the key effects that I reported in those studies were replicated (either in the same paper or in a subsequent paper). The ones that are likely to be bogus are the small, minor, unanticipated effects that I never bothered to replicate. But some of the major effects have not yet been replicated and might also be false positives.

If you look at my more recent papers, you will see that I am now making a concerted effort to address this problem. For example, one recent paper used the negative area measure described in chapter 9, which is less sensitive to the choice of measurement window, and this paper also used permutation statistics to control the Type I error rate (Sawaki et al., 2012). Another paper included two experiments that were nearly identical, but differed in the brightness of the stimuli (Zhang & Luck, 2009). The time windows used in the first experiment were guided by the observed waveforms, whereas the time windows used in the second experiment were based on the windows used in the first experiment (but shifted in time to reflect the fact that latencies are earlier for brighter stimuli). A third paper simply measured the amplitude in a series of consecutive 100-ms latency windows and then used time as a factor in the ANOVA (Gamble & Luck, 2011; see figure 10.6).

I hope it is now clear to you that random noise in the data can easily look like a real effect and can be statistically significant, especially if you choose to measure from a particular time window and set of electrode sites on the basis of the effects you see in the data. This is the problem of multiple implicit comparisons. And note that it was exacerbated by the inclusion of an electrode hemisphere factor in the ANOVAs, giving us more opportunities to see a bogus interaction. See box 10.3 for some further thoughts about this problem (and a shocking admission of guilt).

The waveforms shown in figure 10.5B contain some clues that the P1 and P2 "effects" are false positives. First, the P1 effect is very early for a cognitive manipulation, and it does not have the scalp distribution that one typically sees for early visual ERP components. Second, the P2 effect begins near time zero and extends for hundreds of milliseconds, which is a common pattern for bogus effects. These issues were discussed in the section on "Baseline Correction" in chapter 8 (see figures 8.2 and 8.3). As I noted in that chapter, "it is important to look at the baseline and be very suspicious about any effects (differences between groups or conditions) that begin near time zero. . . . In addition, you should be highly suspicious about any effects that begin within 100 ms of stimulus onset unless they reflect differences in the stimuli (e.g., larger sensory responses for brighter stimuli)." It is possible for real effects to show these patterns, but these patterns indicate that the effects are probably false positives (especially if they were unexpected or if you used the waveforms to guide your selection of time windows and electrode sites).

The problem of multiple implicit comparisons leads to a somewhat counterintuitive principle:

The more-is-less principle The more conditions, time points, and electrodes are in your data, the less true statistical power you will have.

For example, if you have data from two conditions and two electrode sites, and you limit your analyses to a time window of 200–300 ms, you will have very few opportunities for noise to impact your data, and you will have very few implicit or explicit choices to make about how to analyze your data. Consequently, it will be unlikely that a bogus effect will be statistically significant. However, if you have data from 42 conditions and 128 electrode sites, and you look at the entire time period from 50 to 1500 ms poststimulus, there are thousands of opportunities for noise to produce statistically significant effects. Anything you do post hoc to avoid this inflation of the Type I error rate will reduce your statistical power. Thus, having more conditions and more electrode sites may make it more difficult for you to find the truth. This is one of several reasons why having a large number of electrodes can be problematic (see the chapter 5 supplement). See box 10.4 for additional discussion of the more-is-less principle.

Solving the Problem of Implicit Comparisons

There are several ways that you can solve the problem of multiple implicit comparisons and avoid inflating your Type I error rate. There is no one best solution for all studies, so you will need to choose the best approach given the nature of your own research. Whatever you choose, you should provide an explicit justification in your methods or results section. That way, reviewers and readers will know that you are being careful.

The best time to start thinking about this is when you are designing your experiment. You may want to limit the number of conditions and electrode sites to minimize the number of possible comparisons you could make (because of the more-is-less principle). And you may want to design the experiment so that the time windows and electrode sites that you've used previously can be used again in the current study.

A Priori **Hypotheses**   In many cases, you can use prior research rather than the current waveforms to guide your selection of time windows and electrode sites. For example, when my

Box 10.4
When More Is Less

I first encountered the more-is-less principle when I was in graduate school and was conducting spatial cueing experiments. In these experiments, a cue (e.g., an arrow) indicates that a subsequent target is likely to appear in a particular location. The target appears in the cued location on most trials (called *valid trials*) but sometimes appears in an uncued location (called *invalid trials*). The idea is that subjects will focus attention onto the cued location, yielding improved processing (in terms of reaction time, accuracy, and/or ERP amplitudes) when the target appears at the cued location. Many experiments also include *neutral trials*, in which no information is provided about the location of the subsequent target. On these trials, one would expect that attention is broadly or randomly distributed, leading to performance that is somewhere between valid and invalid trials. After conducting several such experiments and reading many papers from other labs, I realized that performance from the neutral trials was sometimes very close to performance on the valid trials, and sometimes closer to performance on the invalid trials. But it did not seem very systematic. Eventually I realized that a little bit of random variation in performance on any of the three trial types could have a fairly substantial effect on exactly where the neutral trials fell relative to the valid and invalid trials. I and other researchers were constantly coming up with post hoc explanations for the patterns we were seeing in specific experiments, but most of these patterns were likely a result of noise (although a few have been shown to be consistent).

This led me to the following realization: The more conditions I included in an experiment, the more likely it was that I found a "weird" result in one of the conditions that was caused by random noise. Consequently, I started narrowing down my experiments to include only the essential conditions. This gave me fewer opportunities to observe random variations, and fewer weird findings to explain. And in retrospect, it gave me fewer opportunities to conduct multiple implicit comparisons, so it reduced my Type I error rate.

Of course, this approach comes at a cost: by including fewer conditions, I am putting on "theoretical blinders," and I may miss interesting results that could be seen by including more conditions. Consequently, when I am trying a completely new type of experiment, I go ahead and include lots of conditions, and I assume I will need to conduct follow-up experiments to replicate the findings of this experiment. But when I am testing a specific hypothesis in a well-developed paradigm, I focus on the smallest number of conditions that can test the hypothesis.

students conduct an N2pc experiment with highly salient target stimuli, I know from previous experience that they should measure N2pc amplitude from the lateral posterior electrode sites from approximately 175 to 275 ms. I don't have much more to say about this approach, but don't take this to mean that it isn't a useful approach. In a large proportion of experiments, this is by far the best approach.

This approach obviously can't work when you're trying a completely new experimental paradigm. When you find yourself in this situation, the best option is often to conduct a follow-up experiment so that you can use the results from the first experiment to guide the choice of time windows and electrode sites (and so that you can demonstrate the replicability of your findings, as recommended in box 10.1).

Functional Localizers   This is a variant of the idea of an *a priori* hypothesis, but it is based on data that you collect at the same time as your main experiment. The idea—which is very popular in functional neuroimaging—is that you can use one very simple and well-understood manipulation to determine the time course and electrode sites of a given effect, and then you can apply this to the comparison that is the main focus of your experiment.

Imagine, for example, that you want to determine whether the N170 component is larger for smiling faces than for frowning faces. This might be a very small effect, and to maximize statistical power it would be useful to know the optimal time window and electrode sites for measuring the voltage. In fact, you might want to determine the optimal time window and electrodes separately for each individual subject. You could do this by including a condition in which you record the ERPs elicited by faces and cars, which is known to elicit a very robust N170 effect (see, e.g., figure 1.2 in chapter 1). You could then make a face-minus-car difference wave and determine the onset and offset time of the N170 effect and the electrode sites at which this effect is present. This would define the measurement window and electrode sites that you would use for measuring the amplitude of the smiling-minus-frowning difference wave. You could do this individually for each subject, or you could measure the face-minus-car difference in the grand average and apply the results uniformly to every subject.

This approach is used very rarely in ERP studies. It has a lot of potential, although it could fail under some circumstances. For example, if the time course of the smiling-versus-frowning difference is not the same as the face-versus-car difference, you won't be selecting the appropriate time window (for a discussion of limitations in neuroimaging studies, see Friston, Rotshtein, Geng, Sterzer, & Henson, 2006). Despite the limitations, it should probably be used in more ERP studies.

Collapsed Localizers   This is a variant of a functional localizer, except that it does not require additional conditions in your experiment. Instead of using a contrast between one set of conditions to determine the window and electrodes that will be used to assess a contrast between a different set of conditions, the data are collapsed across the conditions of interest to determine the window and electrodes.

In our N170 example, we could simply average across the smiling and frowning faces and use the overall N170 to determine the time window and electrode sites that are best for measuring the N170. This window and set of electrodes would then be applied to measure the N170 for the separate smiling and frowning ERPs. An obvious shortcoming is that the N170, when not measured from a difference wave, contains both face-specific and face-nonspecific activity. Consequently, the timing and scalp distribution of the N170 from the collapsed waveform may be quite different from the timing and scalp distribution of the face-specific activity. Nonetheless, there are situations where this can be a reasonable approach.

Now let's consider a different example, in which an oddball paradigm is used with both a patient group and a control group. Imagine that we wanted to ask whether the peak latency of the P3 wave in a rare-minus-frequent difference wave was later in the patients than in the controls. First, we would construct a grand average rare-minus-frequent difference wave that combines all patients and controls. We would then find the latency range and scalp sites that showed the largest P3 wave in this combined difference wave. We could then use this information to define the time window and electrode sites to be used for measuring P3 peak latency in the rare-minus-frequent difference waves from the individual subjects in each group. This approach would minimize bias in a statistical comparison between patients and controls because the differences between patients and controls were not used to select the measurement window and scalp sites used in the comparisons.

This example raises a potential problem with using a combined waveform to determine the window and electrodes. Imagine that the P3 was much larger in the control subjects than in the patients (which is a common finding for many types of patient groups). And imagine that the scalp distribution differed for the patients and controls. The larger amplitude of the P3 in the control subjects would mean that the grand average across groups would be dominated by the control subjects, and the scalp distribution that was selected on the basis of the collapsed data might therefore be more appropriate for the controls than for the patients. This could bias the results in favor of the control subjects. This kind of scenario is relatively rare, so it does not usually preclude the use of this approach. However, if the amplitudes differ across the conditions or groups that you are collapsing, you will want to think carefully about whether this approach might somehow bias your results.

Looking for Condition × Electrode Interactions Imagine that you have *a priori* knowledge that 200–300 ms is a good measurement window, but you don't know which of your 128 electrode sites will contain the effect of interest. You could simply measure the mean amplitude from 200 to 300 ms at all electrode sites and then conduct an ANOVA in which electrode site is a factor. Because the difference between conditions is likely to be large at a subset of the sites and small or even opposite at others, you probably won't see a significant main effect of condition in this ANOVA. Instead, you will be looking for a condition × electrode site interaction. Earlier in the chapter, I discussed some problems that arise with condition × electrode interactions, including the possibility of spurious results owing to the larger number of *p* values and the difficulty of distinguishing between multiplicative and non-multiplicative interactions. However, the former problem arises mainly when the electrode interactions were not predicted, and the latter problem arises mainly when you are trying to draw conclusions about changes in the locations of the underlying generators of the effects. If you are simply trying to conclude that two conditions are different, which will inevitably lead to larger differences at some sites than at others, you can use a condition × electrode site interaction to demonstrate your conditions are significantly different. I have used this approach successfully in several papers. In addition, if this interaction is significant, you are justified in conducting a follow-up analysis to see which electrode sites exhibit a significant main effect of condition.

The main shortcoming of this approach is that the statistical power for seeing an interaction of this nature is usually much lower than the power you would have for seeing a main effect of

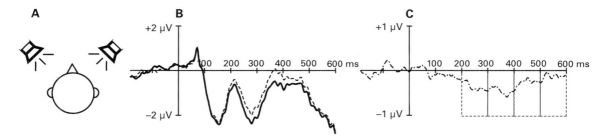

Figure 10.6
Stimuli and data from an experiment designed to find an auditory analog of the N2pc component (Gamble & Luck, 2011). (A) Stimuli. Subjects listened to pairs of auditory stimuli, presented simultaneously in separate speakers. They were asked to detect a particular target sound, which could occur unpredictably in either speaker on a given trial. (B) Grand average ERP waveforms recorded at a cluster of anterior electrode sites (F3, F7, C3, T7, F4, F8, C4, T8), contralateral or ipsilateral to the location of the target. The waveforms are referenced to the average of the mastoids and were filtered offline with a low-pass Gaussian filter with a half-amplitude cutoff at 50 Hz. (C) Difference wave (contralateral minus ipsilateral) from the data shown in panel B. The dashed rectangles show the time windows that were used to measure mean amplitude.

condition if you limited the ANOVA to the electrode sites where the effect is large. The advantage, however, is that it does not require prior knowledge of which sites will have a large effect, allowing you to avoid biasing the results by using the data to help you decide which electrodes to use. But keep in mind that looking for condition × electrode interactions in a post hoc manner can lead to an increase in the Type I error rate.

Looking for Time × Electrode Interactions   An analogous time-based approach can also be used to avoid choosing a specific time window. That is, you can measure the mean voltage from multiple time windows across a broad range, and then include time as a factor in the ANOVA. For example, Marissa Gamble ran a study in my lab in which she was searching for an auditory analog of the N2pc component (figure 10.6). She found a nice contralateral effect for auditory stimuli over anterior electrode sites (which we called *N2ac* for N2-anterior-contralateral) (Gamble & Luck, 2011). We had no strong *a priori* hypothesis about the time window for this effect, because we couldn't assume that it would be the same as the time window of the visual N2pc component. We looked at the data and tried many different measurement windows, but we realized that this was inflating the probability of a Type I error.

To avoid this problem, we simply measured the mean amplitude in consecutive 100-ms windows between 200 and 600 ms, and we conducted an ANOVA in which time window was a factor. We measured N2ac amplitude as the difference in voltage between the contralateral and ipsilateral electrodes (relative to the location of the target) in each of these time windows (figure 10.6C). This difference was significantly greater than zero when we collapsed across the measurement windows (analogous to a significant main effect of contralateral-versus-ipsilateral). In addition, we found an effect of time period on this difference (analogous to an interaction between time window and contralateral-versus-ipsilateral). This justified follow-up analyses of the individual time windows, which indicated that the difference was significantly greater than zero in the 200–300 ms, 300–400 ms, and 400–500 ms windows, but not in the 500–600 ms window.

This approach worked well given that we had no good *a priori* information about what time window to use for measuring the N2ac. That is, despite the fact that we did not inflate the Type I error rate by choosing measurement windows on the basis of the data, the very clear effects were statistically significant. However, part of the reason that it worked in this case was that the effect was present over most of our overall time range. If it had been present only in one of the four 100-ms windows, we would have had relatively low power for detecting the effect (as either a main effect or an interaction with time). If I were to analyze these data today, I would use the *mass univariate approach* described in online chapter 13. In addition, we used the multiple-consecutive-windows approach because these were our first N2ac experiments, so we had no *a priori* information to guide our selection of time windows. I would not take this approach with future N2ac experiments, because I could use our existing N2ac results to guide the choice of the time window for future experiments.

Window-Independent Measures   The problem of defining the measurement window can be very severe in some experiments. An example of this, which was discussed in chapter 9 (see figure 9.2), arises when you try to measure the difference in N2 amplitude between rare and frequent trials in an oddball paradigm. Even if you measure from a difference wave, the N2 may be preceded by a P2 effect and followed by a P3 effect, which may partially cancel the N2 effect (see, e.g., the single subject shown in figure 10.7). Moreover, the latencies of these effects may vary somewhat from subject to subject. If you are measuring mean amplitude, you will need to use a very narrow window to avoid having your N2 measure canceled by P2 or P3 activity in some of the subjects, but the use of a narrow window increases the impact of noise on your measurements. In addition, the results that you get could be influenced by exactly what window you choose, and if you choose the window by looking for the period with the largest effect, this

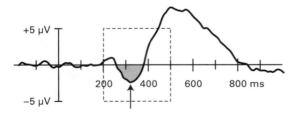

Figure 10.7
Example of how signed area can be used with a broad measurement window. The waveform here is a single-subject difference wave (rare minus frequent), and negative area was used to quantify N2 amplitude. Because only the area below the 0-µV line contributes to the measurement, the measured area would be the same with the broad window shown here (200–500 ms) as with a narrower window (e.g., 250–400 ms). This makes the measurement relatively independent of the choice of time window. The arrow points to the peak voltage.

will inflate your Type I error rate. Although I generally encourage the use of mean amplitude as a measure, this approach can be problematic when the effect you are trying to measure is surrounded by effects on other components, especially when the other components are large. Measuring from difference waves can help, but not if the other components are still present in the difference waves. The best solution, when possible, is to design the experiment so that only one component is present in the difference waves. However, that is not always possible, so you may need a different solution.

One good solution was described in detail in chapter 9; namely, the use of signed area measures. In our oddball experiment, for example, you could quantify the N2 as the area of the region falling below the zero line over a very broad time window (e.g., 200–500 ms; see figure 10.7, along with figures 9.2 and 9.3). This would ordinarily be done from a difference wave, because the zero line is much more meaningful in a difference wave.

A second solution would be to use peak amplitude. Again, this works best in a difference wave that isolates a small number of components. A wide window is possible in this situation (e.g., 200–500 ms), because the positive-going P2 and P3 waves on either side of the N2 make it unlikely that some other negative peak will be measured instead of N2 (see figure 10.7). Chapter 9 described many shortcomings of peak amplitude, but this is one situation in which it may sometimes be superior to mean amplitude.

For both of these solutions, however, you should be aware that the surrounding P2 and P3 components may still have a large impact on your N2 amplitude measurement. That is, these components are likely to overlap with the N2 and will impact the measured amplitude of the N2, even if the overall voltage in this window is negative. The use of signed area or peak amplitude may reduce the cancellation of N2 produced by the P2 and N2 waves, but it will not eliminate it. For example, if you were comparing rare-minus-frequent difference waves between two groups, a larger P3 wave in one group could make it appear that the N2 was smaller (less negative) in that group (see, e.g., figure 2.5F in chapter 2). In addition, both peak amplitude and signed area amplitude are biased: they tend to increase when the signal-to-noise ratio decreases (for an example of how we dealt with this problem when using signed area measures, see Sawaki, Geng, & Luck, 2012).

Correcting for Multiple Comparisons Instead of treating your analysis as a problem of multiple *implicit* comparisons, you could treat it as a problem of multiple *explicit* comparisons. In the oddball experiment shown in figure 10.7, for example, you could measure the voltage at every time point between 200 and 500 ms and do a *t* test comparing each voltage to zero (or comparing the voltage in the rare and frequent conditions at each time point). You could then perform a correction for multiple comparisons. This is called the *mass univariate approach*, because a massive number of individual statistical tests are used. When I wrote the first edition of this book, I didn't seriously consider raising this possibility, because the standard approach to correcting for multiple comparisons (the Bonferroni correction) has ridiculously low statistical power in this situation. However, there have been some important advances in the statistical analysis of ERP data, and it is now possible to explicitly correct for multiple comparisons without a ridiculous loss of power. This approach is very powerful and becoming increasingly common, so I have provided an additional chapter that describes it in some detail (see online chapter 13).

Suggestions for Further Reading

Groppe, D. M., Urbach, T. P., & Kutas, M. (2011). Mass univariate analysis of event-related brain potentials/fields I: A critical tutorial review. *Psychophysiology, 48,* 1711–1725.

Groppe, D. M., Urbach, T. P., & Kutas, M. (2011). Mass univariate analysis of event-related brain potentials/fields II: Simulation studies. *Psychophysiology, 48,* 1726–1737.

Jennings, J. R., & Wood, C. C. (1976). The ε-adjustment procedure for repeated-measures analyses of variance. *Psychophysiology, 13,* 277–278.

Kiesel, A., Miller, J., Jolicoeur, P., & Brisson, B. (2008). Measurement of ERP latency differences: A comparison of single-participant and jackknife-based scoring methods. *Psychophysiology, 45,* 250–274.

Maris, E. (2012). Statistical testing in electrophysiological studies. *Psychophysiology, 49,* 549–565.

McCarthy, G., & Wood, C. C. (1985). Scalp distributions of event-related potentials: An ambiguity associated with analysis of variance models. *Electroencephalography and Clinical Neurophysiology, 62,* 203–208.

Miller, J., Patterson, T., & Ulrich, R. (1998). Jackknife-based method for measuring LRP onset latency differences. *Psychophysiology, 35,* 99–115.

Miller, J., Ulrich, R., & Schwarz, W. (2009). Why jackknifing yields good latency estimates. *Psychophysiology, 46,* 300–312.

Urbach, T. P., & Kutas, M. (2002). The intractability of scaling scalp distributions to infer neuroelectric sources. *Psychophysiology, 39,* 791–808.

Urbach, T. P., & Kutas, M. (2006). Interpreting event-related brain potential (ERP) distributions: Implications of baseline potentials and variability with application to amplitude normalization by vector scaling. *Biological Psychology, 72,* 333–343.

Ulrich, R., & Miller, J. (2001). Using the jackknife-based scoring method for measuring LRP onset effects in factorial designs. *Psychophysiology, 38,* 816–827.

Appendix: Linear Operations, Nonlinear Operations, and the Order of Processing Steps

Overview

ERP data analysis involves many processing steps, including filtering, epoching, artifact rejection, and so forth. One of the most common questions I'm asked in ERP Boot Camps is whether processing step X should be done before or after processing step Y. For example, you may be wondering whether you should filter your data before or after performing artifact rejection or whether you should re-reference your data before or after filtering. The answer depends on whether a given processing step involves a linear or nonlinear operation. The distinction between linear and nonlinear operations is also important for understanding how the jackknife statistical approach works (see chapter 10).

I will define what *linear* and *nonlinear operations* are in a little bit. But first I want to mention why this distinction is important. Linear operations have an important property: They can be performed in any order and the result will be the same. This is just like the fact that addition can be done in any order (e.g., A + B + C gives you the same result as C + B + A). In the context of ERP processing, averaging and re-referencing are linear operations, so you can do them in any order and get the same result (assuming you haven't interposed any nonlinear operations between them). In contrast, the order of operations matters for nonlinear operations. For example, artifact rejection is a nonlinear process, so you will get a different result if you re-reference the data and then perform artifact rejection versus performing artifact rejection and then re-referencing. This is analogous to the combination of addition and multiplication in simple arithmetic (e.g., [A + B] × C is not usually the same as A + [B × C]). By knowing whether all the operations in a set are linear, you can know whether the order of operations matters.

The following sections define the terms *linear* and *nonlinear* and provide specific advice about the optimal order of processing steps in a typical ERP experiment.

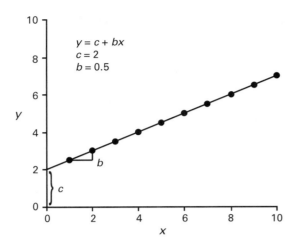

Figure A.1
Example of a linear equation.

Defining Linear and Nonlinear Operations

If you hate math, you can skip this section, because the next section will tell you which common EEG/ERP data analysis procedures are linear and which are nonlinear. However, if you don't mind a tiny bit of math, you should read this section so that you understand what it means for an operation to be linear or nonlinear.

The term *linear* comes from the equation for a line:

$$y = c + bx$$

This equation expresses how each y value is related to each x value when the function is a straight line. This is illustrated in figure A.1. The c value is the y *intercept*, which is the value of y when the line runs through the y axis, which is also the point where x is zero. In this example, c is 2, so the line passes through the y axis at a value of 2. The b value is the slope of the line (the amount that y increases for each one-unit increase in x). The slope is 0.5, so y increases by 0.5 units for each one-unit increase in x. If you know the c and b values for a line, you have everything you need to draw the line. An equation like this might tell you how weight (the y value) tends to increase with height (the x value).

If you've taken a statistics course that covered multiple regression, you know that the equation for a line can be extended to multiple x variables. For example, weight (the y value) could be predicted by a combination of height (which we will call x_1), age (which we will call x_2), and daily caloric intake (which we will call x_3). If these three x variables independently combine to determine someone's weight (along with a constant c, which would represent the minimum possible weight), we could express this relationship with the equation:

$$y = c + b_1x_1 + b_2x_2 + b_3x_3$$

In this equation, the b values are *scaling* or *weighting* factors that indicate the extent to which a given x variable influences the y value. For example, age might have a weaker effect on weight than does daily caloric intake, so it might have a weaker b value.

We can generalize this equation further by adding more pairs of x and b values if we have more factors that influence the y value. This gives us the generalized equation for a line:

$$y = c + b_1x_1 + b_2x_2 + b_3x_3 \ldots + b_Nx_N$$

Any mathematical operation that can be expressed in the form of the general equation for a line is called a linear operation. If a mathematical operation cannot be expressed in this way, it is called a nonlinear operation.

A key feature of a linear operation is that the output value (the y value) depends on the scaled sum of one or more input values (the x values) plus a constant (the c value). The x values are the data points, and the b values are scaling factors that are not determined from the data. For example, the average of three values $(x_1, x_2,$ and $x_3)$ can be expressed as:

$$y = 0 + 1/3x_1 + 1/3x_2 + 1/3x_3$$

In this example, c is zero and each of the b values is 1/3. The x values are the data points that we are averaging together, and each x value is scaled by a b value. Thus, averaging is a linear operation. In averaging and most other ERP processing operations, the c value is zero and can just be ignored.

In a linear function, the scaled data points are simply added together (or subtracted, if a b value is negative). Multiplication can occur between a scaling factor (one of the b values) and a data point (one of the x values), but two data points (x values) cannot be multiplied by each other or divided. If an operation involves combining the scaled x values in any way other than addition or subtraction, then it is not a linear operation. For example, the following equation would not be linear because the x values are multiplied:

$$y = 2 + x_1x_2$$

In addition, if the operation involves a threshold then it is not linear, as in the following equation:

if $x_1 > 0$ then $y = x_1$; otherwise $y = 0$

Artifact rejection involves a threshold, because a given trial is included or excluded depending on whether or not a threshold is exceeded. Therefore, artifact rejection is not a linear operation.

In a linear operation, the b values are independent of the data values. In our example of averaging three points, for example, each b value is 1/3 no matter what the data values are. Of course, we've chosen 1/3 as the scaling value because we are averaging three points together. We would have used 1/4 as the scaling value if we were averaging four points together. Thus, the *b* values are chosen in a manner that depends on the general nature of the data, but they do not depend on the actual observed data values (the *x* values). If a scaling factor is determined from the observed data values, then the process is not linear.

Common Linear and Nonlinear Operations in ERP Processing

Now that we've defined linear and nonlinear operations, let's discuss which of the common procedures we use in ERP research are linear and which are nonlinear.

Averaging

As I mentioned earlier, averaging is a linear operation. There are many types of processes that involve averaging, and all are linear. For example, we average multiple single-trial EEG epochs together to create an averaged ERP waveform, and we also average the values at multiple time points within a time window when we quantify the amplitude of a component with a mean amplitude measurement. Because these two types of averaging are both linear, they can be done in any order with the same result. Consequently, you will get the same result by either (a) measuring the mean amplitude in a specific time window (e.g., 200–300 ms) from the single-trial EEG segments and then averaging these values together, or (b) averaging the EEG segments together into an averaged ERP waveform and then measuring the voltage in the same time window from this averaged waveform.

The same principle applies if you want to take averaged ERP waveforms from multiple conditions and average them together. For example, in a well-counterbalanced oddball experiment with Xs and Os as stimuli, you might have rare Xs and frequent Os in some trial blocks and rare Os and frequent Xs in other trial blocks. You might initially create averaged waveforms for rare Xs, rare Os, frequent Xs, and frequent Os, and you can later make an average of the rare X and rare O waveforms and an average of the frequent X and frequent O waveforms (to reduce the number of factors in the ANOVA, as discussed in chapter 10). Similarly, you might average together the averaged ERP waveforms across multiple nearby electrode sites.

For all of these kinds of averaging, the order of operations does not matter. For example, you could measure the mean amplitude from 200 to 300 ms in the rare X and rare O waveforms and then average these values together, or you could first average the rare X and rare O waveforms together into a single rare stimulus waveform and then measure the mean amplitude from 200 to 300 ms. You will get exactly the same result either way. However, peak amplitude is not a linear measure, so you will typically get one result if you first measure the peak amplitude values from two waveforms and then average these values together and a different result if you average the two waveforms together and then measure the peak amplitude. In most cases, nonlinear measures will be more reliable if measured from cleaner waveforms, so it is usually better to average the waveforms together first and then measure the nonlinear value.

Weighted versus Unweighted Averages

There is a little detail that might cause you to get different results depending on the order in which you perform multiple averaging steps. It's easiest to explain this detail in the context of a concrete example. Imagine that you've performed an oddball experiment with a block of 100 trials in which X was rare and O was frequent and another block of 100 trials in which O was rare and X was frequent. If the rare category occurred on 20% of trials, this will give you 20 rare Xs, 20 rare Os, 80 frequent Xs, and 80 frequent Os. However, after artifact rejection, you might end up with unequal numbers of Xs and Os (just because the blinks are somewhat random). For example, you might have 15 rare Xs and 17 rare Os after artifact rejection. If you combine the EEG epochs for the rare Xs and the rare Os during the initial signal averaging process, you will have an average of 32 trials (15 Xs and 17 Os). Because there were more Os than Xs in the average, the Os will have a slightly larger impact on the averaged waveform than the Xs. However, if you instead make separate averages for the rare Xs and the rare Os and then average these waveforms together, the Xs and Os will have equal impact on the resulting averaged waveform. Because of this slight difference in the number of trials, averaging the single-trial EEG epochs together for the Xs and Os will give you a slightly different result compared to averaging the Xs and Os separately and then averaging these ERP waveforms.

In this example, it won't really matter very much whether you combine the rare Xs and rare Os during the initial averaging process or at a later stage, because the number of Xs is almost the same as the number of Os. However, imagine you wanted to compare the sensory response elicited by Xs with the sensory response elicited by Os, collapsing across rare and frequent. If you combine the EEG epochs for 20 rare Xs with 80 frequent Xs during the initial averaging process, you are giving equal weight to each of the 100 Xs (and therefore greater weight to the category of frequent Xs than to the category of rare Xs). However, if you create separate averaged ERP waveforms for the rare Xs and for the frequent Xs and then average these waveforms together, you are giving equal weight to these two categories (and therefore greater weight to each individual rare X trial than to each individual frequent X trial). These two ways of combining the 100 X trials could lead to substantially different results.

Many ERP analysis systems give you an option for creating *weighted averages* when you take multiple averaged ERP waveforms and average them together. A weighted average gives each individual trial an equal weighting, just as if you had combined the single-trial EEG epochs together during the initial averaging process, but it operates on averaged ERP waveforms. For example, if we first made separate averaged ERP waveforms for the rare Xs and the frequent Xs, and then we averaged these waveforms together using a weighted average, this would give us the same thing as if we combined the 20 rare Xs and 80 frequent Xs during the initial averaging process. In contrast, when we combine two averaged waveforms together without taking into account the number of trials in each average, this is called an *unweighted average*.

You may be wondering which approach is better, using a weighted average or an unweighted average. If you are averaging waveforms together to deal with a theoretically unimportant counterbalancing factor, it usually makes most sense to combine them with a weighted average (so that each trial has equal weight). However, if you are combining waveforms from conditions that differ in a theoretically important way, it usually makes most sense to combine them with an unweighted average (so that each condition has equal weight). If you carefully think through the situation, it should become clear to you which type of averaging is most appropriate (or that it won't matter because the differences in numbers of trials across waveforms are minimal and random).

Difference Waves

Making difference waves is a linear process. For example, the process of making a difference wave between a target waveform (x_1) and a standard waveform (x_2) in an oddball paradigm can be expressed as:

$y = 0 + 1x_1 + -1x_2$ (computed for each point in the waveform)

In this equation, c has a value of zero (so you can ignore it), and the b values are 1 and –1. This may seem like a strange and overly complicated way to express a difference, but it demonstrates that the process of making a difference wave is a linear operation.

Because making difference waves and measuring mean amplitudes are both linear processes, you can combine them in any order. For example, you can measure the mean amplitude from 350 to 600 ms in the target waveform and in the standard waveform and then compute the difference between these two mean amplitudes, or you can compute a rare-minus-frequent difference wave and then measure the mean amplitude from 350 to 600 ms in this difference wave. Either way you will get exactly the same result. However, you would get a different result from these two orders of operations if you measured peak amplitude (or some other nonlinear measure) rather than mean amplitude.

If you are measuring something nonlinear, the results can be enormously different depending on whether you measure before or after making the difference wave. It's difficult to say that one order will be better than the other for most experiments, because it will depend on what you are trying to achieve by measuring a difference. However, if you are making a difference wave to isolate a component (as described in chapters 2 and 4), then you will usually want to measure the component after making the difference wave.

Convolution, Filtering, and Frequency-Based Operations

The most common filters for EEG and ERP data (called *finite impulse response* filters) are based on a mathematical process called *convolution* (defined in online chapter 11). Convolution is a linear process, so these filters are therefore linear. Note, however, that filters in their purest form operate on infinite-duration waveforms, and nonlinear steps are often built into filters to deal with edges at the beginning and end of a finite-duration waveform. Consequently, the order of operations can matter near the edges of an EEG epoch or averaged ERP waveform.

This nonlinear behavior is typically most severe for high-pass filters (see chapter 7 and online chapter 11 for more details). Thus, it is almost always best to perform high-pass filtering on the continuous EEG rather than on epoched EEG data or averaged ERPs. When you filter the continuous EEG, edges are still present, but they are limited to the beginning and end of each trial block and are therefore farther in time from the events of interest. To be especially safe, I recommend recording 10–30 s of EEG data prior to the onset of the stimuli at the beginning of each block and after the offset of stimuli at the end of each block. If you do this, the edge artifacts will not extend into the time period of your stimuli.

Edge artifacts can also occur for low-pass filters, but they are typically very brief, affecting only the very beginning and end of the waveform. Therefore, you can usually apply low-pass filters to epoched EEG or averaged ERP waveforms without any significant problems (assuming that you are using the fairly mild filters recommended in chapter 7). You can also apply these filters to the continuous EEG.

Filters may act strangely if an offset is present that shifts the overall EEG or ERP waveform far above or far below 0 µV. This typically occurs when you record the EEG without a high-pass filter. Baseline correction is typically applied when the data are epoched, and this removes the offset. However, if you are going to apply an offline high-pass filter to the continuous EEG, as I just recommended, you should make sure that the offset voltage is removed prior to filtering. The filtering tool in ERPLAB Toolbox includes an option for this. In other systems, this option may not be part of the filtering system, so it may take you a bit of work to figure out how to do it prior to filtering.

Some EEG/ERP analysis systems (including ERPLAB Toolbox) allow you to use a somewhat different class of filters (called *infinite impulse response* filters) that are not linear. However, when the slope of the filter is relatively mild (e.g., 12 dB/octave), these filters are nearly linear, and the "rules" for using these filters are the same as for truly linear filters (e.g., you should apply high-pass filters only to long periods of continuous EEG, but you can apply low-pass filters to EEG epochs, averaged ERP waveforms, or the continuous EEG).

The Fourier transform and the inverse Fourier transform are linear operations (although, like filters, nonlinear steps may be applied at the edges of finite-duration waveforms). Most time–frequency transformations are also linear (except at the edges).

Baseline Correction

Baseline correction is a linear operation because we are just computing the average of the points from the baseline period (which is a linear operation) and subtracting this average from each point in the waveform (which is also a linear operation). This means that you can perform baseline correction before or after any other linear process and get exactly the same result.

However, artifact rejection is a nonlinear process, so you may not get the same result by performing baseline correction before versus after artifact rejection. This will depend on the nature of the artifact rejection algorithm. If the algorithm uses a simple voltage threshold, then you really need to perform baseline correction prior to artifact rejection (see figure 6.1 in chapter 6). However, the moving average peak-to-peak and step function algorithms are not influenced by baseline correction, so you can apply these algorithms either before or after baseline correction. This does not mean that these artifact rejection algorithms are linear; it just means that they do not take the baseline voltage into account.

Re-referencing

The common methods of re-referencing are linear. For example, the average reference is computed by finding the average across electrode sites (which is a linear process) and then subtracting this value from each site (which is also linear). Changing the reference from one individual site to another is linear (e.g., changing from a Cz reference to a nose reference), as is re-referencing to the average of the mastoids or earlobes (or the average of any other subset of sites). In theory, it would be possible to create a nonlinear re-referencing procedure, but I don't think I've ever seen anyone do this.

Artifact Rejection and Correction

All artifact rejection procedures involve a threshold for determining whether a given trial should be rejected or not. Artifact correction also involves a threshold because a given component is set to zero if it matches a set of criteria for being an artifact. Artifact rejection and correction are therefore nonlinear processes. Consequently, you will need to think carefully about whether to deal with artifacts before or after other processes, such as filtering and re-referencing.

The general principle is that you should apply a given process prior to artifact correction or rejection if it makes the correction/rejection process work better. The most obvious example is that artifacts can be more easily detected and rejected if rejection is preceded by re-referencing the data to create bipolar signals (e.g., under minus over the eyes for VEOG and left minus right or right minus left for HEOG). However, you have to be careful about re-referencing the data prior to most artifact correction procedures because these procedures usually require that all channels have the same, single reference electrode.

In addition, you should typically apply a high-pass filter prior to both artifact correction and artifact rejection. Unless your data are very noisy, you can apply low-pass filtering at some later stage (typically after averaging).

Recall from chapter 6 that trials with blinks or eye movements that change the sensory input should be rejected even if artifact correction is also used. However, this can be complicated to implement. Specifically, it is pointless to apply artifact correction if you have already rejected trials with artifacts, but you can't ordinarily apply artifact rejection after artifact correction because it's impossible to determine which trials have artifacts if you've already performed correction. ERPLAB Toolbox was carefully designed to allow you to combine rejection and correction; if you are using some other system, you may need to contact the manufacturer to figure out how to do this.

Measuring Amplitudes and Latencies

A mentioned earlier, quantifying the magnitude of a component by computing the mean amplitude over a fixed time range is a linear operation. The integral over a time period (see chapter

9) is also a linear measure. Thus, you will get the same results if you use these methods before or after averaging, making difference waves, and so forth.

Area amplitude measures are nonlinear if part of the waveform is negative and part of the waveform is positive during the measurement window. However, they are linear if the waveform is entirely negative or entirely positive during the measurement window (because the area is equal to the integral in this case).

Finding a peak involves comparing different voltage values to determine which one is largest, and there is no way to express this process with a linear equation. Thus, any amplitude or latency measurement procedure that involves finding a peak is nonlinear. More broadly, all latency measurements that I have ever seen are nonlinear. The main implication of this is that, as I described earlier, you will get different results if you measure before versus after averaging waveforms together or computing difference waves.

Recommendations for the Order of Operations in a Typical ERP Experiment

Now that you know which processes are linear and which are nonlinear, you are in a good position to decide what order to use for processing your data. However, it can be a little overwhelming to integrate all this information together, so I've created a simple list that describes the typical order of operations in an EEG/ERP data analysis pipeline. But keep in mind that you may need to change this a bit to reflect the nature of your research and your data. In other words, this list provides a good starting point, but don't just blindly follow it.

Here are the typical steps, in order. Note that the last two steps occur after you have finished recording the data from all subjects. You should apply all of the other steps for each subject, immediately after recording, to verify that there were no technical errors during the recording.

1. High-pass filter the data to remove slow drifts (e.g., half-amplitude cutoff of 0.1 Hz, slope of 12 dB/octave). This should be done on long periods of continuous EEG to avoid edge artifacts, and any offset should be removed if a high-pass filter was not applied during data acquisition. In rare cases, you may also want to apply a mild low-pass filter at this stage (e.g., half-amplitude cutoff of 30 Hz, slope of 12–24 dB/octave).

2. Perform artifact correction, if desired. As described in the online supplement to chapter 6, periods of "crazy" data should be deleted prior to the artifact correction process (but don't delete ordinary artifacts, such as eyeblinks, prior to artifact correction).

3. Re-reference the data, if desired. For example, you may want to re-reference to the average of the mastoids at this point. In addition, you will probably want to create bipolar EOG channels (e.g., lower minus upper and left minus right) at this point to facilitate artifact rejection. You can also re-reference the data again after averaging to see how the waveforms look with different references (as described in chapter 5).

4. Epoch the continuous data to create single-trial EEG segments (e.g., from –200 to +800 ms). In most systems, you will perform baseline correction at this stage (which is essential if you will be using an absolute voltage threshold in the next stage).

5. Perform artifact rejection. Many systems require that you perform artifact after epoching, so I have put this step after epoching. However, it works just as well to perform artifact rejection on the continuous EEG, prior to epoching, if your system allows it.

6. Average the single-trial EEG epochs to create single-subject averaged ERP waveforms.

7. Plot the ERP waveforms to make sure that the artifact correction and rejection processes worked properly. You may want to apply a low-pass filter (e.g., half-amplitude cutoff of 30 Hz, slope of 12–24 dB/octave) before plotting so that you can see the data more clearly. However, I would ordinarily recommend applying the subsequent steps to the unfiltered data.

8. If necessary, average together different trial types if they weren't already combined during the initial averaging process (e.g., to collapse across factors used for counterbalancing).

9. Make difference waves, if desired. Note that averaging across trial types and making difference waves are linear operations, and mild low-pass filtering is either linear or nearly linear, so you can do these steps in any order.

10. If you are averaging multiple waveforms together and/or making difference waves, you should plot the waveforms from each step so that you can verify that the averaging and differencing processes are working properly.

11. Make grand averages across subjects (and possibly leave-one-out grand averages for jackknifing, as described in chapter 10).

12. Measure amplitudes and latencies (from the single-subject ERPs in most cases, but from the leave-one-out grand averages if you are using the jackknife approach). If you are measuring peak amplitude or peak latency, you should usually apply a mild low-pass filter first (e.g., half-amplitude cutoff of 20–30 Hz, slope of 12–24 dB/octave). If you are measuring onset or offset latency, you should usually apply a more severe filter first (e.g., half-amplitude cutoff of 10–20 Hz, slope of 12–24 dB/octave). If you are measuring mean amplitude, integral amplitude, or area amplitude, it is best to avoid low-pass filtering the data (except for the anti-aliasing filter that was applied during the EEG recording; see box 9.2 for the rationale).

Notes

Chapter 2

1. The term *local field potential* is something of a misnomer because even a small electrode tip inside the cortex will pick up distant as well as nearby sources of electrical activity. Moreover, these potentials are also influenced by activity at the reference electrode (which is discussed in chapter 5). However, this term has been around so long that we are probably stuck with it.

2. This is a simplified account. For a more complete account, see Buzsáki et al. (2012).

3. This assumes that the recording electrode is at a sufficient distance from the activated cortex, which will always be true in scalp recordings.

4. In this section, I'm referring to the *absolute voltage*. The distribution of recorded voltage in an actual experiment will depend on the location of the *reference electrode* (see chapter 5 for definitions of these terms).

5. An electrical dipole is accompanied by a magnetic field of proportionate strength, so magnetic field recordings are picking up the same fundamental signal as ERP recordings, but without the spatial blurring caused by the high electrical resistance of the skull. However, there is one small complication: The high resistance of the skull causes a small amount of volume-conducted electrical current to flow parallel to the skull, and this parallel electrical activity is accompanied by a small magnetic field that causes a slight blurring of the magnetic field caused by the main generator dipole. In this manner, the high resistance of the skull causes some blurring of magnetic fields. However, this is minor compared to the blurring of the electrical activity.

6. Fuchs, Wagner, and Kastner (2004) described a technique for estimating the margin of error, but this method is unrealistic when multiple dipoles might be active, and it assumes that the forward model is perfect.

Chapter 5

1. Some ADCs use positive and negative values to reflect positive and negative voltages, unlike our example in which the ADC uses only positive values, with the lower half of the values representing negative voltages. This has no real impact on the ADC's performance.

Chapter 6

1. This framework was originally developed by Jon Hansen at UCSD. He also developed several of the specific artifact rejection algorithms described later in this chapter. Others were developed by Javier Lopez-Calderon, the mastermind behind EPLAB Toolbox.

2. An analogous method is available in BrainVision Analyzer.

3. In theory, a devious experimenter could try a set of artifact rejection parameters, average the data, see if the results conform to the expected pattern, and then keep adjusting the parameters until the results "look right." That would be just plain unethical. However, adjusting the parameters for each subject is not a problem as long as they are adjusted on the basis of success in catching problematic artifacts rather than on the basis of the effects in the averaged ERPs.

Chapter 7

1. The Fourier transform, in its purest form, applies to continuous functions. For discretely sampled data, such as digitized EEG and ERPs, the *discrete Fourier transform* (DFT) is used. The concepts are the same, but the math is a little different. You may also encounter the *fast Fourier transform* (FFT), which is a very efficient way of computing the DFT that can be used when the number of samples is a power of 2 (e.g., 256 points per epoch).

2. This assumes that the roll-off between 20 and 40 Hz is a perfectly straight line, which is only approximately correct.

Chapter 8

1. Note for people who are unfamiliar with American idiom: These terms are not really "advanced"— they are intentionally childish terms that are being used ironically.

2. You may be wondering exactly how to quantify the amplitude of the signal and the amplitude of the noise to compute the SNR. This is actually a fairly complicated topic. For the sake of simplicity, let's just assume that we will quantify the signal as the peak amplitude of the component of interest and the noise as the difference in amplitude between the most positive and most negative points in the prestimulus baseline period.

3. Technically speaking, covariance is used rather than correlation, which means that the matching EEG segment must be large as well as being similar in shape to the template.

4. I am leaving out some details here that are unimportant for understanding the basic idea (but are important for implementation). One detail is that a gradual taper is applied at the edges of the window, so the farthest edges of the window do not contribute as much as the rest of the window.

Chapter 9

1. There are a couple of important details to consider when this procedure is implemented. First, high-frequency noise may cause a point at the edge of the window to be slightly greater than the points on either side. To minimize spurious local peaks like this, the local peak should be defined as having both a greater voltage than the immediately adjacent points and a greater voltage than the average of the 10–20 ms on either side. In addition, you might find cases in which the waveform will not have a local peak anywhere within the measurement (e.g., when the waveform gradually increases or decreases monotonically through the entire measurement window). In such cases, you can just use the simple peak amplitude or the amplitude at the midpoint of the latency range.

2. There usually isn't a time point at which the voltage is exactly 50% of the peak amplitude. However, it's fairly easy to interpolate between the samples and find the 50% point.

3. The sudden onset of a square wave seems unrealistic as a representation of a single-trial ERP. However, this is more realistic than you might think. In choice experiments, individual neurons often start to fire suddenly. Different neurons may start to fire at different times, however, and the local field potential from multiple neurons may therefore ramp up gradually. Thus, the overall ERP may have a gradual onset on each trial, but individual neurons may be well represented by the square wave. Thus, the 50% peak latency measure may do a very good job of representing the average time at which individual neurons make a discrimination.

Chapter 10

1. You may recall from chapter 3 that this method for manipulating task difficulty leads to a physical stimulus confound. I wouldn't recommend using this approach in your own experiments. Despite the confound, however, this experiment provides an adequate example for explaining statistical analysis.

2. The effect of the number of ANOVA factors on the familywise error rate is not as widely appreciated as the effect of the number of separate ANOVAs that are conducted. You can read an interesting blog posting on this issue by Dorothy Bishop at http://deevybee.blogspot.co.uk/2013/06/interpreting-unexpected-significant.html.

3. By "pure effects of brightness," I mean effects that do not interact with other factors.

4. Covariance is closely related to correlation: A correlation between two variables is computed by dividing the covariance between the variables by the average variance of the variables.

5. The Greenhouse–Geisser adjustment tends to be overly conservative, especially at moderate to high levels of nonsphericity. Many statistical packages also include the Hyun–Feldt adjustment, which is more accurate and may be useful when low statistical power is a problem.

6. I have simplified the reasoning behind the jackknife technique a little. A thorough analysis of the reasons why jackknifing works—and a discussion of why it sometimes fails—can be found in a very nice paper from Miller et al. (2009).

Glossary

A note about the abbreviation *ERP*: The abbreviation *ERP* is sometimes used incorrectly in the singular form (mostly by native English speakers) and is sometimes used incorrectly in the plural form (mostly by nonnative English speakers). The key is that *ERP* should usually be written in the plural form when used as a noun and in the singular form when used as an adjective. For example, one conducts an "ERP study" and not an "ERPs study," because in this situation *ERP* is used as an adjective. Conversely, one conducts an experiment "with fMRI and ERPs" not "with fMRI and ERP," because in this situation *ERP* is used as a noun (after all, one would not say that an experiment was conducted "with functional magnetic imaging and event-related potential"). Also, *ERP* stands for *event-related potential*, not *evoked response potential* (which appears to be a miscombination of *evoked response* and *evoked potential*, much as *irregardless* is a miscombination of *regardless* and *irrespective*).

Alpha
In statistics, this is criterion that you have chosen for statistical significance (usually 0.05).

Alternative hypothesis
The hypothesis that the conditions are actually different (i.e., the means of the conditions would differ if we could measure from an infinite number of subjects).

Anti-aliasing filter
A low-pass filter that is specifically designed to avoid aliasing during the digitization process.

Auditory brainstem response (ABR)
These are small and fast ERPs elicited by auditory stimuli such as clicks, occurring within the first 10 ms after stimulus onset and arising from the early stages of the ascending auditory pathway. They are frequently used in clinical audiology. They are also called *brainstem evoked responses* (BERs) or *brainstem auditory evoked responses* (BAERs).

Band-pass filter
A filter that passes an intermediate range of frequencies and attenuates frequencies that are below or above this range. A band-pass filter is equivalent to sequentially applying a low-pass filter and a high-pass filter.

Block of trials
A series of trials, typically followed by a rest break. Also called a *run* or a *trial block*.

Brainstem auditory evoked response (BAER)
See *Auditory brainstem response*.

Brainstem evoked response (BER)
See *Auditory brainstem response.*

CDA
See *Contralateral delay activity.*

Common mode rejection
The ability of a differential amplifier to subtract away the contribution of the common mode voltage (the voltage that is shared between the ground electrode and the other electrodes). It is measured in decibels (dB), and a larger value means that more of the common mode voltage is eliminated.

Component
An underlying neural response that sums together with other neural response to produce the observed waveforms at the scalp electrodes. A given component is often linked with a specific positive or negative peak in the observed waveform, but this link is often tenuous. A component may be defined conceptually as "scalp-recorded neural activity that is generated in a given neuroanatomical module when a specific computational operation is performed" (chapter 2). Similarly, it may be defined as "a subsegment of the ERP whose activity represents a functionally distinct neuronal aggregate" (Donchin, Ritter, & McCallum, 1978, p. 353). It may also be defined operationally as "a set of potential changes that can be shown to be functionally related to an experimental variable or to a combination of experimental variables" (Donchin et al., 1978, p. 353).

Conductance
The ability of a substance to allow current to pass through (the inverse of *resistance*).

Contralateral
On the opposite side. The brain is mainly organized in a contralateral manner, in which one side of the brain receives inputs from the opposite side of space and controls muscles on the opposite side of the body.

Contralateral delay activity (CDA)
A sustained ERP component observed during the delay period of working memory tasks, defined by the difference in voltage between electrodes over the contralateral and ipsilateral hemispheres relative to the objects being maintained in working memory.

Cosine wave
A waveform that reflects the X value of a circle for a given angle of arc on the circle. See *Sine wave* for additional details.

Current
The flow of charges through a conductor.

Decibel (dB)
A logarithmic scale that is most commonly applied to sound waves but is also frequently applied to EEG and ERP waveforms. In the decibel scale, a doubling of power is equal to an increase of 3 dB. If we double the power twice (quadrupling the power), this will be an increase of 6 dB. Because power is amplitude squared, doubling the amplitude is equivalent to quadrupling the power. Therefore, doubling the amplitude is equivalent to a 6 dB increase. A 10-fold increase in power is an increase of 10 dB, and a 10-fold increase in amplitude is an increase of 20 dB.

Dipole
A pair of positive and negative charges, separated by a small distance.

Differential amplifier
An amplifier that amplifies the difference between an active site and a reference site. The active site is initially the electrical potential between the active electrode and the ground electrode, and the reference site is initially the potential between the reference electrode and the ground electrode. The difference between these two electrical potentials (voltages) is amplified, which subtracts away the influence of the ground electrode. This subtraction is not completely perfect, and some small contribution of the ground electrode remains (see *Common mode rejection*).

Differential recording
A recording made by a differential amplifier, in which the ground is subtracted away and the recording represents the electrical potential (voltage) between the active electrodes and a reference electrode. This contrasts with a *single-ended recording*, in which the reference is not subtracted during the recording.

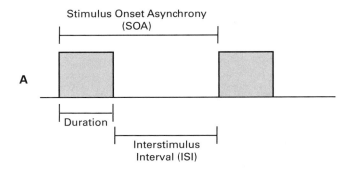

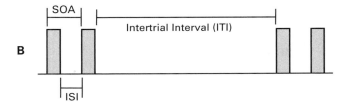

Figure G.1
Terminology for describing the timing of an ERP experiment.

Discrete Fourier transform (DFT)
A version of the Fourier transform that is applied to discretely sampled data (e.g., digitized EEG data and ERPs) rather than to continuous functions.

Downie
Informal (and ironically juvenile) term for a downward-going voltage deflection.

Duration
When describing stimuli, this is the amount of time between the onset of a stimulus and the offset of a stimulus (see figure G.1). When describing ERPs, this is the amount of time between the onset and the offset of a component or effect.

Electrical potential
See *Voltage*.

Electroencephalogram (EEG)
The electrical fields that are generated in the brain and propagate to the surface of the scalp.

Electromyogram (EMG)
Electrical activity generated by a muscle, which can be recorded at a distance by electrodes on the surface of the skin over the muscle.

Event-related magnetic field (ERMF)
The magnetic field that accompanies an event-related potential, as recorded via the magnetoencephalogram. The strength of the ERMF at an given time point is proportional to the strength of the corresponding ERP.

Event-related potential (ERP)
An electrical potential (voltage) that is related to an event (usually a stimulus or a response). This term replaced *evoked potential* for non-obligatory brain potentials in the the late 1960s. The earliest published use of the term "event-related potential" that I could find was by Herb Vaughan, who in a 1969 chapter wrote,

... Since cerebral processes may be related to voluntary movement and to relatively stimulus-independent psychological processes (e.g., Sutton et al., 1967, Ritter et al., 1968), the term "evoked potentials" is no longer sufficiently general to apply to all EEG phenomena related to sensorymotor processes. Moreover, sufficiently prominent or distinctive psychological events may serve as time references for averaging, in addition to stimuli and motor responses. The term "event related potentials" (ERP) is proposed to designate the general class of potentials that display stable time relationships to a definable reference event. (Vaughan, 1969, p. 46)

Related terms include: *evoked potential (EP), evoked response, brainstem evoked response (BER), brainstem auditory evoked response (BAER), auditory brainstem response (ABR), Somatosensory evoked potential (SEP), visual evoked potential (VEP).*

Evoked potential (EP)

An early term, which is still often used for obligatory sensory responses. The idea is that evoked potentials are electrical *potentials* that were *evoked* by stimuli (as opposed to the spontaneous EEG rhythms). This term was replaced in the late 1960s with *event-related potential* for non-obligatory cognitive-related responses. The term *evoked response* is equivalent to *evoked potential*.

Evoked response

See *Evoked potential*.

Evoked response potential (ERP)

This is apparently an accidental miscombination of *evoked response* and *event-related potential* (analogous to combining *irrespective* and *regardless* into *irregardless*). Not recommended.

Excitatory postsynaptic potential (EPSP)

A *postsynaptic potential* in which the net flow of positive charges is inward, bringing the cell closer to the threshold for firing.

Experimentwise error rate

The probability that at least one p value among all the p values in the statistical analyses for a given experiment will be a false positive.

Familywise error rate

The probability that at least one p value among all the p values for a given family of statistical analyses will be a false positive.

Filter

A device or process that removes or attenuates some part of a signal. This term is usually used for devices or processes that attenuate specific frequency bands, but it can be used more broadly. For example, the ADJAR (adjacent response) filter removes the overlapping ERPs from previous and subsequent stimuli.

Fast Fourier transform (FFT)

A computationally efficient method for performing a discrete Fourier transform that can be used when the number of points in the waveform is a power of two (e.g., an ERP waveform consisting of 256 time points).

Fourier transform

A mathematical procedure for decomposing a continuous waveform into the sum of a set of sinusoids of specific frequencies, amplitudes, and phases. The components can also be represented by pairs of sine and cosine waves at each frequency, each with a different amplitude. The phase of the corresponding sinusoid is related to the relative amplitudes of the sine and cosine waves. See also *Discrete Fourier transform* and *Fast Fourier transform*.

Grand average

An average across the single-subject averaged ERP waveforms for all the subjects in an experiment (in within-subject designs) or all the subjects in a group (in between-group designs).

High-pass filter

A filter that passes high frequencies and attenuates low frequencies. Often used to attenuate skin potentials and other slow changes in the EEG.

Impedance
The ability of a substance to keep an alternating current from passing through. A combination of resistance and inductance/capacitance.

Inhibitory postsynaptic potential (IPSP)
A *postsynaptic potential* in which the net flow of positive charges is outward, bringing the cell farther from the threshold for firing.

Interstimulus interval (ISI)
This is the amount of time between the offset of one stimulus and the onset of another stimulus (see figure G.1).

Intertrial interval (ITI)
This is the blank period between the offset of the final stimulus (or the response) in one trial and the onset of the first stimulus in the next trial (see figure G.1). This is equivalent to the *interstimulus interval* in simple experiments with only one stimulus per trial.

Ipsilateral
On the same side. A lesion of the left hemisphere, for example, would be ipsilateral to the left ear and *contralateral* to the right ear.

Lateralized readiness potential (LRP)
An ERP component related to motor preparation that is defined by the difference in voltage between electrodes of the contralateral and ipsilateral electrodes relative to the side of the response.

Line noise
Electrical noise arising from devices powered by the building's main electrical system and reflecting the frequency of this system (either 50 or 60 Hz).

Latency jitter
Trial to trial variations in the timing of one event relative to another event (e.g., variations in the timing of an ERP component relative to a stimulus).

Local field potential (LFP)
An electrical potential recorded from an electrode inside the brain that consists of the summed postsynaptic potentials of nearby neurons.

Low-pass filter
A filter that passes low frequencies and attenuates high frequencies. Often used to avoid aliasing prior to digitization or to attenuate high-frequency noise arising from electronic devices or muscle activity.

LRP
See *Lateralized readiness potential*.

Magnetoencephalogram (MEG)
The magnetic fields that accompany the EEG but are not blurred by the resistance of the skull.

Multi-unit recordings
Recordings of the action potentials from a group of nearby neurons.

Noise
In ERP research, *noise* refers to random variations in the ERP waveform that are unrelated to the brain activity that you are trying to record. For example, electrical activity in the recording chamber that is picked up by the recording electrodes causes variations in the ERP waveform that are not systematically related to the experimental conditions. Alpha oscillations are a form of brain activity that are a source of noise in many experiments, because they cause variations in the ERPs that are unrelated to the experimental manipulations. However, alpha oscillations can also be a signal, the dependent variable that is expected to vary systematically across the experimental conditions (or across subject groups, etc.). It is important to realize that one person's noise may be another person's signal. Any variations in the waveform that are unrelated to the brain activity of interest but are systematic rather than random are not usually considered to be noise, but are instead considered to be confounds.

Notch filter

A filter that attenuates a specific narrow band of frequencies and passes frequencies that are both higher and lower than this band. Often used to attenuate line noise.

Null hypothesis

The hypothesis that two or more conditions are actually equivalent (i.e., the means of the conditions would be exactly the same if we could measure from an infinite number of subjects).

P value

A value returned by a statistical test that we use to decide whether or not to reject the null hypothesis and conclude that there is a real difference between conditions. This is done by comparing the p value with the alpha level (which is typically 0.05). If $p <$ alpha, we conclude that the null hypothesis is false and the alternative hypothesis is true. In theory, this means that we will falsely reject the null hypothesis with a probability of alpha (0.05).

Postsynaptic potential (PSP)

An electrical potential produced in a neuron when a neurotransmitter binds with a receptor, opening or closing ion channels and altering the flow of charged ions into or out of the neuron. Postsynaptic potentials are excitatory (EPSPs) if the net flow of positive charges is inward, bringing the cell closer to the threshold for firing. Postsynaptic potentials are inhibitory (IPSPs) if the net flow of positive charges is outward, bringing the cell farther from the threshold for firing.

Potential

See *Voltage*.

Power

In electrical circuits, power is the ability to get work done, and it is equal to the voltage multiplied by the current. In EEG/ERP research, power is simply the voltage squared (because the current is assumed to be proportional to the voltage). Power can be converted back into voltage by taking the square root, although the sign of the voltage may be incorrect (because, for example $-3^2 = 9$ and sqrt(9) = +3). The term *power* is also used in statistics for a completely different purpose (see *Statistical power*).

Resistance

The ability of a substance to keep a constant current from passing through (the inverse of *conductance*).

Run

See *Block of trials*.

Sine wave

A waveform that reflects the Y value of a circle for a given angle of arc on the circle. This is illustrated in figure G.2. A sine wave has a frequency, which represents how many times the sine wave repeats in a given period of time (e.g., 10 repetitions per sec, or 10 Hz). A sine wave also has an amplitude, which is the height of the peaks. A sine wave also has a phase, which is the horizontal shift in the waveform. A cosine wave is identical to a sine wave, except shifted in phase leftward by 90°. See also *Cosine wave*.

Signal-to-noise ratio (SNR)

A measure of data quality, in which the size of the signal is divided by the size of the noise. For example, if the signal of interest is 10 μV and the noise has a peak-to-peak amplitude of 2 μV, we would say that the SNR is 10:2, which is the same as 5. This means that the signal is 5 times larger than the noise.

Single-ended recording

A recording of the electrical potential (voltage) between an active electrode and a ground electrode, without subtracting the reference electrode. This contrasts with a *differential recording* (see *Differential amplifier*). Some systems (e.g., BioSemi ActiveTwo) initially provide a single-ended recording, and the reference is subtracted offline.

Single-unit recordings

Recordings of a single neuron's action potentials, typically obtained from an electrode just outside the cell body.

SOA

See *Stimulus onset asynchrony* and figure G.1.

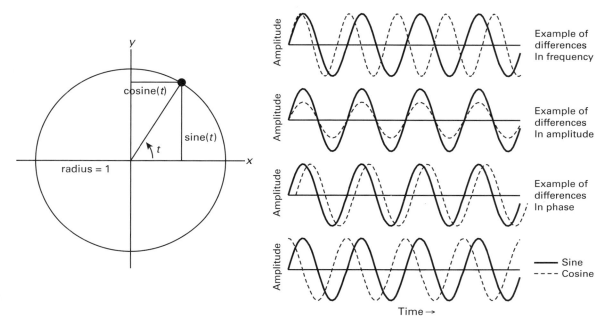

Figure G.2
Illustration of sine and cosine. The left side shows how sine and cosine are the *Y* and *X* values, respectively, of a point on a circle at a given angle (*t*). That is, the sine of angle *t* is the *Y* value of the location of a point at that location on the circle, and the cosine of angle *t* is the *X* value of that point. The right side shows plots of sine and cosine waves (i.e., the value of the sine or cosine function as the angle varies). Sine waves can vary in frequency (number of repetitions per second, which reflects how fast the point is moved around the circle). They can also vary in amplitude (which reflects the radius of the circle). They can also differ in phase (which reflects the degree of rotation of the circle). A cosine wave is simply a sine wave that has been shifted in phase leftward by 90°.

Statistical power
This is the probability that you will obtain a significant *p* value if the null hypothesis is false. If there is a true effect in the infinite population, random variability will sometimes cause the *p* value to be non-significant (a Type II error). The statistical power is a way of determining the likelihood that this will occur. High power means that it is unlikely that we will have a Type II error. Low power means that, even if an effect is present, we are unlikely to achieve statistical significance. Statistical power can be increased by several factors, including increasing the number of subjects in an experiment, increasing the number of trials per subject, etc.

Steady-state auditory evoked response (SSAER)
These are ERPs that are elicited by auditory stimuli that are presented at a constant and relatively fast rate, so that the auditory system begins to oscillate at the stimulus presentation rate (and harmonics of this rate).

Steady-state visual evoked potential (SSVEP)
These are ERPs that are elicited by visual stimuli that are presented at a constant and relatively fast rate, so that the visual system begins to oscillate at the stimulus presentation rate (and harmonics of this rate).

Stimulus duration
See *Duration*.

Stimulus onset asynchrony (SOA)
The interval between the onset of one stimulus and the onset of another stimulus (see figure G.1).

Trial block
See *Block of trials*.

Type I error
This is a false positive in a statistical analysis, in which we reject the null hypothesis (i.e., conclude that the effect is real) when the null hypothesis is actually true (i.e., when there is no true effect). The probability of a Type I error should be the alpha level (e.g., 0.05) if all the assumptions of the statistical test are met. That is, when there is no effect, you will falsely conclude that an effect is present 5% of the time if you use an alpha of 0.05.

Type II error
This is a false negative in a statistical analysis, in which we accept the null hypothesis (i.e., conclude that there is no effect) when the null hypothesis is actually false (i.e., when there is a true effect). Note that accepting the null hypothesis does not ordinarily allow any strong conclusions to be drawn. It is not positive evidence that there is no effect; instead, it just means that we do not have sufficient evidence to reject the null hypothesis. Thus, conventional statistics do not allow you to "prove the null hypothesis."

Uppie
Informal (and ironically juvenile) term for a upward-going voltage deflection.

Visual evoked potential (VEP)
This term is commonly used in clinical contexts to describe largely automatic ERPs elicited by visual stimuli that are used to assess pathology in the visual system, such as demyelination caused by multiple sclerosis. A variant on this term is *visual evoked response* (VER).

Voltage
The potential for current to flow through a conductor. No current will actually flow unless the resistance of the conductor is sufficiently low. If the resistance is low enough, the voltage is essentially the pressure that pushes charges through the conductor (see *Electrical potential*).

References

Adrian, E. D., & Matthews, B. H. C. (1934). The Berger rhythm: Potential changes from the occipital lobes in man. *Brain, 57,* 355–385.

Alcaini, M., Giard, M. H., Thevenet, M., & Pernier, J. (1994). Two separate frontal components in the N1 wave of the human auditory evoked response. *Psychophysiology, 31,* 611–615.

Allison, T., McCarthy, G., Nobre, A., Puce, A., & Belger, A. (1994). Human extrastriate visual cortex and the perception of faces, words, numbers, and colors. *Cerebral Cortex, 4,* 544–554.

American Encephalographic Society. (1994a). Guidelines for standard electrode position nomenclature. *Journal of Clinical Neurophysiology, 11,* 111–113.

American Encephalographic Society. (1994b). Report of the Committee on Infectious Diseases. *Journal of Clinical Neurophysiology, 11,* 128–132.

Anderson, D. E., Vogel, E. K., & Awh, E. (2011). Precision in visual working memory reaches a stable plateau when individual item limits are exceeded. *Journal of Neuroscience, 31,* 1128–1138.

Anderson, D. E., Vogel, E. K., & Awh, E. (2013). A common discrete resource for visual working memory and visual search. *Psychological Science.* Epub ahead of print.

Bach, M. (1998). Electroencephalogram (EEG). In G. K. von Schulthess & J. Hennig (Eds.), *Functional imaging: Principles and methodology* (pp. 391–408). Philadelphia: Lippincott-Raven.

Bastiaansen, M., Mazaheri, A., & Jensen, O. (2012). Beyond ERPs: Oscillatory neuronal dynamics. In S. J. Luck & E. S. Kappenman (Eds.), *The Oxford Handbook of ERP Components* (pp. 31–49). New York: Oxford University Press.

Bentin, S., & Golland, Y. (2002). Meaningful processing of meaningless stimuli: The influence of perceptual experience on early visual processing of faces. *Cognition, 86,* B1–B14.

Bentin, S., Allison, T., Puce, A., Perez, E., & McCarthy, G. (1996). Electrophysiological studies of face perception in humans. *Journal of Cognitive Neuroscience, 8,* 551–565.

Bentin, S., Sagiv, N., Mecklinger, A., Friederici, A., & von Cramon, Y. D. (2002). Priming visual face-processing mechanisms: Electrophysiological evidence. *Psychological Science, 13,* 190–193.

Berg, P., & Scherg, M. (1991a). Dipole modelling of eye activity and its application to the removal of eye artefacts from the EEG and MEG. *Clinical Physics & Physiological Measurement, 12*(Suppl A), 49–54.

Berg, P., & Scherg, M. (1991b). Dipole models of eye movements and blinks. *Electroencephalography and Clinical Neurophysiology, 79,* 36–44.

Berger, H. (1929). Ueber das Elektrenkephalogramm des Menschen. *Archives fur Psychiatrie Nervenkrankheiten, 87,* 527–570.

Bertrand, O., Perrin, F., & Pernier, J. (1991). Evidence for a tonotopic organization of the auditory cortex with auditory evoked potentials. *Acta Oto-Laryngologica, 491,* 116–123.

Brandeis, D., Naylor, H., Halliday, R., Callaway, E., & Yano, L. (1992). Scopolamine effects on visual information processing, attention, and event-related potential map latencies. *Psychophysiology, 29,* 315–336.

Brazdil, M., Roman, R., Falkenstein, M., Daniel, P., Jurak, P., & Rektor, I. (2002). Error processing—evidence from intracerebral ERP recordings. *Experimental Brain Research, 146,* 460–466.

Broadbent, D. E. (1958). *Perception and Communication*. New York: Pergamon.

Bruin, K. J., & Wijers, A. A. (2002). Inhibition, response mode, and stimulus probability: A comparative event-related potential study. *Clinical Neurophysiology*, *113*, 1172–1182.

Brunia, C. H. M., van Boxtel, G. J. M., & Böcker, K. B. E. (2012). Negative slow waves as indices of anticipation: The Bereitschaftspotential, the contingent negative variation, and the stimulus preceding negativity. In S. J. Luck & E. S. Kappenman (Eds.), *The Oxford Handbook of Event-Related Potential Components* (pp. 189–207). New York: Oxford University Press.

Busey, T. A., & Vanderkolk, J. R. (2005). Behavioral and electrophysiological evidence for configural processing in fingerprint experts. *Vision Research*, *45*, 431–448.

Button, K. S., Ioannidis, J. P., Mokrysz, C., Nosek, B. A., Flint, J., Robinson, E. S., et al. (2013). Power failure: Why small sample size undermines the reliability of neuroscience. *Nature Reviews. Neuroscience*, *14*, 365–376.

Buzsáki, G., Anastassiou, C. A., & Koch, C. (2012). The origin of extracellular fields and currents—EEG, ECoG, LFP and spikes. *Nature Reviews. Neuroscience*, *13*, 407–420.

Carmel, D., & Bentin, S. (2002). Domain specificity versus expertise: Factors influencing distinct processing of faces. *Cognition*, *83*, 1–29.

Cheour, M., Leppanen, P. H., & Kraus, N. (2000). Mismatch negativity (MMN) as a tool for investigating auditory discrimination and sensory memory in infants and children. *Clinical Neurophysiology*, *111*, 4–16.

Chun, M. M., & Potter, M. C. (1995). A two-stage model for multiple target detection in rapid serial visual presentation. *Journal of Experimental Psychology. Human Perception and Performance*, *21*, 109–127.

Clark, V. P., Fan, S., & Hillyard, S. A. (1995). Identification of early visually evoked potential generators by retinotopic and topographic analyses. *Human Brain Mapping*, *2*, 170–187.

Clayson, P. E., Baldwin, S. A., & Larson, M. J. (2013). How does noise affect amplitude and latency measurement of event-related potentials (ERPs)? A methodological critique and simulation study. *Psychophysiology, 50*, 174–186.

Coch, D., & Gullick, M. (2012). Event-related potentials and development. In S. J. Luck & E. S. Kappenman (Eds.), *The Oxford Handbook of Event-Related Potential Components* (pp. 475–511). New York: Oxford University Press.

Cohen, M. X. (2014). *Analyzing Neural Time Series Data: Theory and Practice*. Cambridge, MA: MIT Press.

Coles, M. G. H. (1989). Modern mind-brain reading: Psychophysiology, physiology and cognition. *Psychophysiology*, *26*, 251–269.

Courchesne, E., Hillyard, S. A., & Galambos, R. (1975). Stimulus novelty, task relevance and the visual evoked potential in man. *Electroencephalography and Clinical Neurophysiology*, *39*, 131–142.

Csepe, V. (1995). On the origin and development of the mismatch negativity. *Ear and Hearing*, *16*, 91–104.

Curran, T. (2000). Brain potentials of recollection and familiarity. *Memory & Cognition*, *28*, 923–938.

Cuthbert, B. N., Schupp, H. T., Bradley, M. M., Birbaumer, N., & Lang, P. J. (2000). Brain potentials in affective picture processing: Covariation with autonomic arousal and affective report. *Biological Psychology*, *52*, 95–111.

Czigler, I., Balazs, L., & Winkler, I. (2002). Memory-based detection of task-irrelevant visual changes. *Psychophysiology*, *39*, 869–873.

Dale, A. M., & Sereno, M. I. (1993). Improved localization of cortical activity by combining EEG and MEG with MRI cortical surface reconstruction: A linear approach. *Journal of Cognitive Neuroscience*, *5*, 162–176.

Davis, P. A. (1939). Effects of acoustic stimuli on the waking human brain. *Journal of Neurophysiology*, *2*, 494–499.

Davis, H., Davis, P. A., Loomis, A. L., Harvey, E. N., & Hobart, G. (1939). Electrical reactions of the human brain to auditory stimulation during sleep. *Journal of Neurophysiology*, *2*, 500–514.

Dawson, G., Carver, L., Meltzoff, A. N., Panagiotides, H., McPartland, J., & Webb, S. J. (2002). Neural correlates of face and object recognition in young children with autism spectrum disorder, developmental delay, and typical development. *Child Development*, *73*, 700–717.

Dehaene, S., Naccache, L., Le Clec'H, G., Koechlin, E., Mueller, M., Dehaene-Lambertz, G., et al. (1998). Imaging unconscious semantic priming. *Nature*, *395*, 597–600.

Dehaene, S., Posner, M. I., & Tucker, D. M. (1994). Localization of a neural system for error detection and compensation. *Psychological Science*, *5*, 303–305.

Dehaene-Lambertz, G., & Baillet, S. (1998). A phonological representation in the infant brain. *Neuroreport, 9,* 1885–1888.

Delorme, A., & Makeig, S. (2004). EEGLAB: An open source toolbox for analysis of single-trial EEG dynamics including independent component analysis. *Journal of Neuroscience Methods, 134,* 9–21.

Delorme, A., Palmer, J., Onton, J., Makeig, S., & Oostenveld, R. (2012). Independent EEG sources are dipolar. *PLoS ONE, 7,* e30135.

Deutsch, J. A., & Deutsch, D. (1963). Attention: Some theoretical considerations. *Psychological Review, 70,* 80–90.

Dien, J. (1998). Issues in the application of the average reference: Review, critiques, and recommendations. *Behavior Research Methods, Instruments, & Computers, 30,* 34–43.

Di Russo, F., Martinez, A., Sereno, M. I., Pitzalis, S., & Hillyard, S. A. (2002). Cortical sources of the early components of the visual evoked potential. *Human Brain Mapping, 15,* 95–111.

Di Russo, F., Teder-Sälejärvi, W. A., & Hillyard, S. A. (2003). Steady-state VEP and attentional visual processing. In A. Zani & A. M. Proverbio (Eds.), *The Cognitive Electrophysiology of Mind and Brain* (pp. 259–274). San Diego: Academic Press.

Donchin, E. (1981). Surprise!. . .Surprise? *Psychophysiology, 18,* 493–513.

Donchin, E., & Coles, M. G. H. (1988). Is the P300 component a manifestation of context updating? *Behavioral and Brain Sciences, 11,* 357–374.

Donchin, E., & Heffley, E. F., III. (1978). Multivariate analysis of event-related potential data: A tutorial review. In D. Otto (Ed.), *Multidisciplinary Perspectives in Event-Related Brain Potential Research* (pp. 555–572). Washington, DC: U.S. Government Printing Office.

Donchin, E., Ritter, W., & McCallum, W. C. (1978). Cognitive psychophysiology: The endogenous components of the ERP. In E. Callaway, P. Tueting, & S. H. Koslow (Eds.), *Event-Related Brain Potentials in Man* (pp. 349–441). New York: Academic Press.

Drew, T., McCollough, A. W., Horowitz, T. S., & Vogel, E. K. (2009). Attentional enhancement during multiple-object tracking. *Psychonomic Bulletin & Review, 16,* 411–417.

Duncan-Johnson, C. C., & Donchin, E. (1977). On quantifying surprise: The variation of event-related potentials with subjective probability. *Psychophysiology, 14,* 456–467.

Duncan-Johnson, C. C., & Kopell, B. S. (1981). The Stroop effect: Brain potentials localize the source of interference. *Science, 214,* 938–940.

Editorial. (2013). Making methods clearer. *Nature Neuroscience, 16,* 1.

Ehrlichman, R. S., Maxwell, C. R., Majumdar, S., & Siegel, S. J. (2008). Deviance-elicited changes in event-related potentials are attenuated by ketamine in mice. *Journal of Cognitive Neuroscience, 20,* 1403–1414.

Eimer, M., & Kiss, M. (2008). Involuntary attentional capture is determined by task set: Evidence from event-related brain potentials. *Journal of Cognitive Neuroscience, 208,* 1423–1433.

Endrass, T., Reuter, B., & Kathmann, N. (2007). ERP correlates of conscious error recognition: Aware and unaware errors in an antisaccade task. *European Journal of Neuroscience, 26,* 1714–1720.

Eriksen, C. W., & Schultz, D. W. (1979). Information processing in visual search: A continuous flow conception and experimental results. *Perception & Psychophysics, 25,* 249–263.

Ester, E. F., Drew, T., Klee, D., Vogel, E. K., & Awh, E. (2012). Neural measures reveal a fixed item limit in subitizing. *Journal of Neuroscience, 32,* 7169–7177.

Falkenstein, M., Hohnsbein, J., Joormann, J., & Blanke, L. (1990). Effects of errors in choice reaction tasks on the ERP under focused and divided attention. In C. H. M. Brunia, A. W. K. Gaillard, & A. Kok (Eds.), *Psychophysiological Brain Research* (pp. 192–195). Amsterdam: Elsevier.

Fischer, C., Luaute, J., Adeleine, P., & Morlet, D. (2004). Predictive value of sensory and cognitive evoked potentials for awakening from coma. *Neurology, 63,* 669–673.

Folstein, J. R., & Van Petten, C. (2008). Influence of cognitive control and mismatch on the N2 component of the ERP: A review. *Psychophysiology, 45,* 152–170.

Ford, J. M., & Hillyard, S. A. (1981). ERPs to interruptions of a steady rhythm. *Psychophysiology, 18,* 322–330.

Ford, J. M., White, P., Lim, K. O., & Pfefferbaum, A. (1994). Schizophrenics have fewer and smaller P300s: A single-trial analysis. *Biological Psychiatry, 35,* 96–103.

Foxe, J. J., & Simpson, G. V. (2002). Flow of activation from V1 to frontal cortex in humans: A framework for defining "early" visual processing. *Experimental Brain Research, 142,* 139–150.

Friederici, A. D., Hahne, A., & Saddy, D. (2002). Distinct neurophysiological patterns reflecting aspects of syntactic complexity and syntactic repair. *Journal of Psycholinguistic Research, 31,* 45–63.

Friston, K. J., Rotshtein, P., Geng, J. J., Sterzer, P., & Henson, R. N. (2006). A critique of functional localisers. *Neuro-Image, 30,* 1077–1087.

Fuchs, M., Wagner, M., & Kastner, J. (2004). Confidence limits of dipole source reconstruction results. *Clinical Neurophysiology, 115,* 1442–1451.

Galambos, R. G. (1996). Robert Galambos. In L. R. Squire (Ed.), *The History of Neuroscience in Autobiography* (Vol. 1, pp. 178–221). Washington, DC: Society for Neuroscience.

Galambos, R., & Sheatz, G. C. (1962). An electroencephalographic study of classical conditioning. *American Journal of Physiology, 203,* 173–184.

Gamble, M. L., & Luck, S. J. (2011). N2ac: An ERP component associated with the focusing of attention within an auditory scene. *Psychophysiology, 48,* 1057–1068.

Ganis, G., Kutas, M., & Sereno, M. I. (1996). The search for "common sense": An electrophysiological study of the comprehension of words and pictures in reading. *Journal of Cognitive Neuroscience, 8,* 89–106.

Gehring, W. J., Coles, M. G. H., Meyer, D. E., & Donchin, E. (1995). A brain potential manifestation of error-related processing. In G. Karmos, M. Molnar, V. Csepe, I. Czigler, & J. E. Desmedt (Eds.), *Perspectives of Event-Related Potentials Research* (pp. 287–296). Amsterdam: Elsevier.

Gehring, W. J., Goss, B., Coles, M. G. H., Meyer, D. E., & Donchin, E. (1993). A neural system for error-detection and compensation. *Psychological Science, 4,* 385–390.

Gehring, W. J., Gratton, G., Coles, M., & Donchin, E. (1992). Probability effects on stimulus evaluation and response processes. *Journal of Experimental Psychology. Human Perception and Performance, 18,* 198–216.

Gehring, W. J., Himle, J., & Nisenson, L. G. (2000). Action monitoring dysfunction in obsessive-compulsive disorder. *Psychological Science, 11,* 1–6.

Gehring, W. J., Liu, Y., Orr, J. M., & Carp, J. (2012). The error-related negativity (ERN/Ne). In S. J. Luck & E. S. Kappenman (Eds.), *The Oxford Handbook of Event-Related Potential Components* (pp. 231–292). New York: Oxford University Press.

Gibbs, F. A., Davis, H., & Lennox, W. G. (1935). The electro-encephalogram in epilepsy and in conditions of impaired consciousness. *Archives of Neurology and Psychiatry, 34,* 1133–1148.

Gratton, G., Coles, M. G. H., & Donchin, E. (1983). A new method for off-line removal of ocular artifacts. *Electroencephalography and Clinical Neurophysiology, 55,* 468–484.

Gratton, G., Coles, M. G. H., Sirevaag, E. J., Eriksen, C. W., & Donchin, E. (1988). Pre- and post-stimulus activation of response channels: A psychophysiological analysis. *Journal of Experimental Psychology. Human Perception and Performance, 14,* 331–344.

Gray, C. M., König, P., Engel, A. K., & Singer, W. (1989). Oscillatory responses in cat visual cortex exhibit inter-columnar synchronization which reflects global stimulus properties. *Nature, 338,* 334–337.

Groppe, D. M., Makeig, S., & Kutas, M. (2009). Identifying reliable independent components via split-half comparisons. *NeuroImage, 45,* 1199–1211.

Guzman-Martinez, E., Leung, P., Franconeri, S., Grabowecky, M., & Suzuki, S. (2009). Rapid eye-fixation training without eyetracking. *Psychonomic Bulletin & Review, 16,* 491–496.

Hagoort, P. (2007). The memory, unification, and control (MUC) model of language. In A. S. Meyer, L. Wheeoldon, & A. Krott (Eds.), *Automaticity and Control in Language Processing* (pp. 243–270). Hove: Psychology Press.

Hagoort, P., Brown, C. M., & Swaab, T. Y. (1996). Lexical-semantic event-related potential effects in patients with left hemisphere lesions and aphasia, and patients with right hemisphere lesions without aphasia. *Brain, 119,* 627–649.

Hajcak, G., & Olvet, D. M. (2008). The persistence of attention to emotion: Brain potentials during and after picture presentation. *Emotion (Washington, D.C.), 8,* 250–255.

Hajcak, G., Holroyd, C. B., Moser, J. S., & Simons, R. F. (2005). Brain potentials associated with expected and unexpected good and bad outcomes. *Psychophysiology*, *42*, 161–170.

Hajcak, G., Wienberg, A., MacNamara, A., & Foti, D. (2012). ERPs and the study of emotion. In S. J. Luck & E. S. Kappenman (Eds.), *The Oxford Handbook of Event-Related Potential Components* (pp. 441–472). New York: Oxford University Press.

Halgren, E., Boujon, C., Clarke, J., Wang, C., & Chauvel, P. (2002). Rapid distributed fronto-parieto-occipital processing stages during working memory in humans. *Cerebral Cortex*, *12*, 710–728.

Hämäläinen, M. S., Hari, R., Ilmonieni, R. J., Knuutila, J., & Lounasmaa, O. V. (1993). Magnetoencephalography—theory, instrumentation, and applications to noninvasive studies of the working human brain. *Reviews of Modern Physics*, *65*, 413–497.

Hansen, J. C., & Hillyard, S. A. (1980). Endogenous brain potentials associated with selective auditory attention. *Electroencephalography and Clinical Neurophysiology*, *49*, 277–290.

Harrison, S. A., & Tong, F. (2009). Decoding reveals the contents of visual working memory in early visual areas. *Nature*, *458*, 632–635.

Heekeren, K., Daumann, J., Neukirch, A., Stock, C., Kawohl, W., Norra, C., et al. (2008). Mismatch negativity generation in the human 5HT2A agonist and NMDA antagonist model of psychosis. *Psychopharmacology*, *199*, 77–88.

Heinze, H. J., Luck, S. J., Mangun, G. R., & Hillyard, S. A. (1990). Visual event-related potentials index focused attention within bilateral stimulus arrays. I. Evidence for early selection. *Electroencephalography and Clinical Neurophysiology*, *75*, 511–527.

Helmholtz, H. (1853). Ueber einige Gesetze der Vertheilung elektrischer Ströme in körperlichen Leitern mit Anwendung auf die thierisch-elektrischen Versuche [On laws of the distribution of electric currents in bodily conductors with application to electrical experiments in animals]. *Annalen der Physik und Chemie, 89*, 211–233, 354–377.

Hesselbrock, V., Begleiter, H., Porjesz, B., O'Connor, S., & Bauer, L. (2001). P300 event-related potential amplitude as an endophenotype of alcoholism—evidence from the collaborative study on the genetics of alcoholism. *Journal of Biomedical Science*, *8*, 77–82.

Hickey, C., Di Lollo, V., & McDonald, J. J. (2009). Electrophysiological indices of target and distractor processing in visual search. *Journal of Cognitive Neuroscience*, *21*, 760–775.

Hillyard, S. A., & Galambos, R. (1970). Eye movement artifact in the CNV. *Electroencephalography and Clinical Neurophysiology*, *28*, 173–182.

Hillyard, S. A., & Münte, T. F. (1984). Selective attention to color and location: An analysis with event-related brain potentials. *Perception & Psychophysics*, *36*, 185–198.

Hillyard, S. A., Hink, R. F., Schwent, V. L., & Picton, T. W. (1973). Electrical signs of selective attention in the human brain. *Science*, *182*, 177–179.

Hillyard, S. A., Vogel, E. K., & Luck, S. J. (1998). Sensory gain control (amplification) as a mechanism of selective attention: Electrophysiological and neuroimaging evidence. *Philosophical Transactions of the Royal Society: Biological Sciences*, *353*, 1257–1270.

Holcomb, P. J., & McPherson, W. B. (1994). Event-related brain potentials reflect semantic priming in an object decision task. *Brain and Cognition*, *24*, 259–276.

Holroyd, C. B., & Coles, M. G. H. (2002). The neural basis of human error processing: Reinforcement learning, dopamine, and the error-related negativity. *Psychological Review*, *109*, 679–709.

Hopf, J.-M., Luck, S. J., Boelmans, K., Schoenfeld, M. A., Boehler, N., Rieger, J., et al. (2006). The neural site of attention matches the spatial scale of perception. *Journal of Neuroscience*, *26*, 3532–3540.

Hopf, J.-M., Luck, S. J., Girelli, M., Hagner, T., Mangun, G. R., Scheich, H., et al. (2000). Neural sources of focused attention in visual search. *Cerebral Cortex*, *10*, 1233–1241.

Hopf, J.-M., Vogel, E. K., Woodman, G. F., Heinze, H.-J., & Luck, S. J. (2002). Localizing visual discrimination processes in time and space. *Journal of Neurophysiology*, *88*, 2088–2095.

Ikui, A. (2002). A review of objective measures of gustatory function. *Acta Oto-Laryngologica*, *546*(Suppl), 60–68.

Ille, N., Berg, P., & Scherg, M. (2002). Artifact correction of the ongoing EEG using spatial filters based on artifact and brain signal topographies. *Journal of Clinical Neurophysiology*, *19*, 113–124.

Isreal, J. B., Chesney, G. L., Wickens, C. D., & Donchin, E. (1980). P300 and tracking difficulty: Evidence for multiple resources in dual-task performance. *Psychophysiology, 17*, 259–273.

Jasper, H. H. (1958). The ten-twenty electrode system of the International Federation. *Electroencephalography and Clinical Neurophysiology, 10*, 371–375.

Jasper, H. H., & Carmichael, L. (1935). Electrical potentials from the intact human brain. *Science, 81*, 51–53.

Javitt, D. C., Steinschneider, M., Schroeder, C. E., & Arezzo, J. C. (1996). Role of cortical N-methyl-D-aspartate receptors in auditory sensory memory and mismatch negativity generation: Implications for schizophrenia. *Proceedings of the National Academy of Sciences of the United States of America, 93*, 11962–11967.

Jeffreys, D. A. (1989). A face-responsive potential recorded from the human scalp. *Experimental Brain Research, 78*, 193–202.

Jeffreys, D. A., & Axford, J. G. (1972). Source locations of pattern-specific components of human visual evoked potentials. I: Components of striate cortical origin. *Experimental Brain Research, 16*, 1–21.

Jennings, J. R., & Wood, C. C. (1976). The e-adjustment procedure for repeated-measures analyses of variance. *Psychophysiology, 13*, 277–278.

Jeon, Y. W., & Polich, J. (2003). Meta-analysis of P300 and schizophrenia: Patients, paradigms, and practical implications. *Psychophysiology, 40*, 684–701.

John, L. K., Loewenstein, G., & Prelec, D. (2012). Measuring the prevalence of questionable research practices with incentives for truth telling. *Psychological Science, 23*, 524–532.

Johns, M., Crowley, K., Chapman, R., Tucker, A., & Hocking, C. (2009). The effect of blinks and saccadic eye movements on visual reaction times. *Attention, Perception & Psychophysics, 71*, 783–788.

Johnson, R., Jr. (1984). P300: A model of the variables controlling its amplitude. *Annals of the New York Academy of Sciences, 425*, 223–229.

Johnson, R., Jr. (1986). A triarchic model of P300 amplitude. *Psychophysiology, 23*, 367–384.

Jung, T. P., Makeig, S., Humphries, C., Lee, T. W., McKeown, M. J., Iragui, V., et al. (2000). Removing electroencephalographic artifacts by blind source separation. *Psychophysiology, 37*, 163–178.

Kappenman, E. S., & Luck, S. J. (2010). The effects of electrode impedance on data quality and statistical significance in ERP recordings. *Psychophysiology, 47*, 888–904.

Kappenman, E. S., & Luck, S. J. (2011). Manipulation of orthogonal neural systems together in electrophysiological recordings: The MONSTER approach to efficient neurocognitive assessment. *Schizophrenia Bulletin, 38*, 92–102.

Kappenman, E. S., & Luck, S. J. (2012). ERP components: The ups and downs of brainwave recordings. In S. J. Luck & E. S. Kappenman (Eds.), *The Oxford Handbook of ERP Components* (pp. 3–30). New York: Oxford University Press.

Katayama, J., & Polich, J. (1996). P300 from one-, two-, and three-stimulus auditory paradigms. *International Journal of Psychophysiology, 23*, 33–40.

Kayser, J., Tenke, C. E., & Bruder, G. E. (2003). Evaluating the quality of ERP measures across recording systems: A commentary on Debener et al. (2002). *International Journal of Psychophysiology, 48*, 315.

Keil, A., Bradley, M. M., Hauk, O., Rockstroh, B., Elbert, T., & Lang, P. J. (2002). Large-scale neural correlates of affective picture processing. *Psychophysiology, 39*, 641–649.

Keil, A., Debener, S., Gratton, G., Junghöfer, M., Kappenman, E. S., Luck, S. J., et al. (in press). Publication guidelines and recommendations for studies using electroencephalography and magnetoencephalography. *Psychophysiology*.

Kenemans, J. L., Jong, T. G., & Verbaten, M. N. (2003). Detection of visual change: Mismatch or rareness? *Neuroreport, 14*, 1239–1242.

Keppel, G. (1982). *Design and Analysis*. Englewood Cliffs, NJ: Prentice Hall.

Kiesel, A., Miller, J., Jolicoeur, P., & Brisson, B. (2008). Measurement of ERP latency differences: A comparison of single-participant and jackknife-based scoring methods. *Psychophysiology, 45*, 250–274.

Kiss, M., Driver, J., & Eimer, M. (2009). Reward priority of visual target singletons modulates event-related potential signatures of attentional selection. *Psychological Science, 20*, 245–251.

Klimesch, W., Sauseng, P., & Hanslmayr, S. (2007). EEG alpha oscillations: The inhibition-timing hypothesis. *Brain Research. Brain Research Reviews, 53*, 63–88.

Knoblich, U., Siegle, J. H., Pritchett, D. L., & Moore, C. I. (2010). What do we gain from gamma? Local dynamic gain modulation drives enhanced efficacy and efficiency of signal transmission. *Frontiers in Human Neuroscience, 4,* 185.

Kornhuber, H. H., & Deecke, L. (1965). Hirnpotentialanderungen bei Wilkurbewegungen und passiven Bewegungen des Menschen: Bereitschaftspotential und reafferente potentials. *Pflugers Archiv, 284,* 1–17.

Kramer, A. F. (1985). The interpretation of the component structure of event-related brain potentials: An analysis of expert judgments. *Psychophysiology, 22,* 334–344.

Kreitschmann-Andermahr, I., Rosburg, T., Demme, U., Gaser, E., Nowak, H., & Sauer, H. (2001). Effect of ketamine on the neuromagnetic mismatch field in healthy humans. *Brain Research. Cognitive Brain Research, 12,* 109–116.

Kuefner, D., de Heering, A., Jacques, C., Palmero-Soler, E., & Rossion, B. (2010). Early visually evoked electrophysiological responses over the human brain (P1, N170) show stable patterns of face-sensitivity from 4 years to adulthood. *Frontiers in Human Neuroscience, 3,* 67.

Kutas, M., & Hillyard, S. A. (1980). Reading senseless sentences: Brain potentials reflect semantic incongruity. *Science, 207,* 203–205.

Kutas, M., Hillyard, S. A., & Gazzaniga, M. S. (1988). Processing of semantic anomaly by right and left hemispheres of commissurotomy patients. *Brain, 111,* 553–576.

Kutas, M., McCarthy, G., & Donchin, E. (1977). Augmenting mental chronometry: The P300 as a measure of stimulus evaluation time. *Science, 197,* 792–795.

Kutas, M., van Petter, C. K., & Kluender, R. (2006). Psycholinguistics electrified II (1994–2005). In M. J. Traxler & M. A. Gernsbacher (Eds.), *Handbook of Psycholinguistics* (2nd ed., pp. 83–143). New York: Elsevier.

Leonard, C. J., Kaiser, S. T., Robinson, B. M., Kappenman, E. S., Hahn, B., Gold, J. M., et al. (2012). Toward the neural mechanisms of reduced working memory capacity in schizophrenia. *Cerebral Cortex, 23,* 1582–1592.

Leonard, C. J., Lopez-Calderon, J., Kreither, J., & Luck, S. J. (2013). Rapid feature-driven changes in the attentional window. *Journal of Cognitive Neuroscience, 25,* 1100–1110.

Leuthold, H., & Sommer, W. (1998). Postperceptual effects and P300 latency. *Psychophysiology, 35,* 34–46.

Lien, M. C., Ruthruff, E., Goodin, Z., & Remington, R. W. (2008). Contingent attentional capture by top-down control settings: Converging evidence from event-related potentials. *Journal of Experimental Psychology. Human Perception and Performance, 34,* 509–530.

Lins, O. G., Picton, T. W., Berg, P., & Scherg, M. (1993a). Ocular artifacts in EEG and event-related potentials I: Scalp topography. *Brain Topography, 6,* 51–63.

Lins, O. G., Picton, T. W., Berg, P., & Scherg, M. (1993b). Ocular artifacts in recording EEGs and event-related potentials. II: Source dipoles and source components. *Brain Topography, 6,* 65–78.

Liotti, M., Woldorff, M. G., Perez, R., & Mayberg, H. S. (2000). An ERP study of the temporal course of the Stroop color-word interference effect. *Neuropsychologia, 38,* 701–711.

Lorenzo-Lopez, L., Amenedo, E., & Cadaveira, F. (2008). Feature processing during visual search in normal aging: Electrophysiological evidence. *Neurobiology of Aging, 29,* 1101–1110.

Loveless, N. E., & Sanford, A. J. (1975). The impact of warning signal intensity on reaction time and components of the contingent negative variation. *Biological Psychology, 2,* 217–226.

Luck, S. J. (1998a). Neurophysiology of selective attention. In H. Pashler (Ed.), *Attention* (pp. 257–295). East Sussex: Psychology Press.

Luck, S. J. (1998b). Sources of dual-task interference: Evidence from human electrophysiology. *Psychological Science, 9,* 223–227.

Luck, S. J. (1999). Direct and indirect integration of event-related potentials, functional magnetic resonance images, and single-unit recordings. *Human Brain Mapping, 8,* 115–120.

Luck, S. J. (2012a). Event-related potentials. In H. Cooper, P.M. Camic, D. L. Long, A. T. Panter, D. Rindskopf & K. J. Sher (Eds.), *APA Handbook of Research Methods in Psychology: Volume 1, Foundations, Planning, Measures, and Psychometrics.* Washington, DC: American Psychological Association.

Luck, S. J. (2012b). Electrophysiological correlates of the focusing of attention within complex visual scenes: N2pc and related ERP components. In S. J. Luck & E. S. Kappenman (Eds.), *The Oxford Handbook of ERP Components* (pp. 329–360). New York: Oxford University Press.

Luck, S. J., & Ford, M. A. (1998). On the role of selective attention in visual perception. *Proceedings of the National Academy of Sciences of the United States of America, 95*, 825–830.

Luck, S. J., & Girelli, M. (1998). Electrophysiological approaches to the study of selective attention in the human brain. In R. Parasuraman (Ed.), *The Attentive Brain* (pp. 71–94). Cambridge, MA: MIT Press.

Luck, S. J., & Hillyard, S. A. (1990). Electrophysiological evidence for parallel and serial processing during visual search. *Perception & Psychophysics, 48*, 603–617.

Luck, S. J., & Hillyard, S. A. (1994a). Electrophysiological correlates of feature analysis during visual search. *Psychophysiology, 31*, 291–308.

Luck, S. J., & Hillyard, S. A. (1994b). Spatial filtering during visual search: Evidence from human electrophysiology. *Journal of Experimental Psychology. Human Perception and Performance, 20*, 1000–1014.

Luck, S. J., & Hillyard, S. A. (1995). The role of attention in feature detection and conjunction discrimination: An electrophysiological analysis. *International Journal of Neuroscience, 80*, 281–297.

Luck, S. J., & Hillyard, S. A. (2000). The operation of selective attention at multiple stages of processing: Evidence from human and monkey electrophysiology. In M. S. Gazzaniga (Ed.), *The New Cognitive Neurosciences* (pp. 687–700). Cambridge, MA: MIT Press.

Luck, S. J., & Kappenman, E. S. (2012a). ERP components and selective attention. In S. J. Luck & E. S. Kappenman (Eds.), *The Oxford Handbook of ERP Components* (pp. 295–327). New York: Oxford University Press.

Luck, S. J., & Kappenman, E. S. (Eds.). (2012b). *The Oxford Handbook of Event-Related Potential Components*. New York: Oxford University Press.

Luck, S. J., & Vecera, S. P. (2002). Attention. In S. Yantis (Ed.), *Stevens' Handbook of Experimental Psychology: Vol. 1: Sensation and Perception* (3rd ed., pp. 235–286). New York: Wiley.

Luck, S. J., Fan, S., & Hillyard, S. A. (1993). Attention-related modulation of sensory-evoked brain activity in a visual search task. *Journal of Cognitive Neuroscience, 5*, 188–195.

Luck, S. J., Fuller, R. L., Braun, E. L., Robinson, B., Summerfelt, A., & Gold, J. M. (2006). The speed of visual attention in schizophrenia: Electrophysiological and behavioral evidence. *Schizophrenia Research, 85*, 174–195.

Luck, S. J., Girelli, M., McDermott, M. T., & Ford, M. A. (1997). Bridging the gap between monkey neurophysiology and human perception: An ambiguity resolution theory of visual selective attention. *Cognitive Psychology, 33*, 64–87.

Luck, S. J., Heinze, H. J., Mangun, G. R., & Hillyard, S. A. (1990). Visual event-related potentials index focused attention within bilateral stimulus arrays. II. Functional dissociation of P1 and N1 components. *Electroencephalography and Clinical Neurophysiology, 75*, 528–542.

Luck, S. J., Kappenman, E. S., Fuller, R. L., Robinson, B., Summerfelt, A., & Gold, J. M. (2009). Impaired response selection in schizophrenia: Evidence from the P3 wave and the lateralized readiness potential. *Psychophysiology, 46*, 776–786.

Luck, S. J., Mathalon, D. H., O'Donnell, B. F., Spencer, K. M., Javitt, D. C., Ulhaus, P. F., et al. (2011). A roadmap for the development and validation of ERP biomarkers in schizophrenia research. *Biological Psychiatry, 70*, 28–34.

Luck, S. J., Vogel, E. K., & Shapiro, K. L. (1996). Word meanings can be accessed but not reported during the attentional blink. *Nature, 383*, 616–618.

Luck, S. J., Woodman, G. F., & Vogel, E. K. (2000). Event-related potential studies of attention. *Trends in Cognitive Sciences, 4*, 432–440.

Luu, P., & Tucker, D. M. (2001). Regulating action: Alternating activation of midline frontal and motor cortical networks. *Clinical Neurophysiology, 112*, 1295–1306.

Magliero, A., Bashore, T. R., Coles, M. G. H., & Donchin, E. (1984). On the dependence of P300 latency on stimulus evaluation processes. *Psychophysiology, 21*, 171–186.

Makeig, S. (1993). Auditory event-related dynamics of the EEG spectrum and effects of exposure to tones. *Electroencephalography and Clinical Neurophysiology, 86*, 283–293.

Makeig, S., & Onton, J. (2012). ERP features and EEG dynamics: An ICA perspective. In S. J. Luck & E. S. Kappenman (Eds.), *The Oxford Handbook of ERP Components* (pp. 51–86). New York: Oxford University Press.

Makeig, S., Westerfield, M., Jung, T.-P., Enghoff, S., Townsend, J., Courchesne, E., et al. (2002). Dynamic brain sources of visual evoked responses. *Science, 295*, 690–694.

Mangun, G. R. (1995). Neural mechanisms of visual selective attention. *Psychophysiology, 32*, 4–18.

Marco-Pallares, J., Cucurell, D., Muente, T. F., Strien, N., & Rodriguez-Fornells, A. (2011). On the number of trials needed for a stable feedback-related negativity. *Psychophysiology*, *48*, 852–860.

Mathalon, D. H., Ford, J. M., & Pfefferbaum, A. (2000). Trait and state aspects of P300 amplitude reduction in schizophrenia: A retrospective longitudinal study. *Biological Psychiatry*, *47*, 434–449.

Mathewson, K. E., Gratton, G., Fabiani, M., Beck, D. M., & Ro, T. (2009). To see or not to see: Prestimulus alpha phase predicts visual awareness. *Journal of Neuroscience*, *29*, 2725–2732.

Mazaheri, A., & Jensen, O. (2008). Asymmetric amplitude modulations of brain oscillations generate slow evoked responses. *Journal of Neuroscience*, *28*, 7781–7787.

McCarthy, G., & Wood, C. C. (1985). Scalp distributions of event-related potentials: An ambiguity associated with analysis of variance models. *Electroencephalography and Clinical Neurophysiology*, *62*, 203–208.

McCarthy, G., Nobre, A. C., Bentin, S., & Spencer, D. D. (1995). Language-related field potentials in the anterior-medial temporal lobe: I. Intracranial distribution and neural generators. *Journal of Neuroscience*, *15*, 1080–1089.

McClelland, J. L. (1979). On the time relations of mental processes: An examination of systems of processes in cascade. *Psychological Review*, *86*, 287–330.

McMenamin, B. W., Shackman, A. J., Maxwell, J. S., Bachhuber, D. R., Koppenhaver, A. M., Greischar, L. L., et al. (2010). Validation of ICA-based myogenic artifact correction for scalp and source-localized EEG. *NeuroImage*, *49*, 2416–2432.

McMenamin, B. W., Shackman, A. J., Maxwell, J. S., Greischar, L. L., & Davidson, R. J. (2009). Validation of regression-based myogenic correction techniques for scalp and source-localized EEG. *Psychophysiology*, *46*, 578–592.

Metting van Rijn, A. C., Peper, A., & Grimbergen, C. A. (1990). High-quality recording of bioelectric events: 1. Interference reduction, theory and practice. *Medical & Biological Engineering & Computing*, *28*, 389–397.

Miller, J., & Hackley, S. A. (1992). Electrophysiological evidence for temporal overlap among contingent mental processes. *Journal of Experimental Psychology. General*, *121*, 195–209.

Miller, J., Patterson, T., & Ulrich, R. (1998). Jackknife-based method for measuring LRP onset latency differences. *Psychophysiology*, *35*, 99–115.

Miller, J., Riehle, A., & Requin, J. (1992). Effects of preliminary perceptual output on neuronal activity of the primary motor cortex. *Journal of Experimental Psychology. Human Perception and Performance*, *18*, 1121–1138.

Miller, J., Ulrich, R., & Schwarz, W. (2009). Why jackknifing yields good latency estimates. *Psychophysiology*, *46*, 300–312.

Morgan, C. D., & Murphy, C. (2010). Differential effects of active attention and age on event-related potentials to visual and olfactory stimuli. *International Journal of Psychophysiology*, *78*, 190–199.

Morgan, S. T., Hansen, J. C., & Hillyard, S. A. (1996). Selective attention to stimulus location modulates the steady-state visual evoked potential. *Proceedings of the National Academy of Sciences of the United States of America*, *93*, 4770–4774.

Näätänen, R., Gaillard, A. W. K., & Mantysalo, S. (1978). Early selective-attention effect on evoked potential reinterpreted. *Acta Psychologica*, *42*, 313–329.

Näätänen, R., & Kreegipuu, K. (2012). The mismatch negativity (MMN). In S. J. Luck & E. S. Kappenman (Eds.), *The Oxford Handbook of Event-Related Potential Components* (pp. 143–157). New York: Oxford University Press.

Naatanen, R., & Picton, T. W. (1986). N2 and automatic versus controlled processes. In W. C. McCallum, R. Zappoli, & F. Denoth (Eds.), *Cerebral Psychophysiology: Studies in Event-Related Potentials (EEG Supplement 38)* (pp. 169–186). Amsterdam: Elsevier.

Näätänen, R., & Picton, T. (1987). The N1 wave of the human electric and magnetic response to sound: A review and an analysis of the component structure. *Psychophysiology*, *24*, 375–425.

Nagamine, T., Toro, C., Balish, M., Deuschl, G., Wang, B., Sato, S., et al. (1994). Cortical magnetic and electrical fields associated with voluntary finger movements. *Brain Topography*, *6*, 175–183.

Nieuwenhuis, S., Yeung, N., Holroyd, C. B., Schurger, A., & Cohen, J. D. (2004). Sensitivity of electrophysiological activity from medial frontal cortex to utilitarian and performance feedback. *Cerebral Cortex*, *14*, 741–747.

Norman, D. A. (1968). Toward a theory of memory and attention. *Psychological Review*, *75*, 522–536.

Nunez, P. L. (1981). *Electric Fields of the Brain*. New York: Oxford University Press.

Nunez, P. L., & Srinivasan, R. (2006). *Electric Fields of the Brain* (2nd ed.). New York: Oxford University Press.

Ochoa, C. J., & Polich, J. (2000). P300 and blink instructions. *Clinical Neurophysiology, 111*, 93–98.

Olbrich, S., Jodicke, J., Sander, C., Himmerich, H., & Hegerl, U. (2011). ICA-based muscle artefact correction of EEG data: What is muscle and what is brain? Comment on McMenamin et al. *NeuroImage, 54*, 1–3, discussion 4–9.

Olvet, D. M., & Hajcak, G. (2009). The stability of error-related brain activity with increasing trials. *Psychophysiology, 46*, 957–961.

Oranje, B., van Berckel, B. N., Kemner, C., van Ree, J. M., Kahn, R. S., & Verbaten, M. N. (2000). The effects of a sub-anaesthetic dose of ketamine on human selective attention. *Neuropsychopharmacology, 22*, 293–302.

Osman, A., & Moore, C. M. (1993). The locus of dual-task interference: Psychological refractory effects on movement-related brain potentials. *Journal of Experimental Psychology. Human Perception and Performance, 19*, 1292–1312.

Osman, A., Bashore, T. R., Coles, M., Donchin, E., & Meyer, D. (1992). On the transmission of partial information: Inferences from movement-related brain potentials. *Journal of Experimental Psychology. Human Perception and Performance, 18*, 217–232.

Osterhout, L., & Holcomb, P. J. (1992). Event-related brain potentials elicited by syntactic anomaly. *Journal of Memory and Language, 31*, 785–806.

Osterhout, L., & Holcomb, P. J. (1995). Event-related potentials and language comprehension. In M. D. Rugg & M. G. H. Coles (Eds.), *Electrophysiology of Mind* (pp. 171–215). New York: Oxford University Press.

Paller, K. A., Voss, J. L., & Boehm, S. G. (2007). Validating neural correlates of familiarity. *Trends in Cognitive Sciences, 11*, 243–250.

Pascual-Marqui, R. D., Esslen, M., Kochi, K., & Lehmann, D. (2002). Functional imaging with low-resolution brain electromagnetic tomography (LORETA): A review. *Methods and Findings in Experimental and Clinical Pharmacology, 24*(Suppl C), 91–95.

Pashler, H. (1994). Dual-task interference in simple tasks: Data and theory. *Psychological Bulletin, 116*, 220–244.

Pashler, H., & Wagenmakers, E.-J. (2012). Editor's introduction to the special section on replicability in psychological science: A crisis of confidence? *Perspectives on Psychological Science, 7*, 529–531.

Perez, V. B., & Vogel, E. K. (2012). What ERPs can tell us about working memory. In S. J. Luck & E. S. Kappenman (Eds.), *The Oxford Handbook of Event-Related Potential Components* (pp. 361–372). New York: Oxford University Press.

Pernier, J., Perrin, F., & Bertrand, O. (1988). Scalp current density fields: Concept and properties. *Electroencephalography and Clinical Neurophysiology, 69*, 385–389.

Perrin, F., Pernier, J., Bertrand, O., & Echallier, J. F. (1989). Spherical splines for scalp potential and current density mapping. *Electroencephalography and Clinical Neurophysiology, 72*, 184–187.

Picton, T. W. (1992). The P300 wave of the human event-related potential. *Journal of Clinical Neurophysiology, 9*, 456–479.

Picton, T. W. (2011). *Human Auditory Evoked Potentials*. San Diego: Plural Publishing.

Picton, T. W., & Hillyard, S. A. (1972). Cephalic skin potentials in electroencephalography. *Electroencephalography and Clinical Neurophysiology, 33*, 419–424.

Picton, T. W., Alain, C., Woods, D. L., John, M. S., Scherg, M., Valdes-Sosa, P., et al. (1999). Intracerebral sources of human auditory-evoked potentials. *Audiology & Neuro-Otology, 4*, 64–79.

Picton, T. W., Lins, O. G., & Scherg, M. (1995). The recording and analysis of event-related potentials. In F. Boller & J. Grafman (Eds.), *Handbook of Neuropsychology* (Vol. 10, pp. 3–73). New York: Elsevier.

Pliszka, S. R., Liotti, M., & Woldorff, M. G. (2000). Inhibitory control in children with attention-deficit/hyperactivity disorder: Event-related potentials identify the processing component and timing of an impaired right-frontal response-inhibition mechanism. *Biological Psychiatry, 48*, 238–246.

Plochl, M., Ossandon, J. P., & Konig, P. (2012). Combining EEG and eye tracking: identification, characterization, and correction of eye movement artifacts in electroencephalographic data. *Frontiers in Human Neuroscience, 6*, 278.

Plonsey, R. (1963). Reciprocity applied to volume conductors and the EEG. *IEEE Transactions on Bio-Medical Engineering, 19*, 9–12.

Poldrack, R. A. (2006). Can cognitive processes be inferred from neuroimaging data? *Trends in Cognitive Sciences*, *10*, 59–63.

Polich, J. (2004). Clinical application of the P300 event-related brain potential. *Physical Medicine and Rehabilitation Clinics of North America*, *15*, 133–161.

Polich, J. (2012). Neuropsychology of P300. In S. J. Luck & E. S. Kappenman (Eds.), *Oxford Handbook of Event-Related Potential Components* (p. 159-188). New York: Oxford University Press.

Polich, J., & Comerchero, M. D. (2003). P3a from visual stimuli: Typicality, task, and topography. *Brain Topography*, *15*, 141–152.

Polich, J., & Kok, A. (1995). Cognitive and biological determinants of P300: An integrative review. *Biological Psychology*, *41*, 103–146.

Polich, J., & Lawson, D. (1985). Event-related potentials paradigms using tin electrodes. *American Journal of EEG Technology*, *25*, 187–192.

Polich, J., Eischen, S. E., & Collins, G. E. (1994). P300 from a single auditory stimulus. *Electroencephalography and Clinical Neurophysiology*, *92*, 253–261.

Pontifex, M. B., Scudder, M. R., Brown, M. L., O'Leary, K. C., Wu, C.-T., Themanson, J. R., et al. (2010). On the number of trials necessary for stabilization of error-related brain activity across the life span. *Psychophysiology*, *47*, 767–773.

Potter, M. C. (1976). Short-term conceptual memory for pictures. *Journal of Experimental Psychology. Human Learning and Memory*, *2*, 509–522.

Potts, G. F., O'Donnell, B. F., Hirayasu, U., & McCarley, R. W. (2002). Disruption of neural systems of visual attention in schizophrenia. *Archives of General Psychiatry*, *59*, 418–424.

Pratt, H. (2012). Sensory ERP components. In S. J. Luck & E. S. Kappenman (Eds.), *The Oxford Handbook of ERP Components* (pp. 89–114). New York: Oxford University Press.

Pritchard, W. S. (1981). Psychophysiology of P300. *Psychological Bulletin*, *89*, 506–540.

Pritchard, W. S., Shappell, S. A., & Brandt, M. E. (1991). Psychophysiology of N200/N400: A review and classification scheme. In J. R. Jennings & P. K. Ackles (Eds.), *Advances in Psychophysiology* (pp. 43–106). London: Jessica Kingsley.

Raymond, J. E., Shapiro, K. L., & Arnell, K. M. (1992). Temporary suppression of visual processing in an RSVP task: An attentional blink? *Journal of Experimental Psychology. Human Perception and Performance*, *18*, 849–860.

Reiss, J. E., & Hoffman, J. E. (2006). Object substitution masking interferes with semantic processing: Evidence from event-related potentials. *Psychological Science*, *17*, 1015–1020.

Renault, B., Ragot, R., Lesevre, N., & Remond, A. (1982). Onset and offset of brain events as indices of mental chronometry. *Science*, *215*, 1413–1415.

Rhodes, S. M., & Donaldson, D. I. (2008). Association and not semantic relationships elicit the N400 effect: Electrophysiological evidence from an explicit language comprehension task. *Psychophysiology*, *45*, 50–59.

Ridderinkhof, K. R., Ullsperger, M., Crone, E. A., & Nieuwenhuis, S. (2004). The role of the medial frontal cortex in cognitive control. *Science*, *306*, 443–447.

Ritter, W., Simson, R., Vaughan, H. G., & Friedman, D. (1979). A brain event related to the making of a sensory discrimination. *Science*, *203*, 1358–1361.

Roach, B. J., & Mathalon, D. H. (2008). Event-related EEG time-frequency analysis: An overview of measures and an analysis of early gamma band phase locking in schizophrenia. *Schizophrenia Bulletin*, *34*, 907–926.

Robitaille, N., Grimault, S., & Jolicoeur, P. (2009). Bilateral parietal and contralateral responses during maintenance of unilaterally-encoded objects in visual short-term memory: Evidence from magnetoencephalography. *Psychophysiology*, *46*, 1090–1099.

Rohrbaugh, J. W., Syndulko, K., & Lindsley, D. B. (1976). Brain wave components of the contingent negative variation in humans. *Science*, *191*, 1055–1057.

Rossion, B., & Jacques, C. (2012). The N170: Understanding the time course of face perception in the human brain. In S. J. Luck & E. S. Kappenman (Eds.), *The Oxford Handbook of Event-Related Potential Components* (pp. 115–141). New York: Oxford University Press.

Rossion, B., Collins, D., Goffaux, V., & Curran, T. (2007). Long-term expertise with artificial objects increases visual competition with early face categorization processes. *Journal of Cognitive Neuroscience*, *19*, 543–555.

Rossion, B., Delvenne, J. F., Debatisse, D., Goffaux, V., Bruyer, R., Crommelinck, M., et al. (1999). Spatio-temporal localization of the face inversion effect: An event-related potentials study. *Biological Psychology*, *50*, 173–189.

Rossion, B., Kung, C. C., & Tarr, M. J. (2004). Visual expertise with nonface objects leads to competition with the early perceptual processing of faces in the human occipitotemporal cortex. *Proceedings of the National Academy of Sciences of the United States of America*, *101*, 14521–14526.

Rugg, M., & Curran, T. (2007). Event-related potentials and recognition memory. *Trends in Cognitive Sciences*, *11*, 251–257.

Rüsseler, J., Altenmuller, E., Nager, W., Kohlmetz, C., & Munte, T. F. (2001). Event-related brain potentials to sound omissions differ in musicians and non-musicians. *Neuroscience Letters*, *308*, 33–36.

Sawaki, R., & Luck, S. J. (2010). Capture versus suppression of attention by salient singletons: Electrophysiological evidence for an automatic attend-to-me signal. *Attention, Perception & Psychophysics*, *72*, 1455–1470.

Sawaki, R., & Luck, S. J. (2011). Active suppression of distractors that match the contents of visual working memory. *Visual Cognition*, *19*, 956–972.

Sawaki, R., Geng, J. J., & Luck, S. J. (2012). A common neural mechanism for preventing and terminating attention. *Journal of Neuroscience*, *32*, 10725–10736.

Scherg, M. (1990). Fundamentals of dipole source potential analysis. In F. Grandori, M. Hoke, & G. L. Romani (Eds.), *Auditory Evoked Magnetic Fields and Potentials. Advances in Audiology VI* (pp. 40–69). Basel: Karger.

Schreckenberger, M., Lange-Asschenfeldt, C., Lochmann, M., Mann, K., Siessmeier, T., Buchholz, H. G., et al. (2004). The thalamus as the generator and modulator of EEG alpha rhythm: A combined PET/EEG study with lorazepam challenge in humans. *NeuroImage*, *22*, 637–644.

Schupp, H. T., Junghofer, M., Weike, A. I., & Hamm, A. O. (2003). Attention and emotion: An ERP analysis of facilitated emotional stimulus processing. *Neuroreport*, *14*, 1107–1110.

Schupp, H. T., Junghofer, M., Weike, A. I., & Hamm, A. O. (2004). The selective processing of briefly presented affective pictures: An ERP analysis. *Psychophysiology*, *41*, 441–449.

Serences, J. T., Ester, E. F., Vogel, E. K., & Awh, E. (2009). Stimulus-specific delay activity in human primary visual cortex. *Psychological Science*, *20*, 207–214.

Shapiro, K. L., Arnell, K. M., & Raymond, J. E. (1997). The attentional blink. *Trends in Cognitive Sciences*, *1*, 291–296.

Shapiro, K. L., Raymond, J. E., & Arnell, K. M. (1994). Attention to visual pattern information produces the attentional blink in rapid serial visual presentation. *Journal of Experimental Psychology. Human Perception and Performance*, *20*, 357–371.

Shibasaki, H. (1982). *Movement-Related Cortical Potentials: Evoked Potentials in Clinical Testing* (Vol. 3, pp. 471–482). Edinburgh: Churchill Livingstone.

Simmons, J. P., Nelson, L. D., & Simonsohn, U. (2011). False-positive psychology: undisclosed flexibility in data collection and analysis allows presenting anything as significant. *Psychological Science*, *22*, 1359–1366.

Simson, R., Vaughan, H. G., & Ritter, W. (1977). The scalp topography of potentials in auditory and visual discrimination tasks. *Electroencephalography and Clinical Neurophysiology*, *42*, 528–535.

Singer, W. (1999). Neuronal synchrony: A versatile code for the definition of relations? *Neuron*, *24*, 49–65, 111–125.

Singh, P. B., Iannilli, E., & Hummel, T. (2011). Segregation of gustatory cortex in response to salt and umami taste studied through event-related potentials. *Neuroreport*, *22*, 299–303.

Smulders, F. T. Y., & Miller, J. O. (2012). The Lateralized Readiness Potential. In S. J. Luck & E. S. Kappenman (Eds.), *The Oxford Handbook of Event-Related Potential Components* (pp. 209–229). New York: Oxford University Press.

Soltani, M., & Knight, R. T. (2000). Neural origins of the P300. *Critical Reviews in Neurobiology*, *14*, 199–224.

Squires, N. K., Squires, K. C., & Hillyard, S. A. (1975). Two varieties of long-latency positive waves evoked by unpredictable auditory stimuli. *Electroencephalography and Clinical Neurophysiology*, *38*, 387–401.

Sreenivasan, K. K., Goldstein, J. M., Lustig, A. G., Rivas, L. R., & Jha, A. P. (2009). Attention to faces modulates early face processing during low but not high face discriminability. *Attention, Perception & Psychophysics*, *71*, 837–846.

Stahl, J., & Gibbons, H. (2004). The application of jackknife-based onset detection of lateralized readiness potential in correlative approaches. *Psychophysiology, 41*, 845–860.

Sullivan, L. R., & Altman, C. L. (2008). Infection control: 2008 review and update for electroneurodiagnostic technologists. *American Journal of Electroneurodiagnostic Technology, 48*, 140–165.

Sutton, S., Braren, M., Zubin, J., & John, E. R. (1965). Evoked potential correlates of stimulus uncertainty. *Science, 150*, 1187–1188.

Suwazono, S., Machado, L., & Knight, R. T. (2000). Predictive value of novel stimuli modifies visual event-related potentials and behavior. *Clinical Neurophysiology, 111*, 29–39.

Swaab, T. Y., Ledoux, K., Camblin, C. C., & Boudewyn, M. (2012). Language-related ERP components. In S. J. Luck & E. S. Kappenman (Eds.), *The Oxford Handbook of Event-Related Potential Components* (pp. 397–439). New York: Oxford University Press.

Szücs, A. (1998). Applications of the spike density function in analysis of neuronal firing patterns. *Journal of Neuroscience Methods, 81*, 159–167.

Tallon-Baudry, C., Bertrand, O., Delpuech, C., & Pernier, J. (1996). Stimulus specificity of phase-locked and non-phase-locked 40 Hz visual responses in humans. *Journal of Neuroscience, 16*, 4240–4249.

Tanaka, J. W., & Curran, T. (2001). A neural basis for expert object recognition. *Psychological Science, 12*, 43–47.

Thorpe, S., Fize, D., & Marlot, C. (1996). Speed of processing in the human visual system. *Nature, 381*, 520–522.

Tikhonravov, D., Neuvonen, T., Pertovaara, A., Savioja, K., Ruusuvirta, T., Naatanen, R., et al. (2008). Effects of an NMDA-receptor antagonist MK-801 on an MMN-like response recorded in anesthetized rats. *Brain Research, 1203*, 97–102.

Trainor, L., McFadden, M., Hodgson, L., Darragh, L., Barlow, J., Matsos, L., et al. (2003). Changes in auditory cortex and the development of mismatch negativity between 2 and 6 months of age. *International Journal of Psychophysiology, 51*, 5–15.

Treisman, A. M. (1969). Strategies and models of selective attention. *Psychological Review, 76*, 282–299.

Treisman, A. (1986). Features and objects in visual processing. *Scientific American, 255*, 114–125.

Treisman, A., & Gormican, S. (1988). Feature analysis in early vision: Evidence from search asymmetries. *Psychological Review, 95*, 15–48.

Treisman, A., & Souther, J. (1985). Search asymmetry: A diagnostic for preattentive processing of separable features. *Journal of Experimental Psychology. General, 114*, 285–310.

Tsubomi, H., Fukuda, K., Watanabe, K., & Vogel, E. K. (2013). Neural limits to representing objects still within view. *Journal of Neuroscience, 33*, 8257–8263.

Ulrich, R., & Miller, J. (2001). Using the jackknife-based scoring method for measuring LRP onset effects in factorial designs. *Psychophysiology, 38*, 816–827.

Umbricht, D., Schmid, L., Koller, R., Vollenweider, F. X., Hell, D., & Javitt, D. C. (2000). Ketamine-induced deficits in auditory and visual context-dependent processing in healthy volunteers: Implications for models of cognitive deficits in schizophrenia. *Archives of General Psychiatry, 57*, 1139–1147.

Urbach, T. P., & Kutas, M. (2002). The intractability of scaling scalp distributions to infer neuroelectric sources. *Psychophysiology, 39*, 791–808.

Urbach, T. P., & Kutas, M. (2006). Interpreting event-related brain potential (ERP) distributions: Implications of baseline potentials and variability with application to amplitude normalization by vector scaling. *Biological Psychology, 72*, 333–343.

Usher, M., & McClelland, J. L. (2001). The time course of perceptual choice: The leaking, competing accumulator model. *Psychological Review, 108*, 550–592.

van Boxtel, G. J. M., & Böcker, K. B. E. (2004). Cortical measures of anticipation. *Journal of Psychophysiology, 18*, 61–76.

van Boxtel, G. J., van der Molen, M. W., Jennings, J. R., & Brunia, C. H. (2001). A psychophysiological analysis of inhibitory motor control in the stop-signal paradigm. *Biological Psychology, 58*, 229–262.

van Dijk, H., van der Werf, J., Mazaheri, A., Medendorp, W. P., & Jensen, O. (2010). Modulations in oscillatory activity with amplitude asymmetry can produce cognitively relevant event-related responses. *Proceedings of the National Academy of Sciences of the United States of America, 107*, 900–905.

Vanrullen, R., Busch, N. A., Drewes, J., & Dubois, J. (2011). Ongoing EEG phase as a trial-by-trial predictor of perceptual and attentional variability. *Frontiers in Psychology, 2*, 60.

van Turennout, M., Hagoort, P., & Brown, C. M. (1998). Brain activity during speaking: From syntax to phonology in 40 milliseconds. *Science, 280*, 572–574.

Vaughan, H. G., Jr., Costa, L. D., & Ritter, W. (1968). Topography of the human motor potential. *Electroencephalography and Clinical Neurophysiology, 25*, 1–10.

Verleger, R. (1988). Event-related potentials and cognition: A critique of the context updating hypothesis and an alternative interpretation of P3. *Behavioral and Brain Sciences, 11*, 343–427.

Verleger, R. (1997). On the utility of P3 latency as an index of mental chronometry. *Psychophysiology, 34*, 131–156.

Verleger, R., Jaskowsi, P., & Wauschkuhn, B. (1994). Suspense and surprise: On the relationship between expectancies and P3. *Psychophysiology, 31*, 359–369.

Vidal, F., Hasbroucq, T., Grapperon, J., & Bonnet, M. (2000). Is the 'error negativity' specific to errors? *Biological Psychology, 51*, 109–128.

Vogel, E. K., & Luck, S. J. (2000). The visual N1 component as an index of a discrimination process. *Psychophysiology, 37*, 190–203.

Vogel, E. K., & Luck, S. J. (2002). Delayed working memory consolidation during the attentional blink. *Psychonomic Bulletin & Review, 9*, 739–743.

Vogel, E. K., & Machizawa, M. G. (2004). Neural activity predicts individual differences in visual working memory capacity. *Nature, 428*, 748–751.

Vogel, E. K., Luck, S. J., & Shapiro, K. L. (1998). Electrophysiological evidence for a postperceptual locus of suppression during the attentional blink. *Journal of Experimental Psychology. Human Perception and Performance, 24*, 1656–1674.

Vogel, E. K., McCollough, A. W., & Machizawa, M. G. (2005). Neural measures reveal individual differences in controlling access to working memory. *Nature, 438*, 500–503.

Vogel, E. K., Woodman, G. F., & Luck, S. J. (2005). Pushing around the locus of selection: Evidence for the flexible-selection hypothesis. *Journal of Cognitive Neuroscience, 17*, 1907–1922.

Volman, V., Behrens, M. M., & Sejnowski, T. J. (2011). Downregulation of parvalbumin at cortical GABA synapses reduces network gamma oscillatory activity. *Journal of Neuroscience, 31*, 18137–18148.

Voss, J. L., Lucas, H. D., & Paller, K. A. (2012). More than a feeling: Pervasive influences of memory processing without awareness of remembering. *Cognitive Neuroscience, 3*, 193–207.

Vul, E., Harris, C., Winkielman, P., & Pashler, H. (2009). Puzzlingly high correlations in fMRI studies of emotion, personality, and social cognition. *Perspectives on Psychological Science, 4*, 274–290.

Wada, M. (1999). Measurement of olfactory threshold using an evoked potential technique. *Rhinology, 37*, 25–28.

Walter, W. G., Cooper, R., Aldridge, V. J., McCallum, W. C., & Winter, A. L. (1964). Contingent negative variation: An electric sign of sensorimotor association and expectancy in the human brain. *Nature, 203*, 380–384.

Wastell, D. G. (1977). Statistical detection of individual evoked responses: An evaluation of Woody's adaptive filter. *Electroencephalography and Clinical Neurophysiology, 42*, 835–839.

Weinberg, A., & Hajcak, G. (2010). Beyond good and evil: The time-course of neural activity elicited by specific picture content. *Emotion (Washington, D.C.), 10*, 767–782.

West, R., & Alain, C. (1999). Event-related neural activity associated with the Stroop task. *Brain Research. Cognitive Brain Research, 8*, 157–164.

West, R., & Alain, C. (2000). Effects of task context and fluctuations of attention on the neural activity supporting performance of the Stroop task. *Brain Research, 873*, 102–111.

Wilding, E. L., & Ranganath, C. (2012). Electrophysiological correlates of episodic memory processes. In S. J. Luck & E. S. Kappenman (Eds.), *The Oxford Handbook of Event-Related Potential Components* (pp. 373–395). New York: Oxford University Press.

Willems, R. M., Özyürek, A., & Hagoort, P. (2008). Seeing and hearing meaning: ERP and fMRI evidence of word versus picture integration into a sentence context. *Journal of Cognitive Neuroscience, 20*, 1235–1249.

Woldorff, M. (1993). Distortion of ERP averages due to overlap from temporally adjacent ERPs: Analysis and correction. *Psychophysiology*, *30*, 98–119.

Woldorff, M. G., Gallen, C. C., Hampson, S. A., Hillyard, S. A., Pantev, C., Sobel, D., et al. (1993). Modulation of early sensory processing in human auditory cortex during auditory selective attention. *Proceedings of the National Academy of Sciences of the United States of America*, *90*, 8722–8726.

Woldorff, M. G., Hackley, S. A., & Hillyard, S. A. (1991). The effects of channel-selective attention on the mismatch negativity wave elicited by deviant tones. *Psychophysiology*, *28*, 30–42.

Woldorff, M. G., Hillyard, S. A., Gallen, C. C., Hampson, S. A., & Bloom, F. E. (1998). Magnetoencephalographic recordings demonstrate attentional modulation of mismatch-related neural activity in human auditory cortex. *Psychophysiology*, *35*, 283–292.

Woodman, G. F. (2012). Homologues of human ERP components in nonhuman primates. In S. J. Luck & E. S. Kappenman (Eds.), *Oxford Handbook of ERP Components* (pp. 611–625). New York: Oxford University Press.

Woodman, G. F., & Luck, S. J. (1999). Electrophysiological measurement of rapid shifts of attention during visual search. *Nature*, *400*, 867–869.

Woodman, G. F., & Luck, S. J. (2003a). Dissociations among attention, perception, and awareness during object-substitution masking. *Psychological Science*, *14*, 605–611.

Woodman, G. F., & Luck, S. J. (2003b). Serial deployment of attention during visual search. *Journal of Experimental Psychology. Human Perception and Performance*, *29*, 121–138.

Woody, C. D. (1967). Characterization of an adaptive filter for the analysis of variable latency neuroelectric signals. *Medical & Biological Engineering*, *5*, 539–553.

Worden, M. S., Foxe, J. J., Wang, N., & Simpson, G. V. (2000). Anticipatory biasing of visuospatial attention indexed by retinotopically specific alpha-band electroencephalography increases over occipital cortex. *Journal of Neuroscience*, *20*, RC63.

Yeung, N. (2004). Relating cognitive and affective theories of the error-related negativity. In M. Ullsperger & M. Falkenstein (Eds.), *Errors, Conflicts, and the Brain. Current Opinions on Performance Monitoring* (pp. 63–70). Leipzig: MPI of Cognitive Neuroscience.

Yeung, N., Bogacz, R., Holroyd, C. B., Nieuwenhuis, S., & Cohen, J. D. (2007). Theta phase resetting and the error-related negativity. *Psychophysiology*, *44*, 39–49.

Yeung, N., Cohen, J. D., & Botvinick, M. M. (2004). The neural basis of error detection: Conflict monitoring and the error-related negativity. *Psychological Review*, *111*, 931–959.

Yonelinas, A. P., & Parks, C. M. (2007). Receiver operating characteristics (ROCs) in recognition memory: A review. *Psychological Bulletin*, *133*, 800–832.

Yuval-Greenberg, S., Tomer, O., Keren, A. S., Nelken, I., & Deouell, L. Y. (2008). Transient induced gamma-band response in EEG as a manifestation of miniature saccades. *Neuron*, *58*, 429–441.

Zhang, W., & Luck, S. J. (2008). Discrete fixed-resolution representations in visual working memory. *Nature*, *453*, 233–235.

Zhang, W., & Luck, S. J. (2009). Feature-based attention modulates feedforward visual processing. *Nature Neuroscience*, *12*, 24–25.

Index